Ecological
Engineering Design

Ecological Engineering Design

Restoring and Conserving Ecosystem Services

Marty D. Matlock
and
Robert A. Morgan

WILEY

John Wiley & Sons, Inc.

This book is printed on acid-free paper. ∞

Copyright © 2011 by John Wiley & Sons, Inc. All rights reserved.

Published by John Wiley & Sons, Inc., Hoboken, New Jersey

Published simultaneously in Canada

No part of this publication may be reproduced, stored in a retrieval system, or transmitted in any form or by any means, electronic, mechanical, photocopying, recording, scanning, or otherwise, except as permitted under Section 107 or 108 of the 1976 United States Copyright Act, without either the prior written permission of the Publisher, or authorization through payment of the appropriate per-copy fee to the Copyright Clearance Center, 222 Rosewood Drive, Danvers, MA 01923, (978) 750-8400, fax (978) 646-8600, or on the web at www.copyright.com. Requests to the Publisher for permission should be addressed to the Permissions Department, John Wiley & Sons, Inc., 111 River Street, Hoboken, NJ 07030, (201) 748-6011, fax (201) 748-6008, or online at www.wiley.com/go/permissions.

Limit of Liability/Disclaimer of Warranty: While the publisher and the author have used their best efforts in preparing this book, they make no representations or warranties with respect to the accuracy or completeness of the contents of this book and specifically disclaim any implied warranties of merchantability or fitness for a particular purpose. No warranty may be created or extended by sales representatives or written sales materials. The advice and strategies contained herein may not be suitable for your situation. You should consult with a professional where appropriate. Neither the publisher nor the author shall be liable for any loss of profit or any other commercial damages, including but not limited to special, incidental, consequential, or other damages.

For general information about our other products and services, please contact our Customer Care Department within the United States at (800) 762-2974, outside the United States at (317) 572-3993 or fax (317) 572-4002.

Wiley also publishes its books in a variety of electronic formats. Some content that appears in print may not be available in electronic books. For more information about Wiley products, visit our web site at www.wiley.com.

Library of Congress Cataloging-in-Publication Data:
Matlock, Marty D.
  Ecological engineering design : restoring and conserving ecosystem services / Marty D. Matlock, Robert A. Morgan.
    p. cm.
  Includes bibliographical references and index.
  ISBN 978-0-470-34514-6 (hardback); ISBN 978-0-470-87574-2 (ebk); ISBN 978-0-470-87575-9 (ebk); ISBN 978-0-470-87576-6 (ebk); ISBN 978-0-470-95167-5 (ebk); ISBN 978-0-470-95191-0 (ebk); ISBN 978-0-470-94999-3 (ebk)
  1. Ecological engineering. 2. Sustainable design. 3. Environmental engineering. I. Morgan, Robert A. II. Title.
  GE350.M38 2010
  628–dc22
                                                            2010030991

Printed in the United States of America

10 9 8 7 6 5 4 3 2 1

Contents

1. **SUSTAINABLE HUMAN-DOMINATED ECOSYSTEMS / 1**

    Introduction / 1
       Axioms of Ecological Engineering / 2
       Sustainable Design Principles / 3
    Global Population Dynamics—The Forcing
       Function / 4
       Global Fertility Rate Trajectories / 5
       Changing Global Demographics / 6
    Human-Dominated Earth / 8
       Increasing Demands for Ecosystem Services / 8
       Human Impacts through Urbanization / 9
       Land Use Change / 11
       Agricultural Production / 13
       Water Resource Demands / 14
       Lessons from the First Green Revolution / 16
    Structure of This Book / 17
    References / 18

2. **ECOSYSTEM SERVICES / 22**

    Introduction / 22
       Origin of Ecosystem Services / 22
       The Value of Ecosystem Services / 24
       Classifying Ecosystem Services / 24
       The Millennium Ecosystem Assessment / 28
       Why Biodiversity Matters / 35
       Ecosystem Services, Land Use, and
          Biodiversity / 37
    Further Readings / 39
    References / 39

3. DESIGNING ECOSYSTEM SERVICES / 42

Design Challenges and Needs / 42
    Current Design Methods Deficiencies / 43
    Ecosystem Services Design Ethics / 46
    Legitimacy and the Design Process / 48
The Design Process / 50
    Synthesis / 53
    The Ecotechnology Design Team / 54
    Defining the Appropriate Management Structure / 55
    Analysis and Deliberation / 56
    Mapping Ecosystem Services Processes / 56
    Defining Priorities / 58
    Setting Design Goals / 59
    Implementing Design Goals / 60
    Assessing Ecosystem Services Design / 61
Further Readings / 62
References / 62

4. DEFINING PLACE: BIOMES AND ECOREGIONS / 64

Introduction / 64
    Biogeographical Realms / 65
Biomes / 66
Ecoregions / 72
    Bailey's Ecoregions / 72
    Omernik's Ecoregions / 73
    Olson's Ecoregions / 76
Other Land Classification Systems / 78
    Climate Change and Ecoregions / 79
    Land Use Change and Ecoregions / 80
References / 81

5. DEFINING PLACE: THE WATERSHED / 83

Introduction / 83
    Watershed Services / 84
    Watershed Characteristics: Physical Description / 84
    Watershed Hydrologic Characteristics / 92
    Watershed Water Quality Characteristics / 99

Contents   vii

      Watershed Human Impacts / 100
      Summary of Watershed Characteristics / 103
Further Readings / 104
References / 104

6. DEFINING PLACE: THE SITE / 106

Introduction / 106
      Physical Characterization / 106
      Hydrological Characterization / 107
      Biological Characterization / 119
      Climatological Characterization / 124
Summary / 125
Further Readings / 127
References / 127

7. DEFINING PLACE: SOILS AS A LIVING ORGANISM / 129

Introduction / 129
      Morphology / 130
      Soil Physics / 136
      Soil Fertility / 139
      Soil Ecology / 141
Summary / 143
Further Readings / 143
References / 144

8. FUNDAMENTAL PRINCIPLES OF ECOLOGY FOR DESIGN / 145

Introduction / 145
      Fundamental Principles of Ecology / 148
Organisms and Place / 149
      Adaptation Processes / 150
      Responses to Environmental Variation / 152
Landforms and Ecosystem Function / 154
      Patches, Corridors, and Connectivity / 154
      Ecotones and Edge Effects / 156
      Landform Metrics / 158
Further Readings / 160
References / 160

9. ENERGY AND MASS FLOW THROUGH ECOSYSTEMS / 162

Introduction / 162
Energy Flow through Ecosystems / 164
    Energy Balance in the Biosphere / 164
    Emergy as a Unit of Analysis / 168
Trophic Levels / 169
    Energy Density / 169
    Primary Production / 170
    Designing Trophic Levels / 173
Mass Flow through Ecosystems / 175
    Hydrologic Cycle / 176
    Carbon Cycle / 178
    Nitrogen Cycle / 181
    Phosphorus Cycle / 183
References / 184

10. DESIGNING COMMUNITY STRUCTURE / 187

Introduction / 187
    Hierarchical Processes / 187
    Types of Restoration Design / 188
Biotic Interactions / 190
    Community Interactions / 190
    Competition / 191
    Consumption / 192
    Commensalism / 192
Metapopulations / 193
    Species-Area Relationship / 193
    Minimum Viable Populations / 194
    Minimum Viable Metapopulations / 195
Regional Processes / 195
    Species Pool / 196
    Dispersal / 196
    Colonization Sequence / 197
    Dispersion / 197
Environmental and Habitat Impacts / 198
    Abiotic Filters / 198
    Disturbance Regimes / 199
    Habitat Heterogeneity / 200
References / 201

11. ECOSYSTEM CONTROL AND FEEDBACK SYSTEMS / 202

Introduction / 202
Population Control Processes / 204
    Reproductive Strategies / 204
    Survivorship / 205
    Growth Rates / 206
Community Control Processes / 207
    Plants and Nutrients / 208
    Resource Competition / 209
Feedback Processes / 210
    Atmospheric Feedback Loops / 211
    Soil Feedback Loops / 212
    Consumer Feedback Loops / 214
Designing Ecosystem Complexity / 215
    Self-Organization / 217
References / 220

12. STREAM RESTORATION DESIGN / 222

Introduction / 222
Assessment / 223
Hydrology / 227
Sedimentology / 233
Geomorphology / 235
Habitat / 238
Connectivity / 240
Riparian Corridor / 241
Construction / 242
Summary / 242
Further Readings / 243
References / 243

13. DESIGNING ECOSYSTEM SERVICES BY LANDFORM / 245

Introduction / 245
    Ecosystem Services Design Process / 245
Agricultural Lands / 247
Forests / 251
Grasslands / 253
Wetlands / 256

Urban Areas / 260
References / 265

14. GREEN INFRASTRUCTURE DESIGN / 267

Introduction / 267
The Green Infrastructure Network / 268
    Green Infrastructure Planning / 271
        The Tools of Green Infrastructure / 272
        Scale Matters / 275
The Sustainable Cities Initiative / 275
    United Nations World Urban Forum / 276
    ICLEI: Local Governments for
       Sustainability / 278
Summary / 280
Further Readings / 281
References / 281

15. LOW IMPACT DEVELOPMENT / 282

Introduction / 282
Hydrology / 284
Initial Steps / 287
Minimizing Change to Pre-development CN / 287
Maintaining or Increasing t_c / 289
Integrated Management Practices / 290
    Bioretention / 290
    Dry Wells / 291
    Rain Barrels and Cisterns / 291
    Vegetated or Grassed Swales / 292
    Infiltration Trenches / 292
    Tree Box Filters / 292
    Vegetated or Green Roofs / 293
    Filter Strips / 293
    Rain Gardens / 293
    Water Quality / 294
    Minimization / 295
    Natural Filtration / 295
    Constructed Filtration / 296
    Evaporation / 296
Pollution Prevention / 296
Hydrologic Analysis / 296
Refugia / 300

Ecosystem System Services
  Assessment/Design / 301
    Step 1: Define Project Objectives and
      Goals / 301
    Step 2: Perform Site Evaluation and
      Analysis / 302
    Step 3: Develop LID Control Strategies / 302
    Step 4: Design LID Site or Master Plan / 303
    Step 5: Develop Operation and Maintenance
      Procedures / 304
Summary / 304
Further Readings / 305
References / 306

16. ECOSYSTEM SERVICES DESIGN IN AGRICULTURE AND INDUSTRY / 307

Introduction / 307
Agricultural Sustainability Indicators / 308
    Summary of Sustainability Indicators / 309
    Environmental Indicators for Soil / 309
    Environmental Indicators for Water / 312
    Environmental Indicators for Habitat / 313
    Social and Cultural Indicators / 314
    Economic Indicators / 316
    Field-Scale Indicators / 317
Industrial Sustainability Metrics / 320
    Step One: Scope Selection / 320
    Step Two: Identify Priority Ecosystem
      Services / 321
    Step Three: Analyze Trends in Priority
      Ecosystem Services / 323
    Step Four: Identify Business Risks and
      Opportunities / 324
    Step Five: Develop Strategies for Addressing
      Risks and Opportunities / 325
References / 327

Index / 329

1

Sustainable Human-Dominated Ecosystems

> A thing is right when it tends to preserve the integrity, stability and beauty of the biotic community. It is wrong when it tends otherwise.
> —Aldo Leopold

INTRODUCTION

Ecological engineering has been defined as "designing human society with its natural environment for the benefit of both" (Mitch and Jorgenson, 2004). H. T. Odum described the practice of ecological engineering as "Management that joins human design and environmental self-design, so that they are mutually symbiotic" (Odum, 1988). Ecological engineering as a discipline was born in the twentieth century but is emerging as a leading design discipline in the twenty-first century. The foundations of ecological engineering were based in H. T. Odum's concept of quantitative systems ecology; William Mitch developed the first broadly applied ecological design applications using wetlands for waste treatment (Mitch, 2004). Odum's concepts were as broad as the biosphere and as specific as a photosynthetic cell (Kangas, 2002). However, the danger of having such a broad application range is that ecological engineering could be defined as anything. For the purpose of this book, ecological engineering is defined as the process of designing systems that preserve, restore, and create ecosystem services. More succinctly, ecological engineers design ecosystem services, which are the goods and services humanity extracts from the ecosystem. The purpose of this book is to present a framework for ecological engineering design practitioners.

This framework is founded in the science of ecology and the practice of design; it is interdisciplinary by nature. There are few practitioners who could master the body of knowledge necessary to engage in an ecological engineering design project independently. The distinction between ecological engineering and ecological design is important. Ecological engineering is the design process practiced by engineers with credentials that comply with laws or regulations governing the practice of engineering (established by state professional engineering boards in the U.S.).

Ecological design is performed by practitioners of a number of professions, including restoration ecologists, environmental scientists, landscape architects, and others with explicit training and experience in ecology and design processes. A further distinction is necessary between ecological design and ecological science: Ecologists investigating ecosystem processes are not generally designing ecosystem services; they are using the hypothetical reductionist approach (the scientific method) to characterize some aspect of ecosystem function, structure, or process. This scientific body of knowledge informs ecological design. However, the process of design is a practice separate and distinct from the scientific method and requires instruction and expertise apart from investigative methods. The design method will be described in Chapter 3.

Axioms of Ecological Engineering

The notion that human beings can design incredible processes such as ecosystems has been described as the height of hubris by some; ecosystems are too complex, the argument goes, and our knowledge too incomplete. In reality, we design ecosystems every time we start a bulldozer or tractor, every time we change land use or reroute stream flow. We just do not design explicitly, and the consequences are apparent. Designing ecosystem services should be approached with a deep sense of humility and respect for what we do not know. In order to ensure that this philosophy is embodied in the practice of ecological design, we propose the following three axioms of ecological engineering:

1. Everything is connected
2. Everything is changing
3. We are all in this together

The first two axioms are fundamental principles of systems ecology described by H. T. Odum (1988) and are the foundation of ecological design. They are critical for understanding and conceptualizing solutions to the challenges of developing sustainable design strategies.

The interconnectedness of all biotic and abiotic processes throughout the biosphere is demonstrated by the effects of urban land use on almost every aspect of ecosystem function, from climate to hydrology to biodiversity. Everything is changing, and the rate of change is increasing. Changes in the biosphere are being driven by changes in global climate, land use, and human population, among other factors. The third axiom, embodied by the Cherokee cultural ideal *gadugi*, roughly translated as "we are all in this together," is a normative claim that connects ecosystem theory with sustainability. This is the essence of the ecological engineering ethic (see Chapter 3).

Sustainable Design Principles

Sustainability as a concept is difficult to define. For this book, the phrase "sustainable prosperity" more accurately captures the goals of ecological design. The World Commission on Environment and Development (WCED) defined sustainability as "development that meets the needs of the present without compromising the ability of the future generations to meet their own needs" (WCED, 1987). The ethics of sustainability are difficult to define beyond this general framework. Cuello (1997) identified principles of sustainable development that included the following seven elements:

1. Voice for all people in an open and transparent process
2. Respect for the rights of future generations
3. Redefinition of the relationship between the human species and the ecosystems upon which we depend
4. Science-based understanding of the limits of ecosystem services
5. Understanding of the interconnected impacts of activities throughout a production supply chain (all the steps leading to a finished product) and across spatial scales
6. Enhanced self-sufficiency at the community level
7. Pragmatic implementations of practices to test, revise, and adapt to changing conditions

While these principles are aspirational, they can guide formulation of specific goals for sustainability within a sector of the global economy. Sustainable goals for ecological engineering can be formulated to respond to these principles of sustainability (see Chapter 3 on the ethics of ecological engineering design).

GLOBAL POPULATION DYNAMICS – THE FORCING FUNCTION

In 2008, the human population reached 6.8 billion. By 2050, Earth's population will likely reach 9.25 billion (UN, 2009). The population added to Earth in the next 40 years will exceed Earth's total population in 1950. Stated another way, we will be adding the population of Earth circa 1950 to the current population, over the next 40 years. These projections are based upon median estimates of population growth by country, using median fertility and mortality estimates (Figure 1-1). The challenge for ecological engineers is to design a sustainable Earth that supports 9.25 billion people at basal prosperity while preserving biotic diversity, ecosystem integrity, and natural resources. If global fertility rates continue to decline, the human population may reach zero population growth by the mid-twenty-first century, creating an unprecedented opportunity for recovery and optimization of ecosystem services globally (Figure 1-2). This is worth restating: In the next 40 years, for the first time in human history, our population will not be expanding. This rapid rate of change in population growth also will bring great economic and social challenges, as all rapid changes do. Ecological engineering design will provide critical responses to the complex problems and opportunities facing the next generation. These will be as predictable as converting urban areas to forests and agriculture (un-developing), designing estuarine fisheries in submerged urban landscapes, creating affordable and prosperous communities with aging populations, and as unpredictable as the imagination allows.

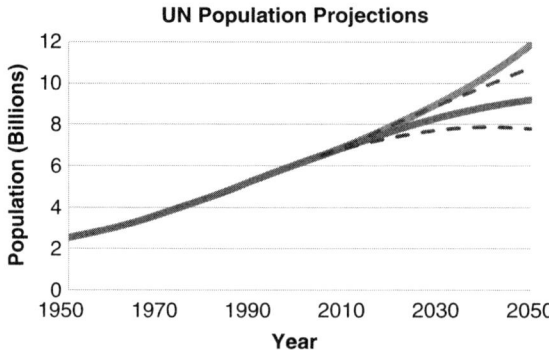

FIGURE 1-1. United Nations estimates of global population growth through 2050. The red line represents no decrease from 2005 fertility rates. The green line represents the median estimate of fertility, bounded by the upper and lower estimates.

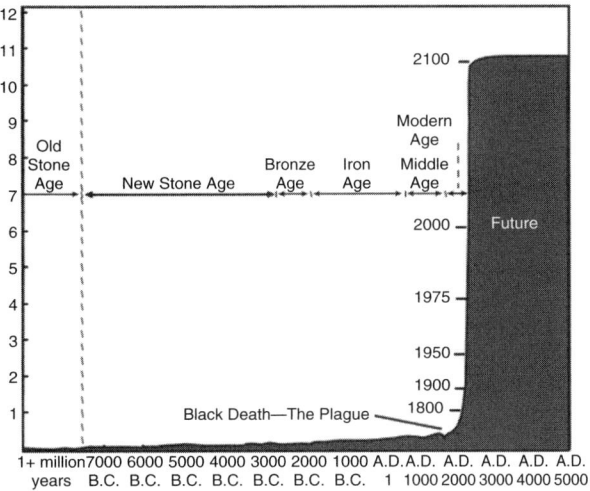

FIGURE 1-2. Population change throughout history. (*Source*: Population Reference Bureau; United Nations, *World Population Projections to 2100* [1998].)

Global Fertility Rate Trajectories

Global fertility rates have been declining over the past decade (UN, 2009). In 1970–1975, world fertility rate (WFR) was 4.5 children per woman, with least developed countries as high as 6.6, less developed at 5.2, and more developed regions at 2.1. In 2000–2005, the WFR was 2.6, with least developed regions at 5.0, less developed at 2.6, and more developed at 1.6. By 2045–2050, those rates are expected to decline even further, with least developed regions reaching 2.4 and less developed countries decreasing to 2.1 children per woman (UN, 2009). The result of changing fertility rates is a projected decrease in the rate of increase in human population through 2050 (Figure 1-3).

These reductions in fertility are a direct consequence of the dramatic reductions in abject poverty around the world, largely driven by increased agricultural productivity within least developed regions. Chronic malnourishment has declined dramatically in the past 40 years (Figure 1-4), and increased prosperity has resulted in dramatic reductions in fertility rates globally (Figure 1-5). Without these reductions, the current fertility rate would give rise to a population of over 12 billion by 2050, with 9.8 billion in less developed regions (Figure 1-1) (UN, 2009). The rate of decrease of fertility throughout the three categories (least, less, and more developed) is expected to decline as birth rates approach replacement rates (approximately 2.1, dependent largely on child mortality).

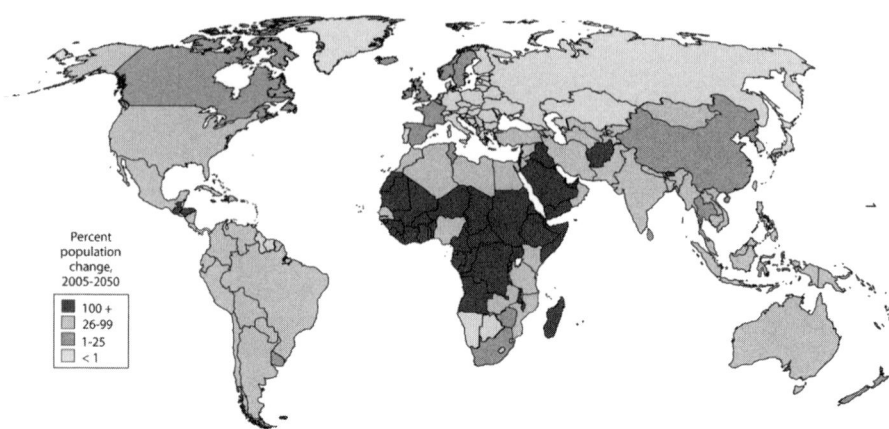

FIGURE 1-3. Percent population change, projected from 2005 to 2050. (*Source*: Population Reference Bureau, *2005 World Population Data Sheet*.)

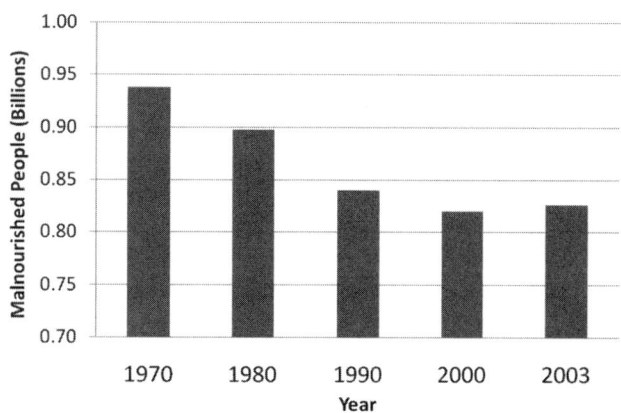

FIGURE 1-4. Proportion and total number of chronically malnourished people. (*Source*: UN Population Data Center.)

Changing Global Demographics

Declining fertility means that the median age of human populations will increase. The median age of human beings was 28 in 2009. That number will likely reach 38 in the next 40 years (UN, 2009). Current populations in less developed regions are very young, in large part due to the devastating impact of HIV/AIDS (UN, 2009). Almost 50 percent of people in less developed countries are less than 24 years old; almost 30 percent are less than 15 years old. In least developed countries,

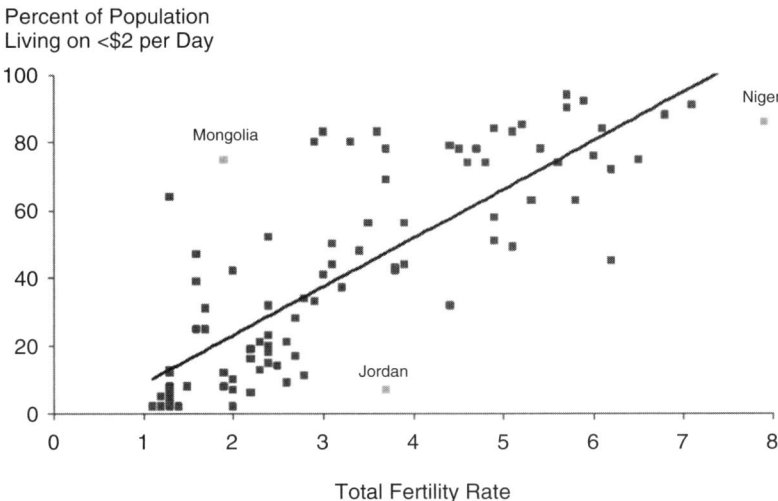

FIGURE 1-5. Relationship of poverty and fertility rates globally. (*Source*: Population Reference Bureau, *Population & Economic Development Linkages 2007 Data Sheet.*)

more than 40 percent are less than 15 years old, largely due to the ravages of HIV/AIDS on the mid-age (25-45 years old) population.

The global percentage of population over 60 will double by 2050. The number of people over 60 in less developed and least developed countries will exceed 1.6 billion. The number of elders on Earth is projected to pass the number of children in 2047 (UN, 2009). The implication for future economic prosperity is significant; economic growth and labor markets, family composition, living arrangements, childhood education support, health-care services, epidemiology, and almost every other facet of economic, social, and political domains will be affected by this age shift. The potential support ratio (PSR) is the ratio of people between the ages of 15 and 65 to those over 65 and represents the potential workers to support the aging sector. The PSR declined from 12 to 9 from 1950 to 2007; by 2050, the PSR will likely reach 4 (UN, 2009). The ability to produce, process, and distribute agricultural products using nonmechanized practices is dependent upon a vibrant and relatively young workforce. That workforce will be in short supply in 40 years, forcing a shift to labor-efficient and mechanized forms of production.

Immigration is a measure of the degree to which political and ecological pressures are disrupting economic and social communities. Population density is unevenly distributed relative to the ecological carrying capacity of a region. This inequity will be increased in the coming decades. The added population will emerge entirely in developing countries (Figure 1-3). Accounting for immigration, the populations of the more developed regions will increase from 1.23 to 1.28 billion (less than 5 percent); without immigration, this population would decrease almost 7 percent (UN, 2009). Currently, 1.5 billion people live in deep poverty (<$1 per day). This disparity motivates an increasing number of people to strive for an increase in their prosperity (security of food and water, security of person, security of opportunity, especially education). Migrations of large human populations across eco-political regions have been central to the human experience. However, increasing resource pressures are resulting in reduced tolerance and opportunity for those immigrants, leading to social and political strife that often results in systemic violence. During the period from 1990 to 2005, the number of international immigrants increased almost 25 percent (UN, 2006). The immigrant population in 2005 was 3 percent of the global population. One immigrant in four lived in North America, and one in three lived in Europe.

HUMAN-DOMINATED EARTH

Increasing Demands for Ecosystem Services

The increased demands for food, feed, fiber, and fuel of 9.25 billion people will be extraordinary, likely creating resource compression and supply chain constriction. Food production will need to increase by at least 50 percent in the next 40 years to meet imminent demand. This increase is not outside the reach of modern agricultural capacity. The increases in production during the beginning of the twenty-first century suggest that the trajectory of growth is in the appropriate scale to meet demand. However, the compounding challenges of competition between food and biofuels for land and water, as well as the uncertain impacts of climate change, make the development of strategies for meeting future demands more difficult to ensure.

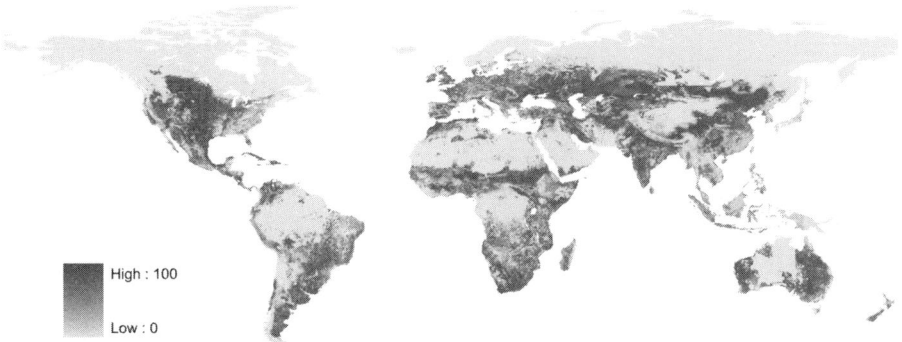

FIGURE 1-6. Global land cover impact from human activities in 2005. (*Source*: Modified from Foley et al., 2005.)

Human beings have changed the biomes of Earth. Agricultural land use (crop-, pasture-, and rangeland) occupies an estimated 40 percent of all of Earth's unfrozen surface area (Figure 1-6, from Foley et al., 2005). Earth's largest terrestrial biome is now agriculture. Human activities currently appropriate between 35 and 50 percent of primary productivity from the landscape (Vitousek, 1994). Ecosystem services are in decline globally (Millennium Ecosystem Assessment, 2005). More than 30 percent of ecosystem services are in serious decline, largely due to habitat loss associated with land use change. Land use pressures are driven by the market for products of primary productivity, especially food, biofuels, timber, and fiber (see Chapter 2).

Human Impacts through Urbanization

The United Nations estimated that 2008 was the threshold year when more than half of the human population (3.3 billion) resided in urban areas (UNFPA, 2007). The urban population has been increasing throughout the twentieth century, from 230 million to over 2.8 billion. This is just the beginning (Figure 1-7). The UN has declared that the

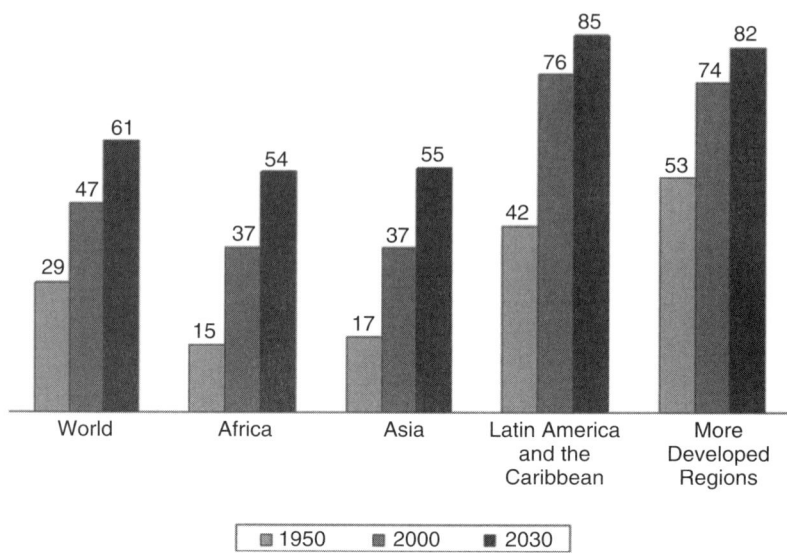

FIGURE 1-7. Global urbanization trends by region (bars represent percent of population). (*Source*: United Nations, *World Urbanization Prospects: The 2003 Revision* [medium scenario], 2004.)

twenty-first century will be the *dawn of the urban millennium*. As humanity transitions to an urban species, design of ecosystem services within the urban landscape becomes critical for our well-being, and perhaps our survival.

The impact of urban landscapes on global ecosystem services is only now being quantified. Evidence is mounting that urban landscapes are causing disruption of solar radiation (global dimming), creating heat island effects, and altering weather processes at unprecedented scales. Land-atmosphere-biosphere interactions cause mesoscale circulation anomalies that can result in large-scale weather pattern changes (Jin et al., 2005). Grimm et al. (2009) described the ecological impact of urbanization as "hot spots that drive environmental change at multiple scales." The landscape-scale impacts of urbanization include patch fragmentation and diversity, changes in hydrology, alterations of biogeochemical cycles, and ultimately reduced biodiversity (Grimm et al., 2009). The site impact of urbanization can best be characterized as paving; soils are encapsulated or otherwise damaged, hydrology is altered to eradicate infiltration and interflow, and plant communities are diminished or destroyed (Pauchard et al., 2006).

From a human perspective, urbanization represents a particularly difficult challenge because the impoverished urban populations are disenfranchised from the ecosystem that could sustain their biological needs. In addition, the loss of social networks, increased crime, and deterioration of social infrastructure for education, employment, and health care create a nightmare scenario for humanity in the twenty-first century (Moore et al., 2003). Integrated design of human habitats, especially urban landscapes, will be critical for enhancing the prosperity of humanity.

Land Use Change

Land use changes have historically occurred along a predictable transition line from forested to intensive agricultural production (Figure 1-8). Over 10 million square kilometers of forest biome have been destroyed by human activities through land use conversion (Foley et al., 2005). That process took centuries in pre-industrialized society. However, tropical forest land use change is now occurring at an increasing rate (Lepers et al., 2005). Forest land use change occurred predominantly in the tropics in the analysis period between 1990 and 2000, with deforestation hotspots concentrated in the Amazon basin and Southeast Asia (Figure 1-9, Lepers et al., 2005).

Dryland areas in Asia were also identified as hotspots of land use change. This land use change was associated with expansion

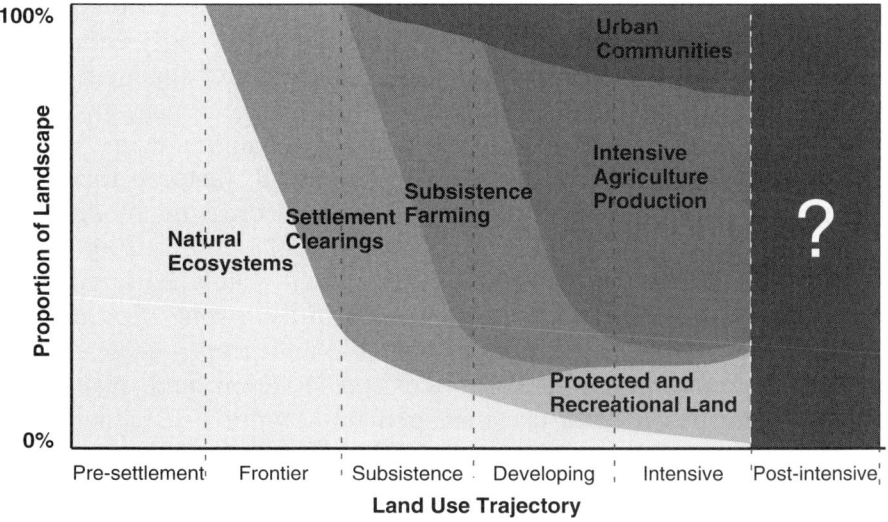

FIGURE 1-8. General transition of land use from human activities (*Source*: Modified from Foley et al., 2005.)

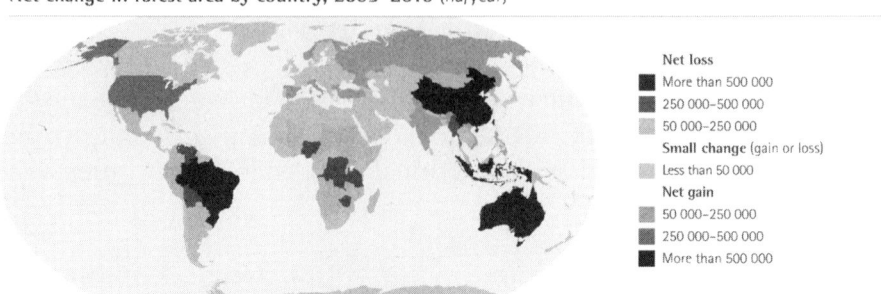

FIGURE 1-9. Food and Agriculture Organization of the United Nations. Global Forest Resources Assessment 2010. (*Source*: FAO. Global Forest Resources Assessment 2010: Main Report. Food And Agriculture Organization Forestry Paper 163, FAO, Rome, Italy. 2010.)

of croplands, which were noted across all continents (Lepers et al., 2005). Croplands increased most extensively during the last decade of the twentieth century in Southeast Asia, but also in Bangladesh, the Indus Valley, the Great Lakes region of Eastern Africa, and in the Amazon basin (Figure 1-9). Cropland area decreased in North America (predominantly the Southeastern U.S.) and Eastern China (Lepers et al., 2005).

Increased demand for agricultural products will result in increased pressure on land use change. Land use change is not the product of poverty; rather, it is the vehicle for prosperity for most economies dependent upon primary productivity. Land use change occurs in response to economic opportunities, with some constraints from governmental and other institutional factors (Lambin et al., 2001). These are global drivers, moderated by local factors. Land use change will likely have the largest effect in decreasing biodiversity for terrestrial ecosystems in this century (Sala et al., 2000). The impact of land use change on biodiversity will likely be larger than that of climate change. Human prosperity in the twenty-first century may come at the cost of infinitely valuable and irreplaceable species and ecosystems. Ecosystem services preservation and restoration must be priorities in land use management (Figure 1-10). Increasing productivity on current agricultural lands will reduce the pressures of land use change, and thus will conserve biodiversity.

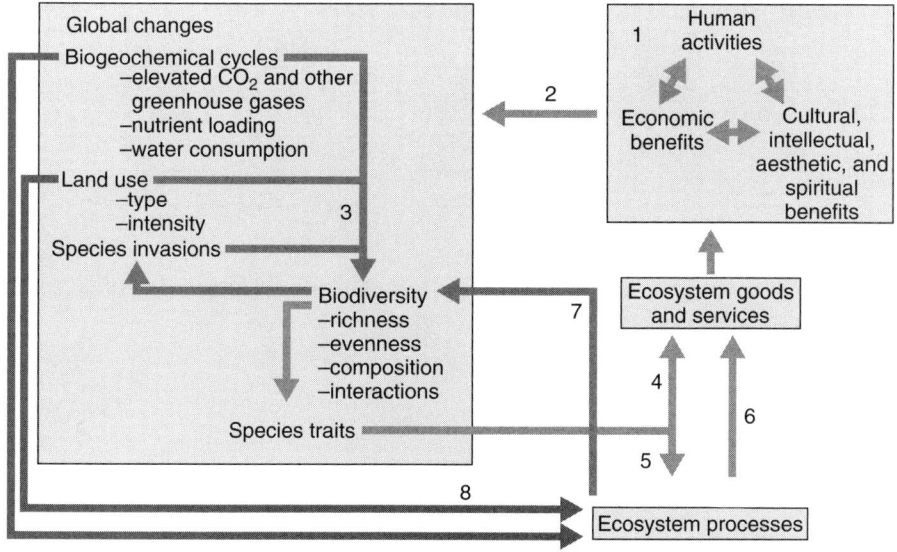

FIGURE 1-10. The consequences of land use change on ecosystem services. (*Source*: Chapin et al., 2000.)

Agricultural Production

Global food production has been increasing, both in terms of efficiency (kilograms produced per hectare) and effectiveness (proportion delivered to market). In 2004, global cereal production (aggregation of maize, rice, wheat, barley, sorghum, millets, oats, rye, triticale, buckwheat, fonio, and quinoa) was almost 9 percent higher than in 2000 (FAO, 2009). During that same period, meat production rose almost 11 percent to 260 million metric tonnes. Production of fruits and vegetables increased more than 14 percent to 1.38 billion metric tonnes. Fertilizer production increased 12 percent from 2000 to 2004, to almost 148 million metric tonnes globally. Overall per capita agricultural production was up 5.3 percent from 2000 to 2004, with cow milk as the highest-value commodity (FAO, 2009). Global annual yields for maize, wheat, and rice crops were at record highs in 2007 (Table 1-1). Global cereal production reached 2,274 million metric tonnes in 2008, the highest production rated in history. However, in 2009 global total grain production dropped three percent to 2,208 million metric tonnes,

TABLE 1-1 Global Maize, Wheat, and Rice Yields for 2007

Grain	Area Cultivated (ha)	Production (Mt)	Yield (Mt/ha)
Maize	142,331,335	637,444,480	3.41
Wheat	204,614,529	549,433,727	2.75
Rice	153,324,898	588,563,933	3.37
Total	500,270,762	1,775,442,140	—

Source: FAO 2009.

triggering concerns about the ability to continue to expand global food supplies (FAO, 2009).

Water Resource Demands

Water is the first need for human survival. An estimated 30 percent of people currently live in areas of chronic water stress (Vörösmarty et al., 2000; Millennium Ecosystem Assessment, 2005). Water resource demands have two major facets: water quality and water quantity. Water quality issues are predominantly pathogens and mineral (salt) content. The disease burden from water, sanitation, and hygiene issues is estimated to be 4 percent of all deaths and 5.7 percent of the total disease burden in disability-adjusted life years (DALYs) occurring worldwide (Table 1-2).

The rate at which water is cycled across the landscape affects the concentration of pollutants. As water resources become increasingly scarce, a given volume of water is cycled more frequently as it moves through the hydrologic cycle. Salinity of many water resources in arid systems is rising due to diversion of flows for irrigation and other uses (Reynolds et al., 2007). The irrigated water collects salts as it

TABLE 1-2 Annual Deaths and DALYs Attributed to Waterborne Infections

Disease	Deaths	DALYs
Schistosomiasis	14,000	1,932,000
Trachoma	0	1,239,000
Ascariasis	3,000	505,000
Trichuriasis	2,000	481,000
Hookworm disease	7,000	1,699,000
Total	26,000	5,856,000
Daily	71	16,044
Diarrhea	2,213,000	82,196,000
Daily Total:	6,063	225,195
Daily Childhood:	5,400	200,000

Source: Prüss et al. 2002, Green et al. 2009.

flows across tilled soils, evaporating (leaving more salts behind) and infiltrating into groundwater (Williams, 1999).

Water resource allocation will be an increasingly contentious issue in the twenty-first century. More than 70 percent of fresh water globally is appropriated for irrigation for agricultural production. In less developed countries, as much as 90 percent of freshwater resources are used for agriculture. Freshwater consumption worldwide has more than doubled since World War II and is expected to rise another 25 to 40 percent by 2030 (Foley et al., 2005). Hoekstra and Chapagain (2008) suggest that minimum water rights should be elevated to a human right to potable water before any other allocation is made. More than 2.5 billion people live in arid and semi-arid areas (mean annual rainfall between 25 and 500 mm); these regions will become increasingly stressed as populations increase and pressures on finite water resources continue to grow (Figure 1-11).

Meeting the water demands for 9.25 billion people will require allocation of fresh water for basic human consumption, agricultural production, biofuels production, municipal sanitation, industrial use, and other applications. There is a growing concern that humanity has passed peak water, or the point of maximum production/utilization of water resources (Gleick, 2009). Humanity uses 26 percent of evapotranspiration and 56 percent of accessible terrestrial runoff (Postel et al., 1996). Globally, 20 percent of freshwater fish are in danger of

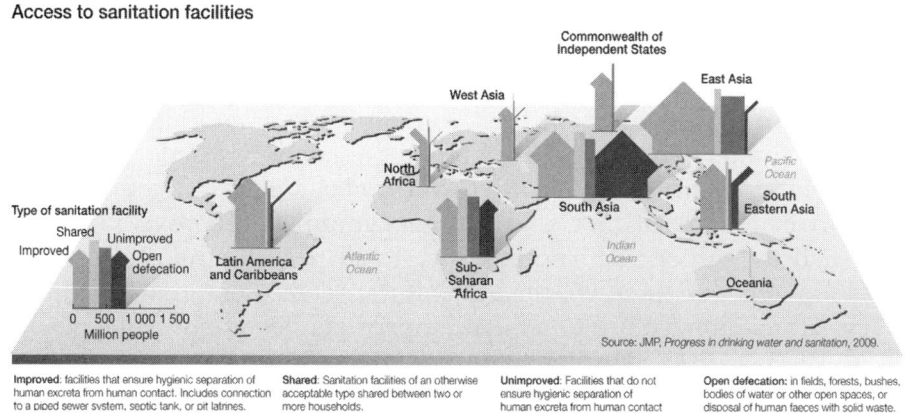

FIGURE 1-11. Geographic distribution of populations without access to clean water. (*Source*: Modified from UNEP/GRID-Arendal, 2010.)

extinction or are already extinct, and 47 percent of all listed endangered species in the U.S. are freshwater species (Jackson et al., 2001). Increasing water consumption will decrease biodiversity. There will be global demand for ecological engineering design solutions to water scarcity.

Lessons from the First Green Revolution

The range and scale of global resource challenges can be daunting. However, in order to maintain an optimistic perspective, we must remember where we came from and what we have accomplished. In 1950, an estimated 70 percent of the human population was chronically malnourished. Hunger was the human condition. Even in the United States, hunger was prevalent in urban and rural areas. The notion that Earth could produce enough food for 3 billion people was outside the realm of imagination for many world leaders. The dominant prosperity ethic was the lifeboat ethic: Only the strong or lucky survive—the rest are tossed overboard (violating the third axiom).

For over 5,000 years prior to 1930 farmers used beasts of burden (predominantly oxen and horses) to till the soil. Farms were relatively small, as a consequence, since a man and a team of oxen could not cultivate much more than 320 acres in a season. Mechanization changed the equation. With the advent of petrochemical-powered agricultural tillage and harvesting machinery, a single farmer could cultivate thousands of acres in a season. In addition, soil fertility and pest management improved dramatically, predominantly through exogenous fertilization and chemical pest control. Within a generation, the human condition in the United States and Europe changed.

The first Green Revolution (GR1) resulted in unprecedented increases in food production. Dr. Norman Borlaug and his colleagues working for the Ford and Rockefeller Foundations used genetically improved crops (predominantly dwarf wheat and rice), combined with irrigation, fertilizers, and mechanization, to improve yields in Asia and the Americas. Through integrated pest control, cultivation, and crop genetics research, the agricultural science community tripled wheat production from 2 to 6 metric tonnes per hectare over the period from 1960 to 2000 (IFPRI, 2002). The development and adoption of high-yield variety (HYV) rice, sorghum, millet, maize, cassava, and beans soon followed. Rice yields more than doubled globally since the introduction of the IR8 variety in 1966 (Cantrell and Hettel, 2004). The doubling of production of cereal grains was achieved with less than 5 percent increase in cultivated area (Tride, 1994).

A series of droughts in India in the early 1960s had threatened to plunge the country into mass starvation, but the increases in production beginning in 1966 ameliorated the impact of the drought. Rural economies flourished with increased production; per capita incomes in Asia doubled between 1970 and 1995 (IFPRI, 2002). The number of poor in rural India dropped from 65 percent in 1965 to 34 percent in 1993. Within 20 years of Norman Borlaug's acceptance of the Nobel Peace Prize for leading GR1, humanity was being fed. By 1990, only 16 percent of humanity was chronically malnourished. An army of committed men and women who spent their lives improving the prosperity of the poorest on the planet revolutionized the human condition in the middle of the twentieth century. They redesigned the twentieth century and put the whole of humanity on the path to prosperity, with the hope of basal prosperity for even the poorest.

Remember Axiom 1: Everything is connected. Prosperity is directly correlated with fertility (Figure 1-5) (FAO, 2009). Fertility rates are declining globally in large part because, as Dr. Norman Borlaug explained, human prosperity is directly correlated with population. "Development is the best contraceptive," Karan Singh, Indian ambassador to the U.S., said in 1970. Fertility rate declines that will result in a stable population in 2050 suggest that humanity can control its appetite and begin reducing its consumptive impact on natural resources.

The cost of this explosion in prosperity has been land use change, increased sediment and nutrient pollution, and chemical pollution from pesticides. Economies of scale resulting from twentieth-century markets for commodity crops have resulted in consolidation of farms, depopulation of rural lands, and loss of indigenous cultivars and production knowledge. These costs are serious, but the opportunities to reduce human suffering remain great, and the negative impacts can be ameliorated with appropriate design and implementation of ecosystem services.

STRUCTURE OF THIS BOOK

The practice of ecological design is by its nature sustainable design, though only a subset of that larger field. This book is structured to provide a narrative for approaching design of ecosystem services, with very explicit guidelines for major areas of practice. However, this is not a comprehensive work; it is unlikely that such a compilation is possible, given the breadth and scope of the discipline.

This book is organized into sixteen chapters. The first three chapters (including this one) contextualize the design process in ecological engineering. Following this chapter, on the challenges facing humanity, Chapter 2 provides a description of ecosystem services, and Chapter 3 offers a set of guidelines for practicing ecosystem design. The following four chapters provide definitions and characteristics of geographic and ecological scale. Chapters 8, 9, 10, and 11 provide a formal body of knowledge necessary for competent ecological design. Chapter 8 reviews the fundamentals of ecology and their relevancy to the design process. Chapter 9 defines mass and energy flows at multiple scales, and describes how they should be characterized in the design process. Chapter 10 provides an overview of community structure processes, and Chapter 11 describes the concept and practice of incorporating self-organization into ecological design.

Chapters 12 and 13 present a set of explicit design guidelines for non-built ecosystems. These include stream (Chapter 12) and landform (Chapter 13) ecosystem services design. Chapters 14, 15, and 16 provide a set of explicit design guidelines for built ecosystems. These chapters provide specific design elements for urban systems, including green infrastructure (Chapter 14), low impact development (Chapter 15), and agricultural systems (Chapter 16). Tools for measuring these processes are too dynamic to characterize in a printed textbook, and therefore are summarized in a supplemental web-based section titled "Ecological Engineering Tools" at www.wiley.com/ecoeng/.

> We shall never achieve harmony with land, any more than we shall achieve absolute justice or liberty for people. In these higher aspirations, the important thing is not to achieve but to strive.
>
> —Aldo Leopold

References
Berthelsen, J., Anatomy of a rice crisis, *Global Asia*, 3(2): 26–31, 2008.
Cantrell, R.P., Hettel, G.P., New challenges and technological opportunities for rice-based production systems for food security and poverty alleviation in Asia and the Pacific, FAO Rice Conference, *FAO*, Rome, Italy, February 12–13, 2004.
Chapin, F. Stuart III, Erika S. Zavaleta, Valerie T. Eviner, Rosamond L. Naylor, Peter M. Vitousek, Heather L. Reynolds, David U. Hooper, Sandra Lavorel, Osvaldo E. Sala, Sarah E. Hobbie, Michelle C. Mack, and Sandra Díaz, Consequences of changing biodiversity, *Nature*, 405: 234–242, May 11, 2000.

Cuello Nieto, C., Toward a holistic approach to the ideal of sustainability, *Philosophy and Technology*, 2(2): 41–48, 1997.

FAO, FAO Statistics Division: Crop Prospects and Food Situation, 2009, www.fao.org/docrep/012/ai484e/ai484e04.htm (accessed October 5, 2009).

Fernandes, S.D., N.M. Trautmann, D.G. Streets, C.A. Roden, and T.C. Bond, Global biofuel use, 1850–2000, *Global Biogeochemical Cycles*, 21(2), GB2019, 2007.

Foley, J.A., R. DeFries, G.P. Asner, C. Barford, G. Bonan, S.R. Carpenter, F.S. Chapin, M.T. Coe, G.C. Daily, H.K. Gibbs, J.H. Helkowski, T. Holloway, E.A. Howard, C.J. Kucharik, C. Monfreda, J.A. Patz, I.C. Prentice, N. Ramankutty, and P.K. Snyder., Global consequences of land use, *Science*, 309: 570–574, 2005.

Gleick, P., *The World's Water*, Island Press, New York, NY, 2009.

Green, Sean T., Mitchell J. Small, and Elizabeth A. Casman, Determinants of national diarrheal disease burden, *Environmental Science & Technology*, 43(4): 993–999, 2009.

Grimm, N., S. Faeth, N. Golubiewski, C. Redman, J. Wu, X. Bai, and J. Briggs, Global change and the ecology of cities, *Science*, 319(5864): 756–760, 2009.

Hoekstra, A., and A. Chapagain, *Globalization of Water*, Blackwell Publishing, Oxford, UK, 2008.

IFPRI, Achieving sustainable food security for all by 2020: Priorities and responsibilities. International Food Policy Research Institute, Washington, DC, 2002.

Jackson, Robert B., Stephen R. Carpenter, Clifford N. Dahm, Diane M. McKnight, Robert J. Naiman, Sandra L. Postel, and Steven W. Running, Water in a changing world, *Ecological Applications*, 11(4): 1027–1045, 2001.

Jin, M., R. Dickinson, and D. Zhang, The footprint of urban areas on global climate as characterized by MODIS. *Journal of Climate*, 18: 1551–1565, 2005.

Kangas, P. *Ecological Engineering: Principles and Practice*. Lewis Publishers, New York, NY, 2004.

Kareiva, P., S. Watts, R. McDonald, and T. Boucher, Domesticated nature: Shaping landscapes and ecosystems for human welfare, *Science*, 316: 1866–1869, 2007.

Lambin, E.F., et al., The causes of land-use and land-cover change: Moving beyond the myths, *Global Environmental Change*, 11(4): 261–269, 2001.

Lepers, E., E. Lambin, A. Janetos, R. DeFries, F. Achard, N. Ramankutty, and R. Scholes, A synthesis of information on rapid land cover change for the period 1981–2000, *Bioscience*, 55(2): 115–124, 2005.

Millennium Ecosystem Assessment, Ecosystems and Human Well-Being: Synthesis, Island Press, Washington, DC, 2005, www.millenniumassessment.org/documents/document.300.aspx.pdf.

Mitsch, W. and S. Jorgensen. *Ecological Engineering and Ecosystem Restoration*. John Wiley and Sons, New York, NY, 2004.

Moore, M., P. Gould, and B. Keary, Global urbanization and impact on health, *International Journal of Hygiene and Environmental Health*, 206: 269–278, 2003.

Odum, H.T., Self-Organization, transformity, and information, *Science*, 242(4882): 1132, 1998.

Pauchard, A., M. Aguayo, E. Peña, and R. Urrutia, Multiple effects of urbanization on the biodiversity of developing countries: The case of a fast-growing metropolitan area (Concepción, Chile), *Biological Conservation*, 127: 272–281, 2006.

Postel, Sandra L., Gretchen C. Daily, and Paul R. Ehrlich, Human appropriation of renewable fresh water. *Science*, 271(5250): 785, 1996.

Prüss, Annette, David Kay, Lorna Fewtrell, and Jamie Bartram, Estimating the burden of disease from water, sanitation, and hygiene at a global level, *Environmental Health Perspectives*, 110(5), 2002.

Reynolds, James F., D. Mark Stafford Smith, Eric F. Lambin, B.L. Turner II, Michael Mortimore, Simon P.J. Batterbury, Thomas E. Downing, Hadi Dowlatabadi, Roberto J. Fernández, Jeffrey E. Herrick, Elisabeth Huber-Sannwald, Hong Jiang, Rik Leemans, Tim Lynam, Fernando T. Maestre, Miguel Ayarza, and Brian Walker, Global desertification: Building a science for dryland development, *Science*, 316(5826): 847, 2007.

Sala, Osvaldo E.F. Stuart Chapin III, Juan J. Armesto, Eric Berlow, Janine Bloomfield, Rodolfo Dirzo, Elisabeth Huber-Sannwald, Laura F. Huenneke, Robert B. Jackson, Ann Kinzig, Rik Leemans, David M. Lodge, Harold A. Mooney, Martín Oesterheld, N. LeRoy Poff, Martin T. Sykes, Brian H. Walker, Marilyn Walker, and Diana H. Wall, Global biodiversity scenarios for the year 2100, *Science*, 287(5459): 1770, 2000.

Tride, D., *Feeding and Greening the World*, CAB International, Wollingford, UK, 1994.

Turner, B.L. II, W.C. Clark, R.W. Kates, J.F. Richards, J.T. Mathews, and W.B. Meyer, *The Earth as Transformed by Human Action: Global and Regional Changes in the Biosphere over the Past 300 Years*, Cambridge University Press with Clark University, Cambridge, UK; New York, NY, 1990.

UN, International Migration Report 2006: A Global Assessment, United Nations Department of Economic and Social Affairs/Population Division 1, New York, NY, 2006.

UN, United Nations Millennium Development Goals, United Nations, New York, NY, 2007, www.un.org/millenniumgoals/environ.shtml.

UN, United Nations Population Prospects, United Nations Department of Economic and Social Affairs/Population Division, New York, NY, 2009, http://esa.un.org/unpp/.

UNEP/GRID-Arendal. Access to sanitation facilities. UNEP/GRID-Arendal Maps and Graphics Library. 2010 March. Available at: http://maps.grida.no/go/graphic/access-to-sanitation-facilities. Accessed January 16, 2011.

UNFPA, State of the World Population, United Nations Population Fund, New York, NY, 2007, www.unfpa.org.

Vitousek, P.M., Beyond global warming: Ecology and global change. *Ecology*, 75(7): 1861–1876, 1994.

Vörösmarty, Charles J., Pamela Green, Joseph Salisbury, and Richard B. Lammers, Global water resources: Vulnerability from climate change and population growth, *Science*, 289(5477): 284, 2000.

WCED, *Our Common Future*, Oxford University Press, Oxford, UK, 1987.

Williams W.D., Salinisation: A major threat to water resources in the arid and semi-arid regions of the world, *Lakes & Reservoirs: Research and Management*, 4(3-4): 85–91, 1999.

2

Ecosystem Services

> Paradise never did exist, at least not in human experience.
> —Robert Ricklefs, *Economy of Nature*

INTRODUCTION

Ecological engineers design ecosystem services. Ecosystem services are the goods, functions, and processes we derive from the biosphere (Costanza et al., 1997). Ecosystem services are the flow of energy, materials, and information from natural capital, or the stock of materials or information that exists at a point in time and space (Costanza et al., 1997). Obviously, ecological services did not originate with ecosystem engineers; they are the cumulative product of over 6 billion years of geomorphologic and evolutionary adaptation to an ever-changing landscape. Therefore, ecological engineers are designing systems that replicate, restore, or at least mimic ecosystems that are much more complex than we will likely ever understand. The best we can hope for is to design systems that enhance ecosystem services beyond current design and management practices. Ecological engineers must be consummate observers, adaptors, and inventors.

The design of ecosystem services demands an understanding of the connectedness and interaction among ecosystem structures, functions, and landforms. This chapter will introduce the most commonly used constructs of ecosystem services, including those of Daily et al. (1997), Costanza et al. (1997), and the Millennium Ecosystem Assessment report. The chapter then provides a review of the status of the categories globally, with a focus on their relationships to terrestrial and aquatic morphology. Finally, guidelines for designing ecosystem services are introduced that will frame the remainder of the book.

Origin of Ecosystem Services

The concept of ecosystem services is rooted in the emerging understanding of the role of humanity in the vast interconnected community of life on Earth. Plato observed the impact of human activity on the landscape, describing loss of arable lands, soils, forests, and even bees (Daily, 1997). George P. Marsh synthesized the observations of a lifetime in human-dominated Europe in the mid-nineteenth century in *Man and Nature* (1864). Marsh wrote in this landmark book:

> ...to indicate the character and, approximately, the extent of the changes produced by human action in the physical conditions of the globe we inhabit; to point out the dangers of imprudence and the necessity of caution in all operations which, on a large scale, interfere with the spontaneous arrangements of the organic or the inorganic world; to suggest the possibility and the importance of the restoration of disturbed harmonies and the material improvement of waste and exhausted regions; and, incidentally, to illustrate the doctrine, that man is, in both kind and degree, a power of a higher order than any of the other forms of animated life, which, like him, ar (sic) nourished at the table of bounteous nature.

While the vestiges of Man the Conqueror still resonated in the pages, Marsh observed with prescient clarity the interrelationship of society with the land. Marsh, the U.S. ambassador to Italy from 1861 until his death in 1882, observed the change in fertility and productivity of the Italian countryside from the "Empire of the Caesars" to his present day:

> The decay of these once flourishing countries is partly due, no doubt, to that class of geological causes, whose action we can neither resist nor guide, and partly also to the direct violence of hostile human force; but it is, in a far greater proportion, either the result of man's ignorant disregard of the laws of nature, or an incidental consequence of war, and of civil and ecclesiastical tyranny and misrule.

Ecological engineering brings these "laws of nature" into the design process. No more can human "ignorant disregard of the laws of nature" be accepted as normal practice in designing Earth. At the beginning of the twenty-first century, the role of humankind in dominating Earth has reached a crescendo.

The notion that the human footprint on Earth is becoming too large to be sustainable is now commonly accepted (Sanderson et al., 2002). The founders of the modern conservation movement, characterized by President Theodore Roosevelt, Gifford Pinchot, and John Muir, certainly understood the interdependence of humanity and the larger ecosystem. This understanding was further articulated by Aldo Leopold, Rachel Carson, Edward Abbey, and E. O. Wilson (see Recommended Readings). However, the concept of ecosystem services began to be applied as a formal sustainability construct in 1997 with Costanza et al.'s "The value of the world's ecosystem services and natural capital" (Costanza et al., 1997), and Gretchen Daily's edited volume *Nature's Services: Societal Dependence on Natural Ecosystems* (Daily, 1997).

The Value of Ecosystem Services

Costanza et al. (1997) estimated the value of ecosystem services for 16 biomes, using current economic value methods, to be on average $33 trillion per year, with a global gross national product of $18 trillion per year. This work was seminal in establishing the debate on ecosystem services valuation, because it created a debate on valuing nonmarket assets that has been critical in driving innovation in market-based ecosystem services solutions.

Many economists point out that this value is nonsensical because it was 1.8 times the gross world product (GWP) in 1997. The GWP in 2008 was $62.25 trillion, illustrating the difficulty of assessing the market value of nonmarket products. Costanza et al. (1997) argued that only a fraction of ecosystem services is accounted for in private goods traded in existing markets. They further acknowledged that there is "no necessary relationship between the valuation of natural capital service flows, even on the margin, and aggregate spending, or GNP, in the economy." That, in fact, is why ecosystem services, managed as a nonmarket common resource (a resource held in the public good, managed by an agent of the public, usually a government), are in tragic decline globally.

Ecosystem services such as gas regulation, disturbance regulation, waste treatment, and nutrient cycling represented almost 75 percent of total value; while a carbon market for gas regulation is emerging, the ability of markets to mediate global processes is still an open question. According to Costanza et al. (1997), the most important biomes in providing ecosystem services were marine coastal systems (30 percent of total); wetlands provided 12 percent of ecosystem service value, by far the densest provider of ecosystem services per unit area.

Classifying Ecosystem Services

Ecosystem services have been characterized using different taxonomies, all of them arbitrary to a certain extent. Ecosystems are continuums, so segregation is largely just a function of definition and context. This book utilizes the taxonomy of the Millennium Ecosystem Assessment (Table 2-1) for most applications; however, this

TABLE 2-1 Ecosystem Services Categories and Status

Category of Ecosystem Services		Subcategory	Status
Provisioning	Food	Crops	Substantial increases
		Livestock	Substantial increases
		Capture Fisheries	Declining due to overharvest
		Aquaculture	Substantial increases
		Wild Foods	Declining due to overharvest
	Fiber	Timber	Regionally variable, globally neutral
		Cotton, Hemp, Silk	Regionally variable, globally neutral
		Wood Fuel	Declining due to overharvest
	Genetic Resources		Lost through extinction
	Biochemical and Pharmaceuticals		Lost through extinction
	Fresh Water		Overextracted for agriculture, industry, municipal use
Regulating	Air Quality		Declining globally
	Climate	Global	Energy increasing due to carbon increasing in atmosphere
		Regional	Declining due to multiple impacts
	Water Flow		Regionally variable, largely correlated with land use change
	Erosion		Declining due to increased vegetative cover disturbance
	Water Treatment		Declining due to altered hydrologic regimes
	Disease		Regionally variable, largely correlated with ecosystem change
	Pest		Declining due to pesticide use
	Pollination		Declining due to chemical pesticide use and ecosystem change
	Natural Hazard		Declining due to loss of natural buffers, especially coastal and riparian zones
Supporting	Soil Formation	These services are not directly used by human beings	Decreasing due to erosion, salinization
	Photosynthesis		
	Primary Production		Increasing, especially in forests and cultivated systems
	Nutrient Cycling		Increasing due to increased N, P, K and other nutrients in the biosphere
	Water Cycling		

(*continues*)

TABLE 2-1 (*continued*)

Category of Ecosystem Services	Subcategory		Status
Cultural	Spiritual and Religious Values		Rapid decline in sacred places due to encroachment
	Aesthetic Values		Declining in quantity and quality of nonhuman-dominated landscapes
	Recreation and Ecotourism		Regionally variable, more people accessing more spaces but those spaces degraded
	Education		

Source: MEA report, UN, 2005.

taxonomy was developed for global-scale assessments. For site assessments, a more spatially explicit taxonomy is necessary.

It is informative to review the different classification schemes at a global scale in order to appreciate the evolution of thinking regarding ecosystem services. Costanza et al. (1997) classified the entire inventory of ecosystem services in the biosphere into 17 categories (Table 2-2). These categories were based largely on the criteria for valuation applied by the research team, and thus were biased towards market-similar services. This publication served the critical function of framing a discourse on ecosystem services valuation, and remains to this day the landmark work for ecosystem services valuation.

Since 1995, much progress has been made in refining classes of ecosystem services. de Groot et al. (2002) sorted 23 ecosystem functions into four functional categories (Table 2-3):

1. *Regulation functions*: The capacity to regulate essential ecological processes and life support systems through biogeochemical cycles and other biospheric processes.
2. *Habitat functions*: Providing refuge and reproduction habitat for wild plants and animals, thus contributing to the (in situ) conservation of biological and genetic diversity and evolutionary processes.
3. *Production functions*: Photosynthesis and nutrient uptake by autotrophs converts energy, carbon dioxide, water, and nutrients into a wide variety of carbohydrate structures that are then used by secondary producers to create an even larger variety of living biomass.
4. *Information functions*: An essential "reference function" that contributes to the maintenance of human health by providing opportunities for reflection, spiritual enrichment, cognitive development, recreation, and aesthetic experience.

TABLE 2-2 Ecosystem Services Categories and Functions According to Costanza et al. 1997

Number	Ecosystem Service	Ecosystem Functions	Examples
1	Gas regulation	Regulation of atmospheric chemical composition	CO_2/O_2 balance, O_2 for UVB protection, and SO_2 levels
2	Climate regulation	Regulation of global temperature, precipitation, and other biologically mediated climatic processes at global or local levels	Greenhouse gas regulation, DMS production affecting cloud formation
3	Disturbance regulation	Capacitance, damping, and integrity of ecosystem response to environmental fluctuations	Storm protection, flood control, drought recovery, and other aspects of habitat response to environmental variability mainly controlled by vegetation structure
4	Water regulation	Regulation of hydrological flows	Provisioning of water for agricultural (such as irrigation) or industrial (such as milling) processes or transportation
5	Water supply	Storage and retention of water	Provisioning of water by watersheds, reservoirs, and aquifers
6	Erosion control and sediment retention	Retention of soil within an ecosystem	Prevention of loss of soil by wind, runoff, or other removal processes, storage of silt in lakes and wetlands
7	Soil formation	Soil formation processes	Weathering of rock and the accumulation of organic material
8	Nutrient cycling	Storage, internal cycling, processing, and acquisition of nutrients	Nitrogen fixation, N, P, and other elemental or nutrient cycles
9	Waste treatment	Recovery of mobile nutrients and removal or breakdown of excess or xenic nutrients and compounds	Waste treatment, pollution control, detoxification
10	Pollination	Movement of floral gametes	Provisioning of pollinators for the reproduction of plant populations
11	Biological control	Trophic-dynamic regulations of populations	Keystone predator control of prey species, reduction of herbivory by top predators

(*continues*)

TABLE 2-2 (*continued*)

Number	Ecosystem Service	Ecosystem Functions	Examples
12	Refugia	Habitat for resident and transient populations	Nurseries, habitat for migratory species, regional habitats for locally harvested species, or overwintering grounds
13	Food production	That portion of gross primary production extractable as food	Production of fish, game, crops, nuts, fruits by hunting, gathering, subsistence farming or fishing
14	Raw materials	That portion of gross primary production extractable as raw materials	The production of lumber, fuel, or fodder
15	Genetic resources	Sources of unique biological materials and products	Medicine, products for materials science, genes for resistance to plant pathogens and crop pests, ornamental species (pets and horticultural varieties of plants)
16	Recreation	Providing opportunities for recreational activities	Ecotourism, sport fishing, and other outdoor recreational activities
17	Cultural	Providing opportunities for non-commercial uses	Aesthetic, artistic, educational, spiritual, and/or scientific values of ecosystems

The first two categories, regulating and habitat functions, are the foundation of the other two. However, markets rarely value the ecosystem services in those functions; they are generally considered commons, and thus protected by institutional law, if at all.

The Millennium Ecosystem Assessment

In 2000 Kofi Annan, then secretary general of the UN, called for the Millennium Ecosystem Assessment in a report to the UN General Assembly titled *We the Peoples: The Role of the United Nations in the 21st Century* (UN, 2005). The Millennium Ecosystem Assessment (MEA) began in 2001, concluded in 2005, and was performed by more than 1,500 scientists around the globe. The objective of the MEA was "to assess the consequences of ecosystem change for human well-being and to establish the scientific basis for actions needed to

TABLE 2-3 Typology of Ecosystem Services by Function Class

Regulating Functions	Production Functions
1 Gas regulation	14 Food
2 Climate regulation	15 Raw materials
3 Disturbance regulation	16 Genetic resources
4 Water regulation	17 Medicinal resources
5 Water supply	18 Ornamental
6 Soil retention	
7 Soil formation	
8 Nutrient regulation	
9 Waste treatment	
10 Pollination	
11 Biological control	
Habitat Functions	**Information Functions**
12 Refugium function	19 Aesthetic information
13 Nursery function	20 Recreation
	21 Cultural and artistic information
	22 Spiritual and historic
	23 Science and education

Source: deGroot et al. 2002.

enhance the conservation and sustainable use of ecosystems and their contributions to human well-being" (UN, 2005). The central premise of the MEA is that all the constituents of well-being, including security, basic materials for prosperity (good life), health, good social relations, and even freedom of choice and action, are directly dependent upon ecosystem services (Figure 2-1). These ecosystem services and the functions that drive them are dependent in turn upon biodiversity, defined roughly as the number, relative abundance, composition, and interactions of species within a geographic area (Figure 2-2).

The MEA defined four groups of ecosystem services with subordinate categories: Provisioning, Regulating, Supporting, and Cultural Services. The MEA framework represents the international consensus on ecosystem services taxonomy, and is the framework for ecosystem services adopted for this book. The subcategories of services within each category are similar to the 17 categories described by Costanza et al. (1997). The MEA evaluated the status of each subcategory (Table 2-4). The MEA divided the biosphere into six major biomes; in Chapter 4 we use the World Wildlife Fund (WWF) framework of 14 terrestrial biomes. The results were disconcerting; over

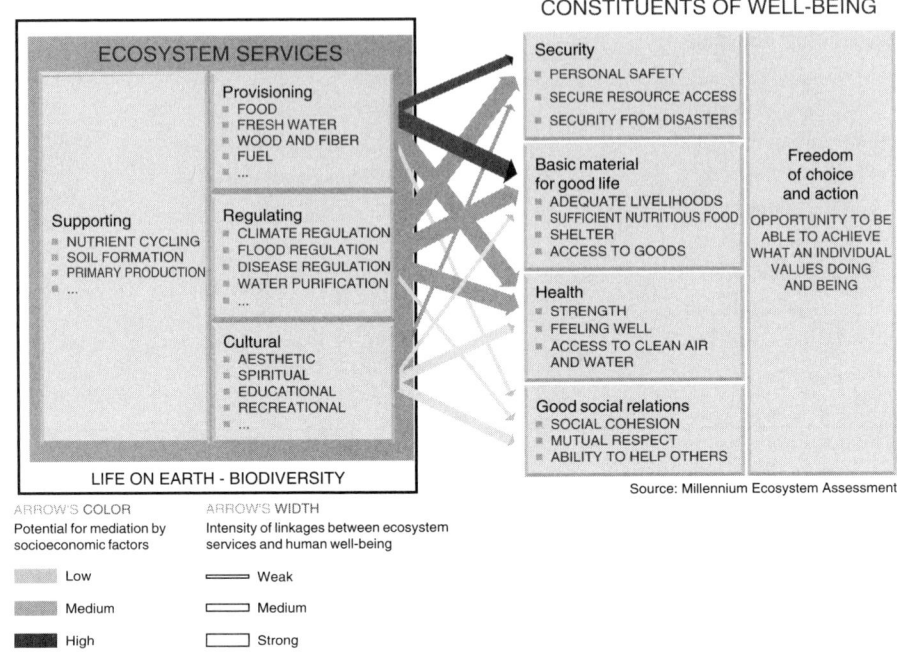

FIGURE 2-1. Relationship between constituents of well-being and ecosystem services. (*Source*: MEA report, UN, 2005.)

the past 50 years, human beings have changed Earth's ecosystem in unprecedented ways. The major findings were:

- More land was converted to cropland in the 30 years after 1950 than in the 150 years between 1700 and 1850.
- 20 percent of the world's coral reefs were lost and 20 percent degraded in the last several decades.
- 35 percent of mangrove area has been lost in the last several decades.
- The amount of water in reservoirs quadrupled since 1960.
- Withdrawals from rivers and lakes doubled since 1960.
- 5–10 percent of the area of five of the six major biomes was converted between 1950 and 1990.
- For the six major biomes, more than two-thirds of the area of two biomes and more than half of the area of four others had been converted to cropland by 1990.
- Since 1960, flows of biologically available nitrogen in terrestrial ecosystems doubled.
- Since 1960, flows of biologically available phosphorus tripled.
- The distribution of species on Earth is becoming more homogenous.

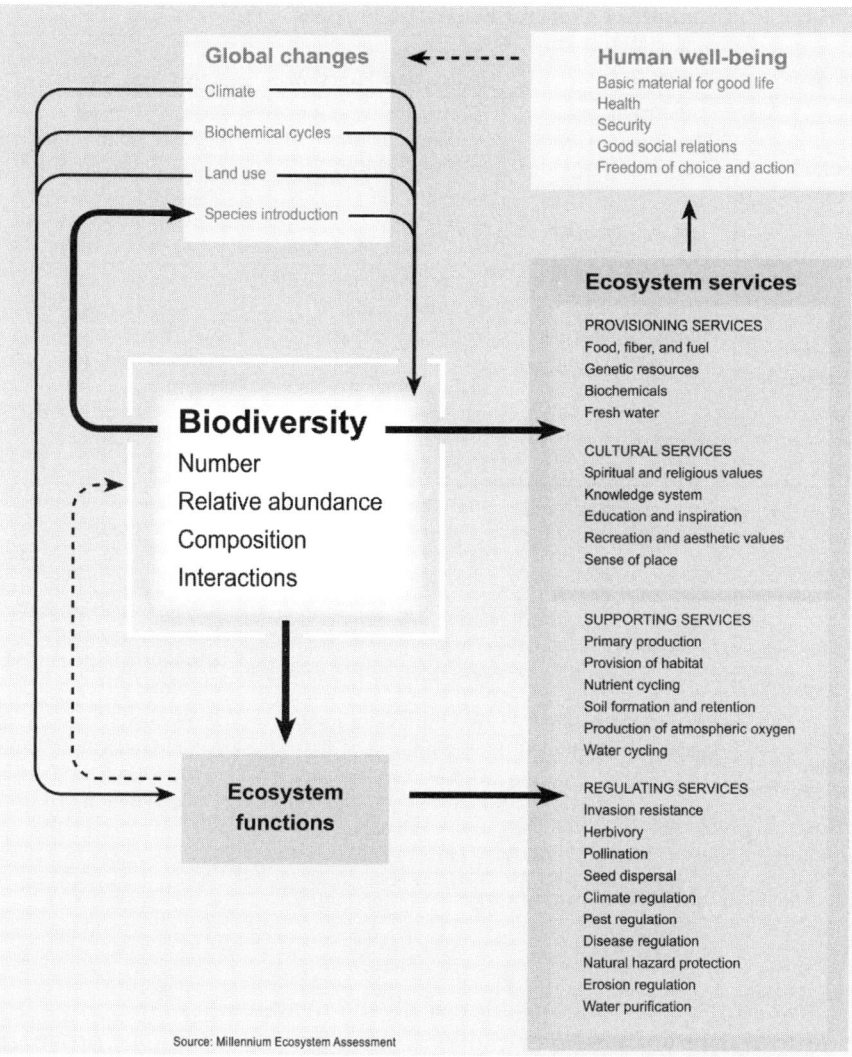

FIGURE 2-2. Impact of biodiversity on ecosystem services. (*Source*: MEA report, UN, 2005.)

- The population size or range (or both) of the majority of species across a range of taxonomic groups is declining.
- Human beings have increased the species extinction rate by as much as 1,000 times over background rates typical over the planet's history (*medium certainty*).
- 10–30 percent of mammal, bird, and amphibian species are currently threatened with extinction (*medium to high certainty*).

TABLE 2-4 Examples of Cultural Ecosystem Services

Biome	Cultural Landscapes	Examples of Ecosystem Functions	Ecosystem State
Humid Tropical	Salina landscape (Densu Delta, Ghana)	Habitat for thousands of wetland species Fishing (primary economy for 20 communities) Salt industry	Ramsar wetland: 6,700 hectares tidal influences extend upstream for some 10 km heavily populated with urban estate development
Humid Temperate	Hedgerow landscapes (e.g., France, United Kingdom, Germany)	Protection against soil erosion Wood production Grassland farming Habitat/corridors for native species Natural pest control Recreation	Regionally distinctive types, regarding patterns, plant compositions, materials, and management Threats: agricultural intensification and abandonment
Semi-arid Boreal	Prairie pothole landscape (Canada)	Farmland Hunting ("duck factory") Biodiversity	Mosaic of 4 million small wetlands; 51 percent of all North American breeding ducks Threats: agricultural activities (pesticides, nutrients)
Warm Desert	Farm-based wildlife landscapes (Namibia)	Wildlife-based rural development Biodiversity (including elephant and endangered black rhinoceros) Tourism	More than 75 percent of African wildlife is found in these landscapes Threats: hunting

Source: Modified from MEA Report, UN 2005.

These impacts are the product of all the changes human beings have made to the biosphere over the past 15,000 years. Human prosperity derives directly from the land. The prosperity of GR1 can be maintained and enhanced if we can design ecosystem services explicitly rather than continue the wanton exploitation and destruction of the land upon which we depend.

Of the 24 ecosystem services evaluated, 60 percent (14 services) were degraded or being used unsustainably (UN, 2005). The warnings are clear, yet very little concerted action is being taken to address these losses. A review of the specific losses by class of ecosystem service illustrates the inequities of the consequences of loss of those services.

Provisioning Services. Provisioning services are the products obtained from ecosystems, including food, fiber, feed, fuel (wood and biofuels), genetic resources, biochemicals, ornamental resources, and fresh water. These ecosystem services are the foundation of most economies of the world and have been explicitly designed for over 10,000 years. The conversions of land use and increased use of nutrients and water on the landscape have dramatically improved the human condition. Never in human history have so many people lived so well. The labor force for agricultural production represents 22 percent of the world's population; considering age demographics, that is half the world's total labor force. Think about that—half of all working people around the globe are engaged in agricultural production. Agricultural production is a critical resource for prosperity beyond nourishment; agriculture accounts for 24 percent of GDP in low-income developing countries (UN, 2005). The MEA estimated market value for the provisioning services globally:

- Food production: $980 billion per year
- Timber industry: $400 billion per year
- Marine fisheries: $80 billion per year
- Marine aquaculture: $57 billion per year
- Recreational hunting and fishing: >$75 billion per year in the United States alone

While all of humanity depends on these provisioning services, the well-being of the poorest among us is directly tied to their viability. According to the MEA, approximately 50 percent (6 out of 12) of the provisioning services evaluated are being degraded or used unsustainably (Table 2-1). This loss of ecosystem services threatens significant harm to human well-being and could result in near-term loss of wealth in those countries that can least afford it (UN, 2005).

Regulating Services. Regulating services are the benefits obtained from the regulation of ecosystem processes, and include regulation of air quality, climate, water, erosion, water purification, disease, pests, pollination, and natural hazards. The ability of ecosystems to regulate processes that affect human affairs is perhaps the most undervalued of ecosystem services classes, and next to provisioning services the most designable. Invasive species resistance, or the ability of indigenous communities to out-compete invading species, is critical for maintaining provisioning services and disease control. However, this service is a product of species composition, arrangement of landscape units, and species richness and diversity. Loss of these elements reduces ecosystem resilience to pressures from invasive species. Pollination is a critical regulating service that depends on the functional composition of the pollinator assemblage. Species richness, density, and distribution across the landscape affect the process of pollination. Loss of pollinator species through habitat destruction, nonspecific pesticide use, and invasive disease results in decreased pollination effectiveness, and thus loss of fruit production. Similarly, climate regulation and carbon sequestration are directly affected by the genetic diversity of plants on the landscape. Pest and disease control is governed by genetic diversity of crops, richness of plant, invertebrate, and vertebrate species in an area, and spatial distribution of landscape units to control and reduce disease vectors (UN, 2005).

Supporting Services. Supporting services are those that are necessary for production of all other ecosystem services. These include soil formation, photosynthesis (net oxygen generation), primary production (net carbon fixation), nutrient cycling, and water cycling. The global or even regional extent of these supporting services is so integrated with the landscape that they are generally outside of the direct control, and therefore design, of human beings. They occur at temporal or spatial scales beyond our ability to manage. However, at the site level, the elements of supporting services are within the control of ecological designers. The amount of photosynthesis of a landscape is dependent upon plant density and species richness, which are explicitly designable. Stability of primary production is dependent upon genetic diversity, species richness, and functional composition of the plant assemblage. The provisioning of habitat is directly correlated with habitat diversity (spatial distribution, size, and shape of landscape units), functional composition of vegetation, and species richness. Nutrient cycling across the landscape is a function of elements within the flow path of those cycles; phosphorus from agricultural lands moves across the surface bound to sediment in runoff, and accumulates in the riparian wetlands and woodlands near a stream, thus contributing to photosynthesis and primary production in those systems. Removing the wetlands and woodlands leads to phosphorus loading directly to streams, increasing eutrophication. These are designable elements on the landscape.

Cultural Services. Cultural services are often thought of in the context of indigenous and tribal peoples. However, all cultures are directly and indirectly influenced by, and often defined by, ecosystem services. The range of cultural services includes identity, heritage, spirituality, inspiration, aesthetics, and recreation and tourism (MEA, 2005). Cultural identity is the connection with place. All cultures are the product of place. Place is characterized by the interaction of human and other species within ecosystems. Cultures in coastal villages are often defined by fisheries, Midwestern U.S. communities by farming. The cultures that evolved in these places contained human generational knowledge about living within those ecosystems, including risk management and sustainable production. The loss of this knowledge through estrangement from place has had consequences throughout the world, and may be the leading cause of overexploitation and loss of ecosystem services.

Cultural heritage derives from the unique features of ecosystems that connect us to our history and ancestry. These often include geographic features such as mountains or bays, biological features such as a grove of trees, or seasonal features such as the harvest season, or the emergence of tree frogs or wild onions. We celebrate these events and places through our religious and civic holidays. We celebrate these places through our iconic imagery: Mount Fuji in Japan, Kilimanjaro in Tanzania; the deserts of North America, North Africa, and Australia; coral reefs of Hawaii and Florida. More locally, we define our economies by these cultural ecosystem services (Table 2-4).

Spiritual ecosystem services are found through the sacred places we observe, and reflected in the sacred places we build. Mountains have often been associated with spiritual power; however, groves of trees are equally often represented as places to connect to deities (Hughes and Chandran, 1998). Similarly, places provide inspiration to humanity through verbal art, writing, performing arts, fine arts, design and fashion, and electronic media (UN, 2005; Chapter 16). Aesthetic services are reflected in people's preference for nonbuilt over built environments, preferences for savanna (Figure 1-6) ecosystems as landforms, and preferences for wild over cultivated landscapes (Kaplan and Kaplan, 1989).

Activities such as bird watching, camping, fishing, hunting, swimming, surfing, hiking, canoeing, and the whole range of outdoor recreational activities we engage in are derived from ecosystem services. Recreational tourism, often referred to as ecotourism, has become very profitable in the early twenty-first century. People are visiting places around the world to experience novel cultures, ecosystems, and rituals in increasing numbers; global ecotourism has more than doubled since 1990, accounting for more than 20 percent of international travel (UN, 2005).

Why Biodiversity Matters

Biodiversity is the underlying metric of ecosystem health (Figure 2-2). The suite of ecosystem services provided by a place—which may be as small as a meter of soil or as large as a biome—are all provided by the interactions of the abiotic components of that place with the number of species (diversity) and number of organisms within a species in the place (populations) (Luck et al., 2003). The number of species alone is not an indicator of ecosystem services density, however (O'Connor and Crowe, 2005). In fact, there are clearly key species serving pivotal roles in ecosystem functions. When those species are removed or reduced in density in a place, other species may compensate for their role, but these compensations often result in shifts of other processes, and ultimately shifts in ecosystem function (services). These "idiosyncratic" effects underscore the need to understand the context of ecosystem services, and to better understand the interplay between species being managed and processes being designed (O'Connor and Crowe, 2005).

There are numerous metrics of biodiversity, but all are based on the four metrics of population diversity: richness, size, spatial distribution, and genetic differentiation. Species richness is the number of populations within an area; for this metric, landform should be the unit of assessment (see Chapter 10). Population size is the number of individuals of each population within an area. These are often not directly measurable, but estimates to within an order of magnitude should be made, using observation surveys and other population demographic tools. There is a threshold population size for a given area, below which populations become at risk of extinction in that area. These differ by species, but a set of indicator species for each guild (group of species that exploit the resources of a place in the same way, such as primary producers, detritivores, and predators) should be established for assessment in design and management of ecosystem services. Spatial distribution is a measure of where those individuals reside. Species rarely are distributed uniformly across space—they are inherently clumpy. The more motile the species, the more important pathways and corridors become. The more sessile the species (plants, for example), the more important refuge areas become. Setting aside refuge for a management indicator species of birds that does not allow them access to water, for example, is ineffective. Similarly, designating a patch of ground for refuge for a key pollinator that does not include plants necessary for the pollinator to complete its life cycle is a fail-start. The metrics of species distribution include range, dispersion, and aggregation (Luck et al., 2003). Genetic differentiation is the fourth characteristic of populations that is critical for designing ecosystem services. The most common example of this characteristic is the measurement of diversity of responses of crops to various stresses, including disease, lack of nutrients, and drought. These variations in response are the basis for crop improvement through selective breeding, as they are for natural selection in nonhuman-dominated systems. Species that have homogeneous genotypes are vulnerable to extinction from blight or other episodic stresses.

Defining the boundaries of place within which to consider populations is difficult because of the multiple scales that interact simultaneously in ecosystems. Luck et al. (2003) describe the Service Providing Unit (SPU) of ecosystems as the combination of Evolutionary Units (EUs), Demographic Units (DUs), and Conservation Units (CUs) that interact within a defined boundary (place) to provide a discrete ecosystem service (Figure 2-3). EUs are populations with independent evolutionary dynamics, typically characterized as occupying a unique place and separated from other breeding populations by geographic, behavioral, or physiological barriers. Example evolutionary barriers include islands, mountains, and in some cases, human structures or activities. Examples are roadways that separate reptile communities, fields that segregate plant populations, and urban areas that transect migratory pathways of mammals and birds. Demographic Units are populations with unique demographic characteristics or dynamics. Example DUs are populations of the same species that do not interbreed frequently because of fragmentation of their habitat, but that still have the capacity to interbreed. Examples include agricultural tillage that separates pollinator populations at distances greater than foraging behavior supports (>2 km between patches for native bees) (Kremen et al., 2002, 2004). A Conservation Unit is the smallest unit of a population that meets conservation goals, usually maintaining genetic diversity within a species. Thus, SPUs can be defined as locally as a single species (Figure 2-3a) or as broadly as all the species within an ecoregion (Figure 2-3b). The key to designing ecosystem services is understanding the species that are critical for these services.

Ecosystem Services, Land Use, and Biodiversity

Ecosystem services are the product of place. They are the collection of the interactions of all species within communities within watersheds within ecoregions within biomes. The relationship between land use and ecosystem services is conceptually apparent. As land use changes, the processes on the land change. The cumulative impact of those land use changes on a watershed is becoming apparent as the proportion of human-dominated land uses in watersheds increases. These topics will be addressed in Chapters 5, 6 and 7. The impact of economic policy

38 Ecological Engineering Design

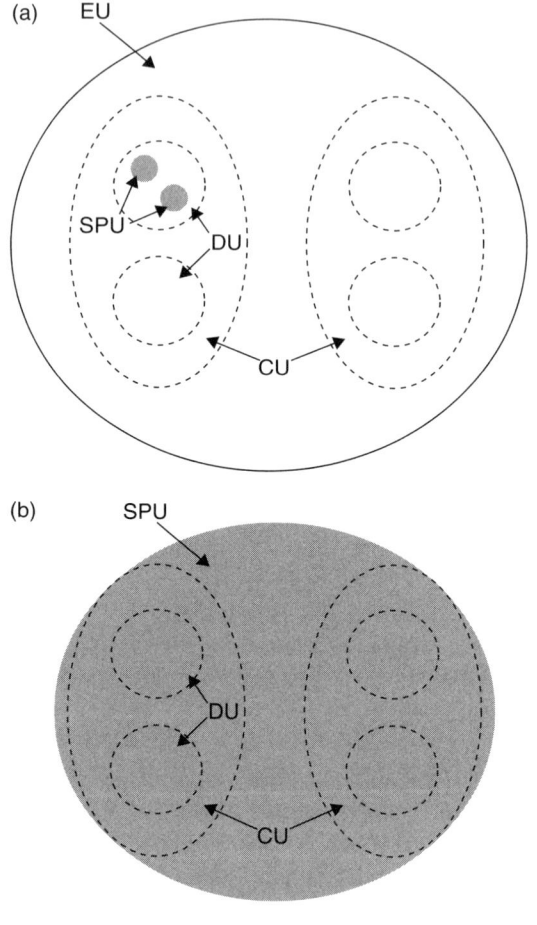

FIGURE 2-3. Biodiversity units for consideration in ecosystem services design. (*Source*: From Luck et al., 2003.)

at global, national, regional, and local scales on land use, and thus ecosystem services, is still poorly understood because the effects are so intricate (Figure 2-4). Thus, the challenge for ecological engineers is to provide designs that anticipate, respond to, and inform policies for land-based prosperity.

Generally speaking, our understanding of ecosystem services processes is inadequately resolved to support a species-level design approach. Instead, ecological engineers use a framework design approach. The framework design approach is described in Chapter 3; the approach is to design critical land forms and conditions necessary

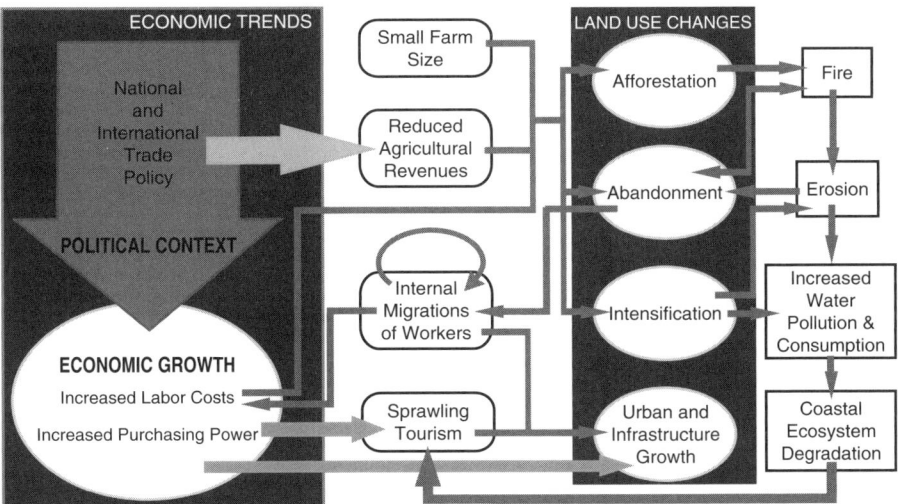

FIGURE 2-4. Economic trends drive land use changes, which drive ecosystem services loss. (*Source*: Modified from MEA report, UN, 2005.)

to support the ecosystem services of an area. This is more explicit than the *Field of Dreams* approach ("If you build it they will come"). However, the challenges of understanding the processes of ecosystem services represent a critical area of research need. The ecological engineering community must contribute to and provide focus for these research activities.

> Engineering is clearly the dominant idea of the industrial age. What I have here called ecology is perhaps one of the contenders for a new order.
>
> —Aldo Leopold

Further Readings
Abbey, E., *Desert Solitaire*, McGraw-Hill Book Co., New York, NY, 1987.
Carson, R. *The Edge of the Sea*. Mariner Books, Boston, MA. 2005.
Leopold, A., *A Sand County Almanac and Sketches Here and There*, Oxford University Press, New York, NY, 1987.
Wilson, E.O., *The Future of Life*, Alfred A. Knopf, New York, NY. 2002.

References
Bjorklund, J., K. Limburg, and T. Rydberg, Impact of production intensity on the ability of the agricultural landscape to generate ecosystem services: An example from Sweden, *Ecological Economics*, 29: 269–291, 1999.

Costanza, R., R. d'Arge, R. de Groot, S. Farber, M. Grasso, B. Hannon, K. Limburg, S. Naeem, R.V. O'Neill, J. Paruelo, R.G. Raskin, P. Sutton, and M. van den Belt, The value of the world's ecosystem services and natural capital, *Nature*, 387: 253–260, 1997.

Daily, G., ed., *Nature's Services: Societal Dependence on Natural Ecosystems*, Island Press, Washington, DC, 1997.

De Groot, R. S., M. A. Wilson, R. M.J. Boumans. A typology for the classification, description and valuation of ecosystem functions, goods and services, *Ecological Economics* 41: 393–408, 2002.

Harris, J., and S. Kennedy, Carrying capacity in agriculture: global and regional issues. *Ecological Economics*, 29: 443–461, 1999.

Hughes, J.B., et al., The loss of population diversity and why it matters, in P.H. Raven, ed., *Nature and Human Society*, National Academy Press, Washington, DC, pp. 71–83, 1998.

Hughes, J.D., and M.D.S. Chandran, Sacred groves around the earth: An overview, in P.S. Ramakrishnan, K.G. Saxena, and U.M. Chandrashekara, eds., *Conserving the Sacred: For Biodiversity Management*, UNESCO, Paris, France/Oxford, UK, and IBH Publishing, New Delhi, India, pp. 69–86, 1998.

Kaplan, S., and R. Kaplan, *The Experience of Nature: A Psychological Perspective*, Cambridge University Press, New York, NY, 1989.

Kremen, C., et al., Crop pollination from native bees at risk from agricultural intensification, *Proceedings of the National Academy of Sciences USA*, 99, pp. 16812–16816, 2002.

Kremen, C., N. Williams, R. Bugg, J. Fay, R. Thorp, The area requirements of an ecosystem service: Crop pollination by native bee communities in California. *Ecology Letters*, 7: 1109–1119, 2004.

Luck, G., G. Daily, and P. Ehrlich, Population diversity and ecosystem services, *Trends in Ecology and Evolution*, 18(7): 331–336, 2003.

Millennium Ecosystem Assessment, Ecosystems and Human Well-Being: Synthesis, Island Press, Washington, DC, 2005, www.millenniumassessment.org/documents/document.300.aspx.pdf.

Marsh, George P., Man and nature; or, Physical geography as modified by human action, C. Scribner, New York, 1864, http://memory.loc.gov/cgi-bin/query/r?ammem/consrv:@field(DOCID+@lit(amrvgvg07)):.

O'Connor, N., and T. Crowe, Biodiversity loss and ecosystem functioning: Distinguishing between number and identity of species, *Ecology* 86(7): 1783–1796, 2005.

Sanderson, E., J.M. Jaiteh, M. Levy, K. Redford, A. Wannebo, and G. Woolmer, The human footprint and the last of the wild, *Bioscience*, 52(10): 891–904, 2002.

Tscharntke, T., A. Klein, A. Kruess, I. Steffan-Dewenter, and C. Thies, Landscape perspectives on agricultural intensification and biodiversity–ecosystem service management. *Ecology Letters*, 8: 857–874, 2005.

UN, We the Peoples: The Role of the United Nations in the 21st Century. United Nations Environment Programme. New York, NY, 2005.

UN, United Nations Millennium Development Goals, United Nations, New York, NY, 2007, www.un.org/millenniumgoals/environ.shtml.

3
Designing Ecosystem Services

> Ecological design at the level of culture resembles the structure and behavior of resilient systems in other contexts in which feedback between action and subsequent correction is rapid, people are held accountable for their actions, functional redundancy is high, and control is decentralized.
> —David Orr, *The Nature of Design*

DESIGN CHALLENGES AND NEEDS

In his groundbreaking book *The Nature of Design* (Orr, 2002), David Orr postulated several pathologies that prevent ecological design from becoming commonplace. Among these were the loss of slow knowledge (cultural wisdom about place), speed (the increased velocity of change across all aspects of our culture), verbicide (the loss of depth of language and therefore frames of reference), technological fundamentalism (the irrational faith in technology for solutions to all problems), and ideasclerosis (the logjam of innovative ecological solutions in the morass of information and communication).

The framing of ecological challenges in the context of ecosystem services addresses many of these pathologies by providing place-based context for design. Ecosystem services are the product of a place on Earth, and are unique to that place. In addition, the process of designing ecosystem services requires a continuity of understanding across time and space in design practices. No structure, development, or community can be designed for ecosystem services outside its ecological context and eco-history. Designing ecosystem services is not high technology or low technology, but rather ecotechnology. This novel framing creates opportunities for innovation at every imaginable scale.

The purpose of this chapter is to define the process that should be employed in designing ecosystem services. This process requires a review of current design methods to learn what does and does not work from traditional practice. The chapter provides an assessment of the design ethic necessary for designing ecosystem services, as part of that analysis. The challenge of designing with complex, multidisciplinary teams is summarized, concluding with design and assessment criteria for ecosystem services.

Current Design Methods Deficiencies

Engineering design as currently practiced is the product of hundreds of years of trial and revision. Engineers use this method because it works; we have been able to send human beings to space and bring them back; we designed a city on the Gulf Coast that functioned below sea level. However, recent failures in both these areas suggest that current engineering design methods must be adapted in order to perform adequately in complex, changing systems. Current design practices often limit the engineer to addressing technical challenges while deferring critical decisions on function and scope to management. This artificial distinction between decision domains has been at the heart of many critical design failures in the past 40 years. The failure of the U.S. Army Corps of Engineers in designing an adequately stable floodwall for New Orleans is an example of the limits of the current design method (Figure 3-1).

FIGURE 3-1. Satellite image of Hurricane Katrina as it reached landfall in Louisiana and Mississippi. (*Source*: NASA, 2005.)

In 2007, the American Society of Civil Engineers conducted a comprehensive review of the failure of the New Orleans hurricane protection system as a result of the impact from Hurricane Katrina on August 29, 2005 (ASCE, 2007). This failure was one of the worst disasters to strike the United States, resulting in more than 1,400 deaths and nearly $30 billion in property damage. The total economic and social costs of this disaster are still accruing, more than five years after the event.

Much of the flooding and subsequent mortality and damage resulted not from Hurricane Katrina, but from "the storm's exposure of engineering and engineering-related policy failures" (ASCE, 2007). The technical design failures included collapse of concrete floodwalls (I-walls), overtopping of floodwalls and levees, and pump station failures (Figure 3-2).

The I-walls failed because the engineers who designed them used a margin of safety that was too low for a critical life-safety structure. Engineers did not consider variability in soil structure across the levee system, and failed to anticipate the impact of I-wall structural deformation on the integrity of the anchoring system. Several earthen levees failed as a result of overtopping. These levees should have been armored to prevent erosion from overtopping.

These technical failures were the result of compartmentalized design decisions that, in turn, resulted in a host of design and management failures, including:

1. Inadequate communication of risk to the affected community by the design community, resulting in a failure of understanding of the threat posed to the community.
2. Inadequate integration of the elements of the hurricane protection system, resulting in systemic weaknesses throughout the network of I-walls, levees, and pump stations.
3. Failure to engage appropriate technical expertise (soil geotextile engineers, meteorologists, systems modelers, risk assessors) to ensure that design criteria were appropriate to mitigate risk to acceptable levels.
4. Failure to confirm design criteria through external quality review, resulting in simple errors such as use of the incorrect datum to measure levee elevation. This single error resulted in some levees being built two to three feet below design elevation, exacerbating the failure event. Similarly, geoscientists and engineers were not consulted to address subsidence intensity within the region.

Designing Ecosystem Services 45

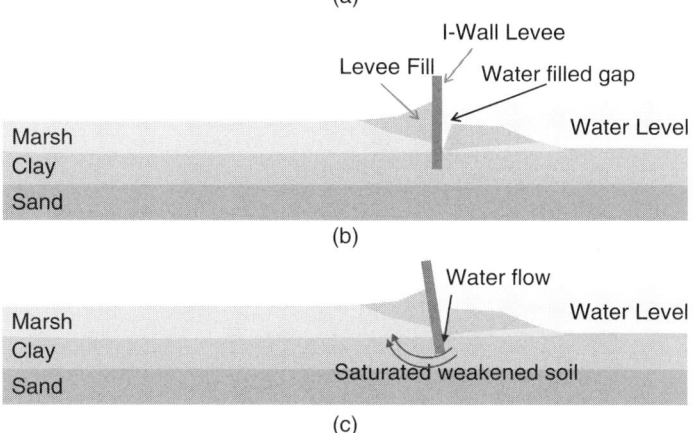

FIGURE 3-2. Photograph and diagrams of New Orleans levees b. Floodwaters caused the I-wall to lean away from the canal, creating a water filled gap between the wall and the soil behind it. c. The water-filled gap allowed the I-wall to be undermined, as water seeped under the barrier and weakened the foundation soil. (*Source*: a The Army Corp of Engineers, Source b and c Adapted from ASCE, The New Orleans Hurricane Protection System: What Went Wrong and Why, The American Society of Civil Engineers. The American Societry of Civil Engineers 2007.)

5. Distribution of responsibility and authority for managing and assessing the hurricane protection system across many agencies, creating barriers to coordination and communication.
6. Failure to integrate funding into a comprehensive priority-based strategy resulted in further disintegration of systemic cohesion.

FIGURE 3-3. Collapse of the Interstate 35W Bridge in Minnesota in 2007, which killed 13 people. [*Source*: Wikipedia Commons File: I35W Collapse - Day 4 - Operations & Scene (95)]

These individual decisions, made over a period of decades, resulted in cascading failure of a critical infrastructure system. This set of cascading failures does not consider the further impact of construction of the Mississippi River Gulf Outlet navigation canal, which created a storm surge channel to the Ninth Ward, which killed more than 1,400 people.

The design failures highlighted by Hurricane Katrina are not that uncommon; on August 1, 2007, an eight-lane bridge spanning the Mississippi River in Minneapolis collapsed, killing 13 people (Figure 3-3). This failure was more typical of engineering design failures, where loss of life is tragic but limited, since most designs do not encompass the level of risk associated with the New Orleans hurricane protection system. However, the risk to life and property, when considered in total, is severe. To many engineers, this was an additional symbol of the failure of the U.S. to maintain, manage, and design infrastructure for transportation, drinking water, wastewater disposal, power, and telecommunications. Orr (2002, p. 14) suggests that ecological failures are design failures. The lessons of Hurricane Katrina suggest that designing for single purposes will not meet the needs of our complex, interconnected society.

Ecosystem Services Design Ethics

Gadugi, the Cherokee cultural ideal of coming together and working for the benefit of the community, embodies the notion of an ecosystem services design ethic. This is an ethic born of the struggle for existence that depends directly on ecosystem services. The recognition that we are all in this together acknowledges the inherent limitations of resources upon which community prosperity depends. The notion that this approach is in some way communist or socialist is simply cynical and short-sighted, and is often derived from the frontier myth of individualism as the dominant resource management strategy. That approach has resulted in much of the loss of ecosystem services documented in Chapter 1, and will ultimately lead to a global decline in human prosperity. Gadugi design principles embody the lessons of Hurricane Katrina; we are all in this together, and must work across boundaries and barriers to achieve our common goals.

The Gadugi principle applied to ecosystem services design ethics is not that dissimilar from the 100-year-old engineering ethic adopted by the American Society of Civil Engineers (ASCE, 2006). These ethics were established as a set of principles and canons (Table 3-1) that address the breadth of common ethical issues that a practicing engineer might face. The ecological engineer, engaged in designing ecosystem services, will face many of these same issues, but will also

TABLE 3-1 The American Society of Civil Engineers Code of Ethics

Fundamental Principles of Engineering Ethics	
Engineers uphold and advance the integrity, honor and dignity of the engineering profession by:	
1	using their knowledge and skill for the enhancement of human welfare and the environment;
2	being honest and impartial and serving with fidelity the public, their employers and clients;
3	striving to increase the competence and prestige of the engineering profession; and
4	supporting the professional and technical societies of their disciplines.
Fundamental Canons of Engineering Ethics	
1	Engineers shall hold paramount the safety, health and welfare of the public and shall strive to comply with the principles of sustainable development in the performance of their professional duties.
2	Engineers shall perform services only in areas of their competence.
3	Engineers shall issue public statements only in an objective and truthful manner.
4	Engineers shall act in professional matters for each employer or client as faithful agents or trustees, and shall avoid conflicts of interest.
5	Engineers shall build their professional reputation on the merit of their services and shall not compete unfairly with others.
6	Engineers shall act in such a manner as to uphold and enhance the honor, integrity, and dignity of the engineering profession and shall act with zero-tolerance for bribery, fraud, and corruption.
7	Engineers shall continue their professional development throughout their careers, and shall provide opportunities for the professional development of those engineers under their supervision.

encounter challenges unique to the complexity and integrated nature of ecosystem design.

The first canon of Civil Engineering Ethics was modified in 1996 to include the admonition to "strive to comply with the principles of sustainable development." This seemingly benign addition created a significant difficulty for enforcing the ethical standard, because the phrase "sustainable development" was defined as "... the challenge of meeting human needs for natural resources, industrial products, energy, food, transportation, shelter, and effective waste management while conserving and protecting environmental quality and the natural resource base essential for future development" (ASCE, 2007). How does an engineer strive to meet human needs for natural resources, et cetera, while also designing a water distribution system, a highway interchange, or a bridge? This type of well-intended hand waving towards sustainable design suffers from lack of explicit criteria for evaluation and assessment.

Ecosystem services design, performed by ecological engineers, provides a quantitatively rigorous method for assessing design impact on sustainability criteria. The ecological engineer therefore requires a more explicit set of ethics from which to benchmark decision making. A set of proposed ethics for ecological engineers is presented in Table 3-2; these have not been vetted through a professional society or adopted as a canon, but represent reasonable responses to the ethical challenges of ecological engineers. These ethics are based on the ASCE Code of Ethics, with modification, and are informed by the failures of current engineering design practices (ASCE, 2006).

Legitimacy and the Design Process

The notion that we are all in this together, a paraphrase of the Gadugi principle of community service, should be incorporated into the fabric of decision making in ecosystem services design, since everything is connected. The design team should therefore meet the criteria for legitimate policy when framing and developing ecosystem services design (Sabatier et al., 2005). These criteria represent a normative framework for design that is similar to the lessons learned from the failure of the New Orleans hurricane protection system. Sabatier et al. (2005) identified four criteria for legitimate policy that apply to the process of ecosystem services design. These four criteria should guide the structure and composition of design teams, and implementation of the design process:

TABLE 3-2 Proposed Ecosystem Services Design Code of Ethics for Ecological Engineers

Fundamental Principles of Ecological Engineering Ethics	
Engineers uphold and advance the integrity, honor, and dignity of the engineering profession by:	
1	Using their knowledge and skill for the enhancement of human welfare and the protection, restoration, and enhancement of ecosystem services
2	Being honest and impartial and serving with fidelity known current and future interests of the public, their employers, and clients
3	Striving to increase the competence and prestige of the engineering profession through strong collaboration across disciplines
4	Supporting the professional and technical societies of their disciplines in developing, implementing, and innovating design standards of practice
Fundamental Canons of Ecological Engineering Ethics	
1	Engineers shall hold paramount the safety, health, and welfare of the public (current and future generations) and shall comply with the principles of ecosystem services design in the performance of their professional duties.
2	Engineers shall perform services only in areas of their competence, being mindful to document and communicate additional competency areas necessary for successful design.
3	Engineers shall issue public statements only in an objective and truthful manner, specifically quantifying risks and uncertainties associated with design decisions.
4	Engineers shall act in professional matters for each employer or client as faithful agents or trustees, and shall avoid conflicts of interest.
5	Engineers shall build their professional reputation on the merit of their services and shall not compete unfairly with others.
6	Engineers shall act in such a manner as to uphold and enhance the honor, integrity, and dignity of the engineering profession and shall act with zero tolerance for bribery, fraud, and corruption, ecosystem exploitation, or wanton destruction of ecosystem services.
7	Engineers shall continue their professional development throughout their careers, and shall provide opportunities for the professional development of those engineers under their supervision through engagement with professionals across disciplines and within ecological engineering.

1. *Autonomy*: Autonomy is the principle that people ought to choose the rules and conditions under which they live. This is central to the notion of freedom in the United States. Ecosystem design teams ought not to prescribe the conditions under which people live, not the engineer in isolation. The legitimate design process must consider that people who are affected by a design must participate in the design itself. If the design of ecosystem services creates constraints, rules, or other impacts on people's lives, they have a stake in the design and must be included as stakeholder in the process.
2. *Procedural Legitimacy*: The way design is conducted matters. The principles of transparency and openness must prevail in even the

most contentious ecosystem services design project. An eco-dictator could proclaim by fiat a policy that might be effective, but it would never be viewed as legitimate from the perspective of autonomy and justice.
3. *Substantive Legitimacy*: The design must work. The danger of complex, collaborative design is that conflict is often resolved through a consensus process or other mediation methods, and the technical efficacy of the design may be lost in the process. This explains much of the failure of the New Orleans hurricane protection system; no single group decided how the system would function, and the emphasis became coexistence, not design competency. Ideally, substantive legitimacy is linked with procedural legitimacy in the design process.
4. *Justice*: The principle of justice simply requires that the conditions of life must be improved by the design, and that a person or group of people must not bear a disproportionate share of the costs of improving the lives of others. The affected parties must consent to the impact of the design.

If the criteria for legitimate ecosystem services design are adhered to, the likelihood that the design process will succeed is increased; the potential for innovation is enhanced, and the acceptance of the design constraints will be facilitated through the process. The challenge for the ecological engineer in this process is to understand the complexity of the community (human and ecological) within which the design resides, and to utilize this complexity to make the design more robust.

THE DESIGN PROCESS

The process of engineering design has often been described as the process of solving complex problems with inadequate information and limited resources. This definition captures several challenges in design: lack of information and lack of resources (generally money). For complex designs such as ecosystem services, it is critical that these limitations be articulated at the outset. The three axioms of ecosystem services design described in Chapter 1 are:

1. Everything is connected.
2. Everything is changing.
3. We are all in this together.

Since everything is connected, designing ecosystem services requires an understanding of the extent and implications of those connections when modifying an ecosystem function. This means that no ecosystem services design process can be explicitly complete at the outset. Every design must be iterative and expansive. Mitch and Jorgensen (2004) defined 19 principles of ecological design. These principles are fundamental to designing ecosystem services (Table 3-3).

The team must also consider rates of change within and adjoining the design area, so competency in land use change within the

TABLE 3-3 Ecological Design Principles

Principle	Description
1	Ecosystem structure and functions are determined by forcing functions of the system.
2	Energy inputs to the ecosystems and available storage of matter are limited.
3	Ecosystems are open and dissipative systems.
4	Attention to a limited number of factors is most strategic in preventing pollution or restoring ecosystems.
5	Ecosystems have some homeostatic capability that results in smoothing out and depressing the effects of strongly variable inputs.
6	Match recycling pathways to the rates of ecosystems to reduce the effect of pollution.
7	Design for pulsing systems wherever possible.
8	Ecosystems are self-designing systems.
9	Processes of ecosystems have characteristic time and space scales that should be accounted for in environmental management.
10	Biodiversity should be championed to maintain an ecosystem's self-design capacity.
11	Ecotones, or transitions zones, are as important for ecosystems as membranes are for cells.
12	Coupling between ecosystems should be utilized wherever possible.
13	The components of an ecosystem are interconnected, interrelated, and form a network, implying that direct as well as indirect effects of ecosystem development need to be considered.
14	An ecosystem has a history of development.
15	Ecosystems and species are most vulnerable at their geographic edges.
16	Ecosystems are hierarchical systems and are part of a larger landscape.
17	Physical and biological processes are interactive. It is important to understand both physical and biological interactions and to interpret them properly.
18	Ecotechnology requires a holistic approach that integrates all interacting parts and processes as far as possible.
19	Information in ecosystems is stored in structures.

Source: Mitch and Jorgensen 2004.

52 Ecological Engineering Design

design team is critical. Finally, the process of ecological design is not like conventional engineering design; this process "is as much about politics and power as it is about ecology" (Orr, 2002).

The standard process of engineering design is not easy to define, as this process is more craft than science. Holtzapple and Reece (2005) define the method in five stages: synthesis, analysis, deliberation, implementation, and assessment. The ecosystem services design method is not significantly different, but has a more iterative structure (Figure 3-4). The cyclical nature of the design process recognizes that in self-organized systems, design is never finished. Much of the difficulty arises from the generally poor understanding of how ecosystems function while under transition pressure, and associated difficulty in predicting outcomes based on design. Thus, designing ecosystem services is an iterative process, with intrinsic adaptive management strategies and feedback loops to ensure that the process is informed (Figure 3-5).

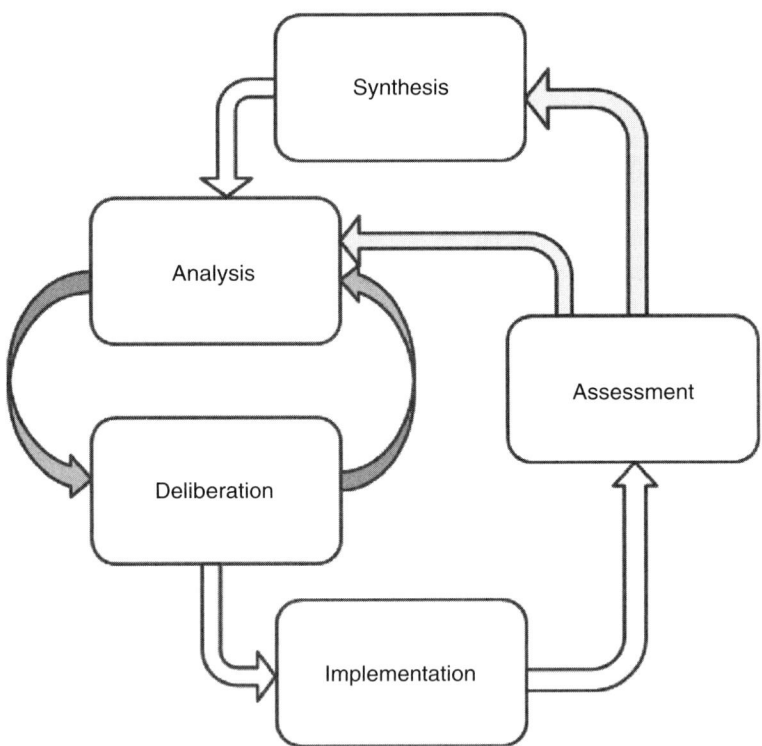

FIGURE 3-4. The ecosystem services design method.

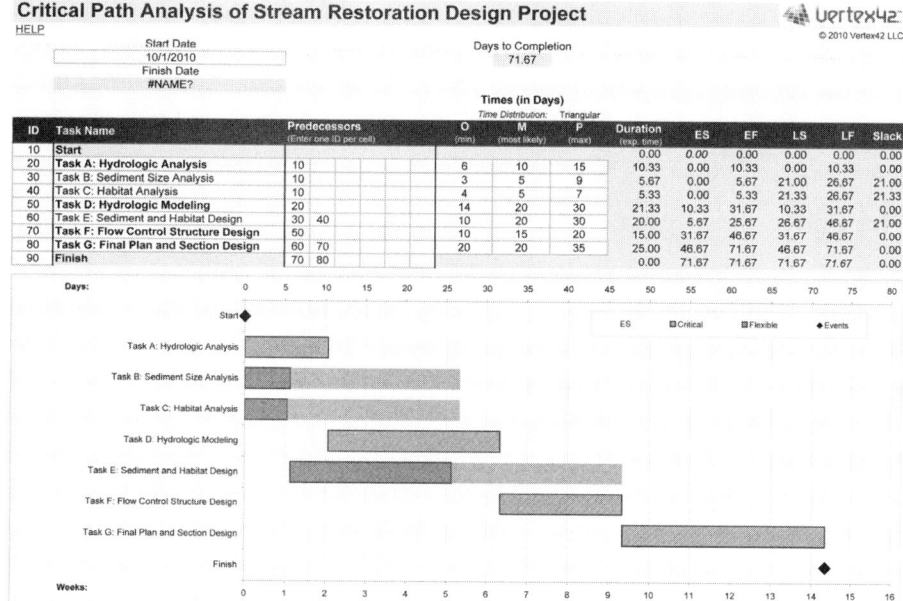

FIGURE 3-5. Example critical path analysis and Gantt charts.

The process of primary design, in which the design team is actively modifying elements of the landscape, is finite. This process is dependent upon an explicit set of assessment criteria that trigger initiation of additional design elements. In practice, the most iterative stages are analysis and deliberation, whereby stakeholders participate in collaborative learning processes to inform design decisions. Design elements should be implemented in phased stages so that each element can be integrated within the context of the larger design, rather than in a simultaneous implementation of elements across spatial and temporal scales.

Synthesis

The entry point into the ecosystem services design cycle is synthesis, in which problems and opportunities are identified, the design team is formulated, design processes are mapped across area and over time increments, and options are evaluated based upon performance criteria and costs. Synthesis is critical to the success of any ecosystem services design process; without a clear formulation of goals, assessment criteria, and performance expectations, any design is at risk of failing to meet a definition of success established at the outset. The first activity of the ecological engineer is problem identification. This process may seem simple, but in most ecosystem services design projects, this proves to be the most problematic stage. The ecological engineer should assemble the ecotechnology team early in the Synthesis process.

The Ecotechnology Design Team

A basic premise of ecosystem services design is that no single person, discipline, or profession possesses adequate breadth and depth of knowledge to be independently competent in the process. No individual or profession encompasses this breadth of knowledge; a design team is necessary for competent ecosystem services design. Leaving aside the stakeholder participation and communications elements, even the technology elements are too complex for a single discipline. The ecological engineer must be competent in the process of design to formulate and organize the ecosystem design team. The ecosystem design team composition is determined by the technology needs of the project and the management structure of the design team. The ecological engineer must be adept at defining and assembling a team of designers with appropriate technological expertise to address the problems at hand. Technology needs within a design team are difficult to assess at the outset of a project. The process of inventorying the technology needs of an ecosystem services design project requires an assessment of scope, scale, and function.

Defining the scope of an ecosystem services design project involves defining the problems to be solved. Remember the first axiom of ecological engineering: EVERYTHING IS CONNECTED. The process of scope definition is a precursor to design, and involves defining connections. This process must adhere to the principles of legitimacy previously described, and therefore must be inclusive and transparent. The lead designer, often an ecological engineer, will have some sense of the scope of a project from the client at the outset. However, the ecological engineer must recognize that any design decisions will affect other processes. Therefore, engaging stakeholders, organized as an Ecosystem Services Advisory Committee (ESAC), at the initial stages of problem definition may lead to innovations at the outset. These include redefinition of a problem that opens alternative solutions, mitigation of conflict through direct communications, and at the minimum establishing relationships with individuals and groups who will be affected by or who are interested in the design. The ESAC should be engaged using a process such as collaborative learning (Daniels and Walker, 1996) to develop common understanding across groups on the connections and causes of problems.

Ecological engineers design on the landscape; the challenges of the particular landscape define the skills and expertise needed to develop competent designs. For terrestrial designs, the most common technology elements will include soils (geotextile as well as ecology), hydrology, plant ecology, and perhaps landscape ecology. The disciplines for these bodies of knowledge include agronomy, botany, ecology, and landscape architecture. If there are stream restoration design elements, an expert in stream geomorphology is necessary. If fish habitat is a concern (and it should be, in almost every stream restoration project), a fisheries ecologist should be on the ecotechnology team. The tension between inclusiveness of expertise and management of costs is perhaps the biggest challenge of the ecological engineer.

Defining the Appropriate Management Structure

Managing complex teams is difficult; egos, competing demands on resources and priorities, and ideological differences can lead to team structure failure. The ideal management structure does not exist for these complex projects. The role of the ecological engineer is to manage the ecotechnology team and interact with the ESAC. The principles of legitimacy must be adhered to throughout this process. Sabatier et al. (2005) found that the single best predictor of the success of complex collaborative teams was trust between participants.

The ecological engineer has very specific management responsibilities beyond design innovation. The project must be completed on schedule and on budget. From the outset, the goals and principles for priority setting must be made clear to the ecotechnology team. Key technical concepts must be articulated in a manner that translates across disciplines and encompasses the values of each discipline in a design context. The mechanisms for appropriate resolution of conflict must be articulated at the outset of the process. The team must accept these goals and apply the principles as a common guiding ethic in decision making throughout the project. The ecological engineer is the final arbitrator of the design process because he/she ultimately is responsible for the integrity of the process. Managing interdisciplinary teams in these complex design environments is ultimately an interpersonal activity.

The ecological engineer must develop skills in communications, organization theory, and systems assessment in order to manage ecosystem services design projects. Standard engineering management tools are appropriate for ecosystem services design projects. These tools include critical path analysis, schedule-task management using Gantt charts, and time-on-task budgeting (Figure 3-5). The ecological engineer should establish a communications framework, a database for design information, criteria for data incorporation, and the process for engaging the ESAC. Finally, the ecological engineer should serve as the authoritative voice for technical issues.

Analysis and Deliberation

Legitimate design of ecosystem services integrates stakeholder participation in each step of the project; the recommended method is analysis and deliberation (A/D) (National Research Council, 1996). Analysis is used to inform design deliberation so that the best information is brought to bear upon the problem to be solved ("getting the design right"). Deliberation is used not only to make a decision, but also to frame the analysis and to empower participants in understanding analytic findings ("getting the right design"). Thus, ecosystem services design decision making is not solely within the discretion of the ecological engineer. All interested and affected parties (stakeholders) decide what information should be considered in the design, what approaches should be used to measure and assess outcomes, and how to evaluate alternative ecosystem services management schemes (Sabatier et al., 2005). Nontechnical stakeholders participate in framing the issues that are salient to the design problem from the outset. Such issues include deciding what information should be considered, what further design elements should be incorporated to reduce uncertainty, and which ecosystem services should be prioritized. New information from the design process, as it moves into implementation and assessment, may stimulate another round of analyses to further inform deliberation. The careful integration of analysis and deliberation in a recursive manner is the most important element of ecosystem services design (see Figure 3-4).

Mapping Ecosystem Services Processes

One of the most effective methods of developing this common understanding is conceptual mapping, whereby the ESAC works with the ecological engineer at the project outset to identify and link all the elements associated with a process of concern (Norton et al., 2009). The group literally maps their understanding of a process as a systems diagram (Figure 3-6). With experience and iteration, the ESAC can

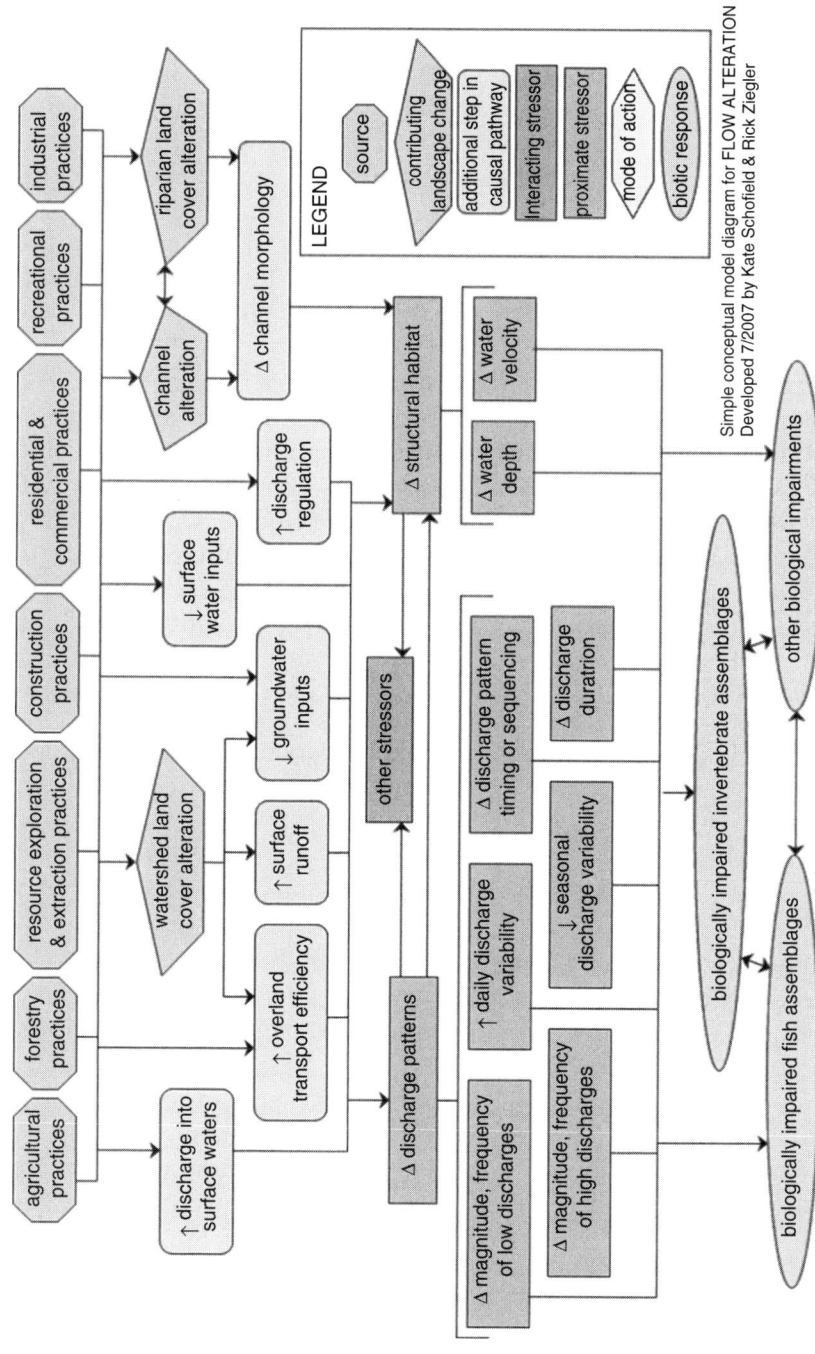

FIGURE 3-6. Conceptual map of flow alteration elements. (*Source:* Norton et al., 2009, created by Schofield and Ziegler, 2007.)

define levels of confidence in knowledge, data, and innovation potential across each element and relationship in a process. This exercise allows the ecological engineer to work with the ESAC to define explicitly the expertise needed to design ecosystem services in a specific context. The group of experts engaged with stakeholders in developing and implementing ecosystem services designs is called the ecotechnology design team. Libraries of conceptual maps are being developed for a number of complex landscape-scale processes; however, the process of developing the map with a group of stakeholders familiar with a location allows the ESAC to identify site-specific relationships. Generic conceptual maps should only be used to initiate the deliberation process.

For smaller projects with minimum stakeholder impact, the ESAC can be composed of the ecological engineer and the project management team. However, the strategy of open mapping of process should be used even for the smallest and narrowest of projects. What will emerge is a pattern of expertise for typical projects within specific ecoregions; over time, the ecological engineer can anticipate ecotechnology team requirements and reduce the transaction costs of assembling the team. However, ecological engineers must understand that the practice of ecosystem services design will never be a cookie-cutter affair. Every project will be unique, nuanced, and difficult. Remember the failures of compartmentalized design in the New Orleans hurricane protection system as a warning to avoid the temptation to make the process more "efficient" by streamlining communications and narrowing participation.

Defining Priorities

Simply put, design priorities are the design variables that matter most, in rank order. Conventional engineering design defines priorities early, and decisions are driven from those priorities. This is a very rational and effective method for designing systems that are predominantly linear or controlled. When designing ecosystem services, characterized by dynamic, complex, and nonlinear feedback mechanisms, priority setting should be iterative in the sense that design alternatives should be evaluated incrementally. This allows emergent properties of the complex system to be recognized, incorporated into the design, and enhanced through design structure.

Priorities should be initially defined by the regulatory and safety considerations of the project. The first canon of ethics is unambiguous: the safety of the public is the primary priority. Within that context, the suite of ecosystem services should be assessed based on scale (site, local, regional), timelines (short, medium, long term), and influence (direct, indirect, and association). Priorities should be defined for each scale, timeline, and landform, within the scope of overall project goals. The effectiveness of design implementation should be measured throughout the project, and for a period after primary intervention, to evaluate success and prioritize incremental follow-up design activities.

Scale in ecosystem services design is difficult to explicitly define. For any integrated system, elements at one scale are influenced by the next higher scale (Lyle, 1999). Simply stated, you cannot understand a site without understanding its location and region. Design at larger than regional scale (subcontinent, for example) is rare, but does happen. These are largely hydrologic diversion projects, and have impact throughout the systems they displace. As critical resources such as water become more scarce, subcontinent-scale planning and design will increase as a critical tool for sustainable prosperity. For ecosystem services design, the typical regional scale is the watershed (eight-digit hydrologic unit code, roughly 1,000 square miles). The location is the subcatchment within a watershed and the jurisdictional boundaries (city limits, zoning areas) that transect the location. The site is the explicit locale of the design elements under consideration. Sites exists within a location, locations within an ecoregion; thus, sites integrate the complex elements of location and region.) Priorities for design are selected at the site scale; goals are set at the location and ecoregional scales.

Setting Design Goals

Design goals are the set of outcomes (quantitative and qualitative) intended from the design process. These outcomes should be measurable, affordable, and designable (the MAD criteria of outcomes). The ESAC should work with the ecotechnology team to develop these goals for each scale, using conceptual mapping to identify critical control variables. The ecological engineer and ecotechnology team should perform an inventory of ecosystem services at all three scales. This inventory should include an assessment of ecosystem services states (Table 3-4).

Design goals should be informed by each ecosystem service's state values. For a given scale (site, local, regional), ecosystem services can be inventoried by landform (temperate deciduous forest, shortgrass meadow, stream riparian zone, etc.). Each individual ecosystem service (ES_i) can be indexed to the historic ecosystem service level for that

TABLE 3-4 States of Ecosystem Services across Scales

Symbol	State	Definition
ES_H	Historic	Suite of ecosystem services based upon pre-industrial human-dominated land use (what was)
ES_C	Current	Suite of ecosystem services present at each scale (what is)
ES_P	Potential	Highest value of ecosystem services that could exist given current conditions and potential land use changes (what could be)
ES_D	Design	Proposed ecosystem services to be designed (what will be)

service (ES_{Xi}/ES_{Hi}), where X is the state of ecosystem service, which represents an ecosystem services loss index (ESLI).

What was does not necessarily inform what could be. The potential ecosystem services that could be designed at a site represents the highest potential value of a site. The sum of individual ecosystem service at a site (ES_i) divided by the potential ecosystem service (ES_{pi}) for all ecosystem services divided by the number of services represents the sustainability status of ecosystem services at each scale. This represents the ecosystem services sustainability index (ESSI) of a given scale:

$$\text{ESSI} = \Sigma(ES_{Xi}/ES_{Pi})/n$$

where ES_X is the state of ecosystem service (Table 3-4), and n is the total number of ecosystem services considered (i). Each state and scale of ecosystem services informs the design goals differently. The index of the current state of ecosystem services provides a benchmark from which all potential designs can be assessed. A site with an ESSI of 0.35 would therefore have a total ecosystem services index of 35 percent of potential conditions. A site with a design ESSI of 0.45 could be said to have a net increase of 10 percent of current condition. The potential ecosystem services index therefore provides a theoretical maximum for ecosystem services, and a benchmark for optimization. A site could not have a design ESSI of greater than one, as the design ESSI approaches one, the site is more sustainable. An ESSI of 0.85 or greater could be reasonably considered sustainable for this system.

Implementing Design Goals

Ecosystem services designs are implemented by element. Elements are physical design components that represent discrete installations. Examples of ecosystem services design elements include vegetative infiltration buffers at the edge of a parking lot, riparian wetland modules in the floodplain of a stream, and terrestrial corridors for animal movement. Elements should be budgeted discretely, based upon unit area or mass (dollars per square foot or cubic meter, for example). Each element provides numerous ecosystem services, but represents an incremental piece of the overall design. Implementation of ecosystem services design requires strategies for sequencing, constructing, and assessing each element through the design strategy.

Sequencing elements is fairly straightforward; there are often seasonal constraints on when an element can be initiated, and often other constraints (permits, property acquisition, etc.) inform sequencing. The tools described previously, especially critical path analysis, should be employed at the element level. Each element should be tied to similar elements across scales. For example, restoring a riparian corridor in an urban stream segment that is channelized and denuded of riparian vegetation upstream and downstream may achieve site goals but not achieve the desired local or regional goals.

Assessing Ecosystem Services Design

The purpose of assessing ecosystem services design is to determine the extent to which design goals are being met, and to provide insights for improvement. Assessment of ecosystem services design goals should be element-specific. Parameters for assessment should be quantitative, relatively easy to measure (remotely sensed when possible), outcomes-based, easy to communicate, and scalable across site, locale, and region. If a stream restoration project has the goal of restoring benthic macro-invertebrate community integrity, measuring observed versus expected species (or similar community metrics) across site, location, and region is a reasonable assessment tool.

Each element must be assessed at adequate frequency and during critical periods to determine effectiveness of design implementation. The assessment criteria for each element should be indexed to a point in time or space for comparison. Typical assessment criteria would be reduction in sediment load to streams, increased habitat for songbirds, decreased nutrient concentrations in discharge waters, and decreased summer critical period temperatures of urban streams. The criteria for triggering design actions must be identified at the outset. There are three assessment thresholds for design actions (Table 3-5). The goals identified at the project outset should be used to define assessment criteria and thresholds of success. If a design element is partially successful, the ecotechnology team should work with the ESAC to review

TABLE 3-5 Thresholds for Design Assessment Actions for Each Element

Threshold	Definition	Action
Success	Element meets or exceeds design goals across most ecosystem services.	Continue assessment
Partial success	Element meets design goals across some ecosystem services, but fails others.	Iterate through A/D
Failure	Element fails to meet design goals across most ecosystem services.	Iterate through synthesis

and revise the design implementation. If a design element fails to meet the goals established by the ESAC, the design team should return to the synthesis process prior to engaging the A/D process (Figure 3-4). The ecological engineer should establish timeframes for meeting each goal in the assessment stage.

> If regional landscape design is a rarity, design at the higher levels is rarer still.
>
> —John Tillman Lyle, *Design for Human Ecosystems*, Island Press, 1999

Further Readings
Holtzapple, M. and D. Reece. *Concepts in Engineering*. McGraw Hill, New York, NY. 2005.
Mitsch, W.J. and S.E. Jørgensen, *Ecological Engineering and Ecosystem Restoration*. John Wiley and Sons, Inc., New York. 2004.
Orr, D., *The Nature of Design*, Oxford University Press, Oxford, UK, 2002.

References
ASCE, American Society of Civil Engineers. *Code of Ethics*. ASCE Press, Reston, VA. 2006.
ASCE, American Society of Civil Engineers. The New Orleans Hurricane Protection System: What went wrong and why. *ERP Report*, ASCE Press, Reston, VA, 2007.
Daniels, S., and G. Walker, Collaborative learning: Improving public deliberation in ecosystem-based management, *Environmental Impact Assessment Review*, 16: 71–102, 1996.
Lyle, J.T. *Design for Human Ecosystems: Landscape, Land Use, and Natural Resources*. Island Press, 1999.
Mitsch, W.J. and S.E. Jørgensen, *Ecological Engineering and Ecosystem Restoration*. John Wiley and Sons, Inc., New York. 2004.
National Research Council, *Understanding Risk: Informing Decisions in a Democratic Society*, National Academy Press, Washington, DC, 1996.

Norton, S., G. Suter, and S. Cormier, Causal Analysis/Diagnosis Decision Information System (CADDIS), U.S. Environmental Protection Agency, Washington, DC, 2009, http://cfpub.epa.gov/caddis/index.cfm.

Orr, D. *The Nature of Design*, Oxford University Press, Oxford, UK, 2002.

Sabatier, P., W. Focht, M. Lubell, Z. Trachtenberg, A. Vedlitz, and M. Matlock, Swimming upstream: Collaborative approaches to watershed management, MIT Press, Cambridge, MA, 2005.

4
Defining Place: Biomes and Ecoregions

> Humankind has not woven the web of life. We are but one thread within it. Whatever we do to the web, we do to ourselves. All things are bound together. All things connect.
>
> —Chief Seattle

INTRODUCTION

Ecological design begins with recognition of the importance of "place" for the reason that Chief Seattle articulated. Every ecosystem services design project has geographic context, or "place." In this context, place is the sum of the biological, physical, and chemical properties of the site, and the interconnectedness of those properties with the larger landscape. Place context includes the geology, soils, hydrology, climate, topography, ecohistory, ethnohistory, and many other characteristics of a site. The ecotechnology team is responsible for understanding the ecological context of a project beyond the site boundaries in order to exploit every opportunity for preserving, conserving, restoring, and enhancing ecosystem services.

Chapters 4 through 7 of this book provide resources for defining place with respect to available ecosystems and ecosystem services. First we look at the largest-scale ecosystems, biogeographical realms, and biomes. Subsequent chapters look more closely at manageable units of watershed, site, and even individual soils. Each of these scales has been explored as a framework for assessment of ecosystems for at least 100 years, with many scholarly works committed to them. A detailed analysis of every construct of geographic location is beyond the scope of this book. The purpose of these chapters is to provide an overview of place, provide guidance for ecological engineers in defining spatial boundaries, and highlight important characteristics that need to be considered in ecological design.

Biogeographical Realms

In the nineteenth century, biologists recognized that natural barriers to movement of animals generally divided the world into biogeographic realms. In 1876, A. R. Wallace published *The Geographic Distribution of Animals* (Wallace, 1876), in which he delineated six realms based on these natural barriers, including:

- Nearctic: North America, Greenland, and related islands
- Palearctic: Eurasia north of the Himalaya plus Africa north of the tropic of Cancer and portions of the Middle East
- Ethiopian or Afrotropical: Tropical and Southern Africa, Madagascar, Tropical Arabia, and a few isolated islands
- Oriental: India and the Malayan peninsula and islands east to Java plus China north of the Himalaya
- Australian: Australia and its islands
- Neotropical: South America plus the Caribbean islands, Central America, and portions of the coast of Mexico

Today, biologists generally recognize Wallace's realms, with minor revisions, as the major biological divisions of the Earth, with the addition made by Udvardy (1975) of the Antarctic and Oceanian (Udvardy, 1975) (Figure 4-1):

- Antarctic: Antarctica, New Zealand, and southern islands
- Oceanian: Borneo and the Pacific islands

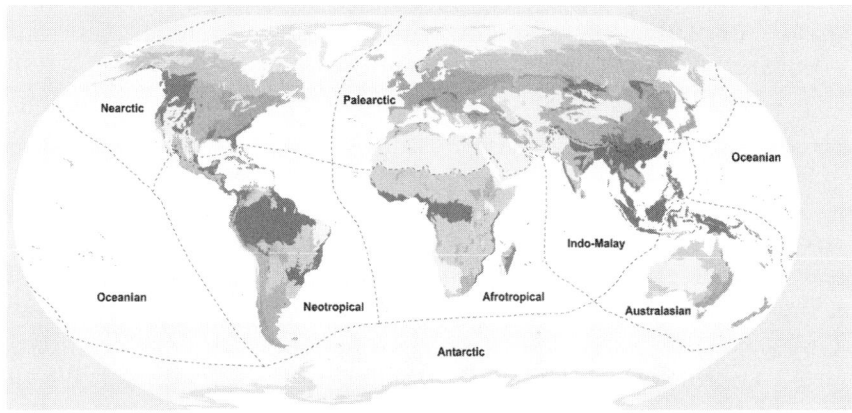

FIGURE 4-1. The biogeographical realms of the Earth. (*Source*: From *The Millennium Ecosystem Assessment report: Current State and Trends,* Figure 4-3, page 839, Appendix A.)

In 2000, then United Nations Secretary General, Kofi Annan called for the completion of The Millennium Ecosystem Assessment (Millennium Ecosystem Assessment, 2005) to assess the consequence of ecosystem change on human well being. The Millennium Ecosystem Assessment contains the current map and descriptions of the biographical realms of the earth (Figure 4-1). Only very slight changes have been made from the work of Wallace and Udvardy.

Habitat is where a particular species prefers to live. The species niche is its role in the ecosystem. Sites within a given biogeographic realm have similar habitats and niches, but the animals that occupy those habitats and fill the various niches are distinct. When a species is transported from one realm to another, it will find an appropriate habitat and niche. The species will then be in competition with the native occupants of that habitat. Because these "invasive" species do not have natural predators, they frequently displace the native species and become nuisances.

BIOMES

Physical barriers define biogeographic realms. Each biogeographic realm contains multiple distinct ecosystems, which are referred to as biomes. Several researchers have developed biome identification systems, which differ depending on their intended usage. Biomes are determined largely by climate. Climate is in turn a function of latitude, position on the continent, and precipitation.

One classification of climate based on vegetation groups is that developed by Wladimir Köppen in 1900 (Kottek et al., 2006). Köppen's classification of the world's climates includes six zones: A—Equatorial, B—Arid, C—Warm temperate, D—Snow, E—Polar, and H—Highlands. A second letter is added to the classification to designate precipitation, and a third for temperature. Thus, a climate may be Cfb, or "warm temperate, fully humid, warm summer." The full classification is given in Table 4-1. The classification system is referred to today as the Köppen-Geiger climate classification because Rudolf Geiger updated the classification and provided maps of the zones in 1951. The Köppen-Geiger classification is widely used today by climatologists. Maps showing the Köppen-Geiger classification for different continents are available on the Internet (National Weather Service, JetStream—Online School for Weather).

Another scheme for classifying biomes based solely on physical parameters is provided by Leslie Holdridge (1947). Holdridge's scheme

TABLE 4-1 Köppen-Geiger Climate Classification System

Main Climate	Description	Precipitation	Description	Temperature	Description
A	Equatorial	W	Desert	h	Hot arid
B	Arid	S	Steppe	k	Cold arid
C	Warm temperate	f	Fully humid	a	Hot summer
D	Snow	s	Summer dry	b	Warm summer
EF	Polar frost	w	Winter dry	c	Cool summer
ET	Polar tundra	m	Monsoonal	d	Extremely continental
H	Highland				

Source: Adapted from Kottek (2006) and the National Weather Service, JetStream–Online School for Weather.

classifies life zones according to three criteria: biotemperature, mean annual precipitation, and potential evapotranspiration ratio (Figure 4-2). Biotemperature is found by adding the mean monthly temperatures for months when that mean is greater than 0°C and dividing by 12. Temperatures below 0 are taken as 0. The rationale of this is that plants

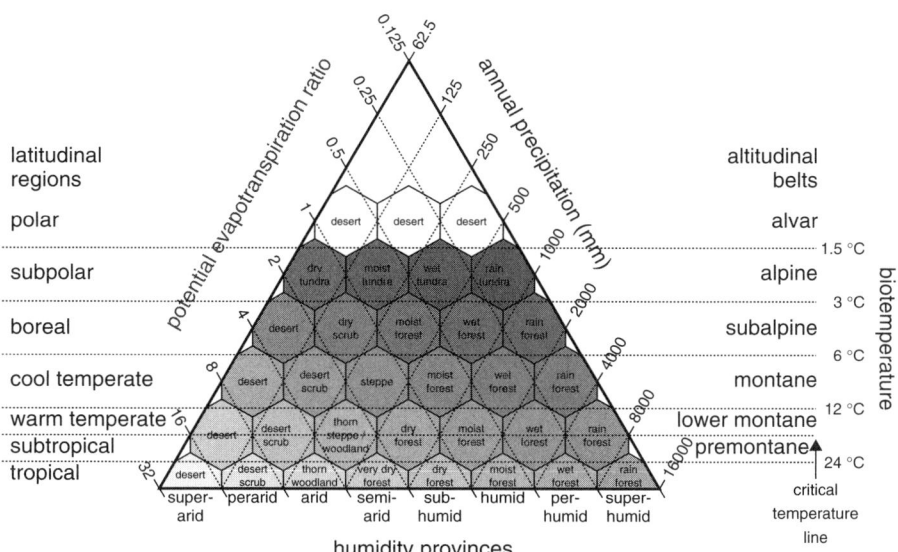

FIGURE 4-2. Holdridge's life zones. Life zones are based on the biotemperature of the region and the annual rainfall. Latitudinal zones are read up the left side of the triangle, and altitudinal zones up the right side. (*Source*: From Halasz, Peter, http://commons.wikimedia.org/wiki/File:Lifezones_Pengo.svg. Adapted from Holdridge, 1947.)

are dormant below 0°C and therefore should not figure into the classification. Mean annual precipitation is self-explanatory. The potential evapotranspiration ratio is the ratio of potential evapotranspiration to precipitation.

In Holdridge's system, latitudinal belts extend north and south from the equator as follows: tropical, subtropical, warm temperate, temperate, cool temperate, cold, frigid, and polar, corresponding to biotemperature. The distinction between subtropical and warm temperate is the presence of a frost line. Subtropical regions do not experience frost. To use the chart, enter from the right by the region's biotemperature, slide across to the left until the diagonal line representing mean annual precipitation is reached, and read the biome on the left of the diagonal. In all, 38 distinct biomes are classified, from polar desert to tropical rainforest.

Holdridge provided for altitudinal belts as well as latitudinal belts. Altitudinal belts correspond to latitudinal belts with similar characteristics. Lower montane corresponds with warm temperate, montane with cool temperate, subalpine with boreal, alpine with subpolar, and alvar with polar.

The advantage of Holdridge's system is that it is based solely on physical parameters. Any two researchers or ecological designers should derive the same biome classification if given equivalent data. The link between biome, temperature, and precipitation is also clearly shown.

Heinrich Walter used a different approach to describing biomes, based on climate zone classification. Walter describes vegetation types in nine worldwide climate zones (Ricklefs, 1996). The classification is based on annual temperature and precipitation. The nine major biomes extending out from the equator are described in Table 4-2. Walter's classification has been widely adopted and is easily understood because the zones correspond with vegetation type.

Robert H. Whittaker (1975) uses yet another approach to defining biomes based on climate. According to Whittaker, when terrestrial locations are plotted according to mean annual temperature and mean annual precipitation, the locations fall approximately in a triangle, with temperature across the bottom decreasing from left to right and temperature increasing from bottom to top. Cold wet regions are rare because water does not evaporate readily at cold temperatures. Therefore, the cold biomes occupy the lower-right portion of the triangle. Tropical regions are uniformly warm, but vary from dry to extremely moist. Therefore those regions occupy the left-hand side of the graph.

TABLE 4-2 Classification of Climate Zones of the World According to Heinrich Walter

Climate Zone	Climate	Vegetation Type
Equatorial	Warm and humid without seasonality	Evergreen tropical rainforest
Tropical	Summer rainy season and cooler "winter" dry season	Seasonal forest, shrub, savanna
Subtropical	Highly seasonal, arid climate	Desert vegetation with considerable exposed surface
Mediterranean	Winter rainy season and summer drought	Drought-adapted, frost-sensitive shrublands and woodlands
Warm temperate	Humid, occasional frost, often with summer rainfall maxima	Temperate evergreen forest, somewhat frost-sensitive
Nemoral	Moderate temperate climate with winter freezing	Frost-resistant, broad-leafed deciduous, temperate forest
Continental	Arid, with warm or hot summer and cold winters	Very frost-resistant grasslands and temperate deserts
Boreal	Cold temperate with cool summers and long winters	Evergreen, frost-hardy, needle-leaved forest or taiga
Polar	Very short, cool summers and long, very cold winters	Treeless tundra, mostly with permafrost soils

Source: Modified from Robert E. Ricklefs, 1996, *The Economy of Nature*.

Whittaker's diagram combines the simplicity of Walter's system with the specificity of Holdrich's. The reliance on temperature and precipitation to classify biomes provides a process that is replicable between researchers. At the same time, Whittaker maintains the system of nine major biomes that Walter uses.

Olson et al. (2001) developed a classification system composed of 14 separate terrestrial biomes for use in global biodiversity conservation. The World Wildlife Fund has adopted Olson et al.'s classification system for implementation of their conservation programs. (Olson et al., 2001) (Figure 4-3). This delineation was adapted from the biogeographic realms of Pielou (1979) and Udvardy (1975). In order to promote coordination and standardization, ecological engineers should use the Olson WWF biome system. A short description of the 14 terrestrial biomes is given in Table 4-3.

TABLE 4-3 The World Wildlife Fund's Biome Classification

Biome	Description	Vegetation Type
Tropical and Subtropical Moist Broadleaf Forests	Annual temperature 20–25°C, rainfall generally >200 cm, no seasonality.	Semi-evergreen and evergreen trees, Nutrient-poor and acidic soils, light limitation on production, the most diverse fauna on earth.
Tropical and Subtropical Dry Broadleaf Forests	Year-round warmth, abundant precipitation but with high variability, seasonal droughts.	Deciduous trees that lose their leaves during the dry season, because of the opening of the forest, thick underbrush develops, highly diverse fauna.
Tropical and Subtropical Coniferous Forests	Tropical temperatures but with little precipitation with moderate variability.	A diverse collection of conifers, thick closed canopy with little underbrush, fungi and ferns frequently cover the ground. The understory is shrubs and small trees. Many migratory birds and butterflies winter here.
Temperate Broadleaf and Mixed Forests	Widely ranging temperature and precipitation.	Deciduous trees mix with evergreens. The forest has four layers: a canopy, lower mature trees, shrub, and ground cover of grass or other herbaceous plants.
Temperate Coniferous Forests	Warm summers and cool winters. Frequently fire-dominated.	Evergreen trees including pine, cedar, redwood, and fir. The forest has a canopy and an understory of shrubs or grasses and forbs.
Boreal Forests/Taiga	Low annual mean temperature and precipitation ranging from 40 to 100 cm per year.	Nutrient-poor soils and poor drainage favor conifers with some deciduous trees mixed in. Ground cover is mostly mosses and lichens.
Tropical and Subtropical Grasslands, Savannas, and Shrublands	Tropical temperatures and rainfall of only 90–150 cm per year.	Grasses dominate with some isolated trees. Large foraging animals are common.
Temperate Grasslands, Savannas, and Shrublands	Widely varying temperature and precipitation, although precipitation is less than temperate broadleaf forests.	Generally devoid of trees except in riparian belts. Grasses dominate the landscape. Fauna includes large grazing animals and their predators, burrowing animals, birds, and insects.
Flooded Grasslands and Savannas	Not climate-based.	Plants and animals adapted to hydric conditions. Frequently large assemblages of migratory and resident water birds.
Montane Grasslands and Shrublands	Tropical, subtropical, and temperate.	Similar to other grasslands and shrublands of the same temperature range.

(*continues*)

TABLE 4-3 (*continued*)

Tundra	Long, frigid winters with extended periods of darkness, very dry. Most precipitation falls as snow in winter.	Communities of sedges and heaths and dwarf shrubs.
Mediterranean Forests, Woodlands and Scrub	Hot and dry summers with cool and moist winters. Most precipitation arrives during winter.	Animals and plants uniquely adapted to the hot summers with little rain. Most plants are fire-adapted and dependent on fire for their persistence.
Deserts and Xeric Shrublands	Less than 25 cm precipitation annually, extreme temperatures.	Woody-stemmed plants and shrubs and desert adapted animals.
Mangroves	Subject to twice-daily tides.	Waterlogged, salty soils of sheltered tropical and subtropical shores. Salt-tolerant trees and many aquatic animal species.

Source: Olson 2001.

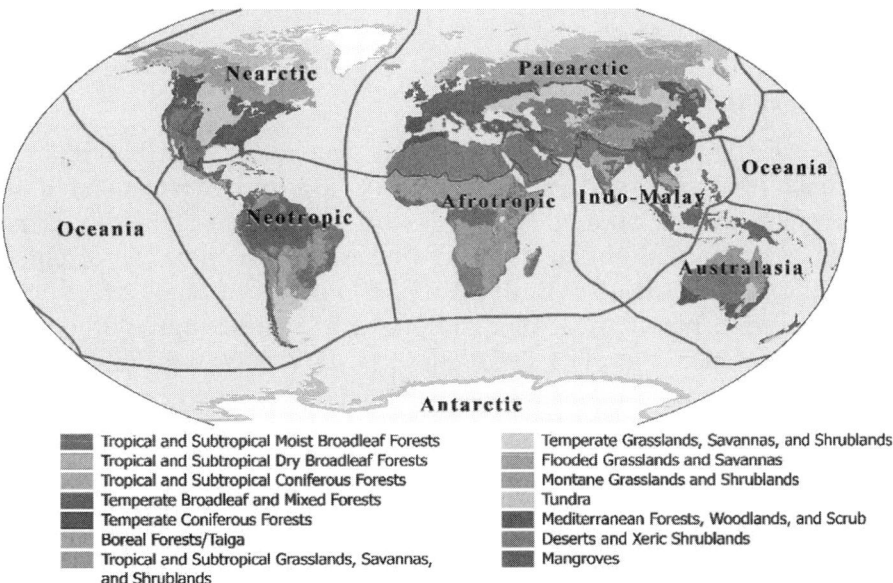

FIGURE 4-3. The 14 major biomes of the world from the World Wildlife Fund. (*Source*: Olson et al., 2001.)

ECOREGIONS

The designation of biomes is at the global scale. Within each biome, similar conditions of climate occur, but the biological communities that fill the similar niches and habitats may be quite dissimilar. Ecological regions, or ecoregions, provide local information regarding specific conditions. Ecoregions are "areas of general similarity in ecosystems and in the type, quality and quantity of environmental resources" (Omernik, 1995). Ecoregions serve as a spatial framework for the research, assessment, management, and monitoring of ecosystems and ecosystem components.

Ecoregions are the comparative scale for ecosystem services design; any design within a site in an ecoregion should be transferable to another site within the same ecoregion, with some obvious site-specific limitations. The knowledge of place allows ecological engineers to utilize locally available materials and design for local conditions. Place also defines what ecological services are available to the ecotechnology team, as well as what must be done to conserve those services.

Three ecoregion delineations are in common use today: Bailey (1998), Omernik (1995), and the Olson World Wildlife Fund (WWF) system (Olson et al., 2001). Bailey (1995), working for the U.S. Forest Service, delineated ecoregions primarily for use by forest managers.

Bailey's Ecoregions

Bailey (1998) used three hierarchical categories for ecoregions: domains, divisions, and provinces. This approach corresponded to forest management strategies, and is still broadly used in the forest industry. Ecological engineers designing for forest industry–oriented projects should include Bailey's ecoregion as a site descriptor. Bailey used a hierarchical description with a three-digit number to differentiate each site. The hundreds digit describes the domain, the tens digit the division, and the ones digit the province.

Domains are continental in scale and are largely climate-defined. The domains relate well to Köppen's climate zones. Four domains are described: polar, humid temperate, dry, and humid tropical. These domains are given identifying numbers of 100, 200, 300, and 400, respectively. Within the domains, there are 15 subcontinental divisions. Divisions exist because of the variety of climate within the provinces. But because of a paucity of data at the smaller scale, divisions are delineated by discontinuities in physiography and/or vegetation physiognomy (Bailey, 1995).

Finally, local microclimate determines specific plant communities that exist in the local area. These major plant formations may be lumped together into fine-scale ecoregions called provinces. There are 34 provinces within Bailey's ecoregions for North America. The Bailey ecoregions for the world may be found at the United States Forest Service website: www.fs.fed.us/land/ecosysmgmt.

Omernik's Ecoregions

Omernik's ecoregions were developed by integrating watersheds with soil type and plant associations, and thus are more commonly used in ecosystem assessment. Omernik's system of ecoregion delineation defines three levels of ecoregions. Level I ecoregions are continental in scale and correspond closely to Bailey's divisions. In North America, there are 15 Level I regions (Figure 4-4). As with biomes, this level of delineation is dependent mostly on climate, especially temperature extremes and precipitation. While Level I ecoregions have similar biotic communities, they span vast geographic areas. For instance, the Great Plains ecoregion extends from West Texas to southeastern Alberta. The climate variability across this large an area makes Level I ecoregions limited in utility for ecosystem services design. They are useful largely for the same reason biomes are useful—they give a general set of characteristics for a site.

Level II ecoregions (Figure 4-5) are more finely delineated than Level I and provide more local information. West Texas is in Level II ecoregion 9.4, south central semi-arid prairies, while southeastern Alberta is in ecoregion 9.3, west central semi-arid prairies. These are both semi-arid prairie ecosystems within the Level I Great Plains ecoregion, suggesting that they have similar ecosystem constraints and processes. An ecological engineer working in one of these ecoregions could translate design at some level to the other, but with clear caution, given the differences in latitude between these semi-arid prairie systems.

Level III ecoregions (Figure 4-6 a and b) provide even more local detail. For example, the south central semi-arid prairies (ecoregion 9.4) contain a number of Level III ecoregions, such as 9.4.1, High Plains, 9.4.3, the Southwestern Tablelands, and 9.4.6, the Edwards Plateau. Each of these ecoregions has very specific characteristics and challenges in ecosystem services design. For example, the Edwards Plateau includes the recharge zone for the Edwards Aquifer, an ecologically sensitive aquifer that supports a number of endangered species, and

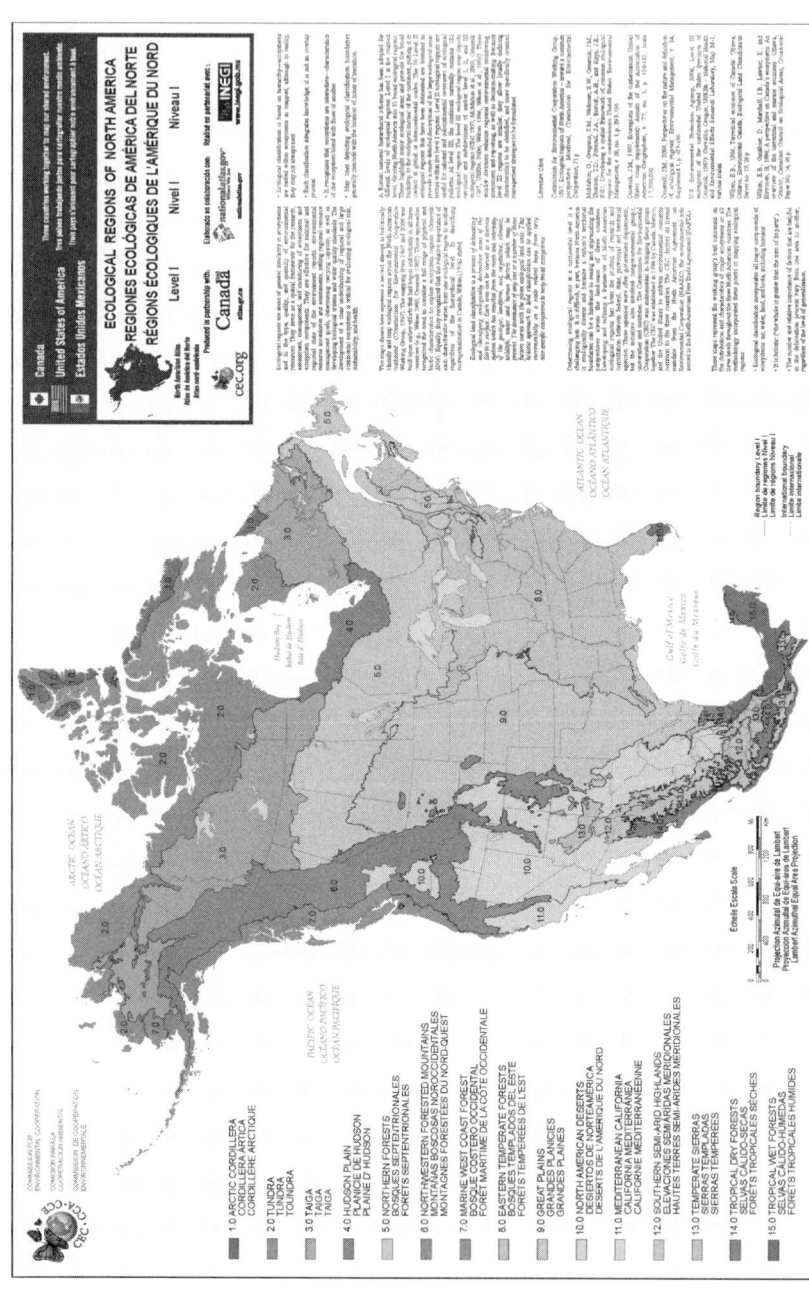

FIGURE 4-4. Omernik's Level I ecoregions of North America. [Environmental Protection Agency (ftp://ftp.epa.gov/wed/ecoregions/cec_na/NA_LEVEL_I.pdf)]

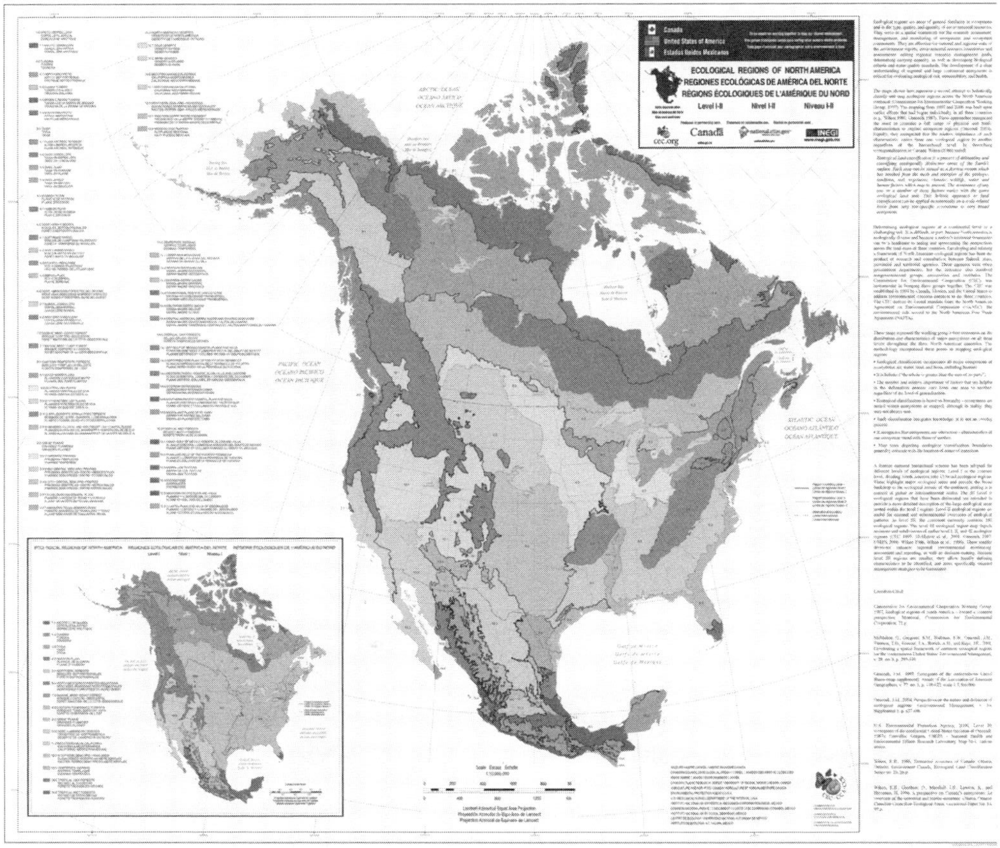

FIGURE 4-5. Level II ecoregions of North America according to Omernik. [Environmental Protection Agency (ftp://ftp.epa.gov/wed/ecoregions/cec_na/NA_LEVEL_II.pdf)]

provides much of the drinking water for the cities of San Antonio and Austin, Texas. Level III ecoregions provide very specific local characteristics that are critical for ecosystem services design.

Olson's Ecoregions

Olson's WWF ecoregion delineation was developed to assess conservation needs at the global scale (Olson et al., 2001). Olson et al. developed this system to reflect the distribution of fauna and flora across the entire biosphere, using a nested framework similar to Omernik's. This approach was designed to provide a framework for comparisons among units, and to characterize representative habitats and species assemblages within ecoregions. This approach supports ecosystem services design explicitly, by providing enhanced comparability across continents.

Olson et al. (2001) delineated global ecoregions from their 14 biomes by adapting the biome systems of Dinerstein et al. (1995) and Ricketts et al. (1999), and incorporating a suite of global and regional map units based on distribution of plants and animals, vegetation type, and region-specific characterizations such as White's (1983)

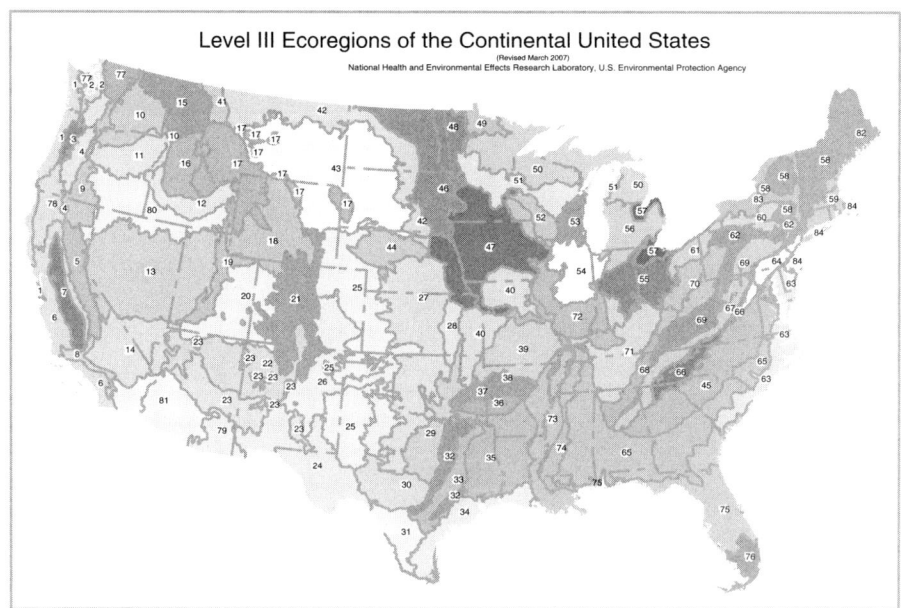

FIGURE 4-6. Omernik's Level III ecoregions for the continental United States. (*Source*: From U.S. EPA Western Ecology Division, 200 S.W. 35 Street, Corvallis, Oregon 97333.)

Omernik Level III Ecoregions Key

1. Coast Range
2. Puget Lowland
3. Willamette Valley
4. Cascades
5. Sierra Nevada
6. Southern and Central California Chaparral and Oak Woodlands
7. Central California Valley
8. Southern California Mountains
9. Eastern Cascades Slopes and Foothills
10. Columbia Plateau
11. Blue Mountains
12. Snake River Plain
13. Central Basin and Range
14. Mojave Basin and Range
15. Northern Rockies
16. Idaho Batholith
17. Middle Rockies
18. Wyoming Basin
19. Wasatch and Uinta Mountains
20. Colorado Plateaus
21. Southern Rockies
22. Arizona/New Mexico Plateau
23. Arizona/New Mexico Mountains
24. Chihuahuan Deserts
25. High Plains
26. Southwestern Tablelands
27. Central Great Plains
28. Flint Hills
29. Cross Timbers
30. Edwards Plateau
31. Southern Texas Plains
32. Texas Blackland Prairies
33. East Central Texas Plains
34. Western Gulf Coastal Plain
35. South Central Plains
36. Ouachita Mountains
37. Arkansas Valley
38. Boston Mountains
39. Ozark Highlands
40. Central Irregular Plains
41. Canadian Rockies
42. Northwestern Glaciated Plains
43. Northwestern Great Plains
44. Nebraska Sand Hills
45. Piedmont
46. Northern Glaciated Plains
47. Western Corn Belt Plains
48. Lake Agassiz Plain
49. Northern Minnesota Wetlands
50. Northern Lakes and Forests
51. North Central Hardwood Forests
52. Driftless Area
53. Southeastern Wisconsin Till Plains
54. Central Corn Belt Plains
55. Eastern Corn Belt Plains
56. Southern Michigan/Northern Indiana Drift Plains
57. Huron/Erie Lake Plains
58. Northeastern Highlands
59. Northeastern Coastal Zone
60. Northern Appalachian Plateau and Uplands
61. Erie Drift Plain
62. North Central Appalachians
63. Middle Atlantic Coastal Plain
64. Northern Piedmont
65. Southeastern Plains
66. Blue Ridge
67. Ridge and Valley
68. Southwestern Appalachians
69. Central Appalachians
70. Western Allegheny Plateau
71. Interior Plateau
72. Interior River Valleys and Hills
73. Mississippi Alluvial Plain
74. Mississippi Valley Loess Plains
75. Southern Coastal Plain
76. Southern Florida Coastal Plain
77. North Cascades
78. Klamath Mountains
79. Madrean Archipelago
80. Northern Basin and Range
81. Sonoran Basin and Range
82. Laurentian Plains and Hills
83. Eastern Great Lakes and Hudson Lowlands
84. Atlantic Coastal Pine Barrens

FIGURE 4-6. (*Continued*)

phytogeographic regions for the Afrotropics (Olson et al., 2001). The goal was to aggregate existing information about biogeography at the highest resolution that (1) matched recognized biogeographic divisions across methods, (2) achieved similar levels of spatial and biogeographic resolution between units, and (3) matched units and

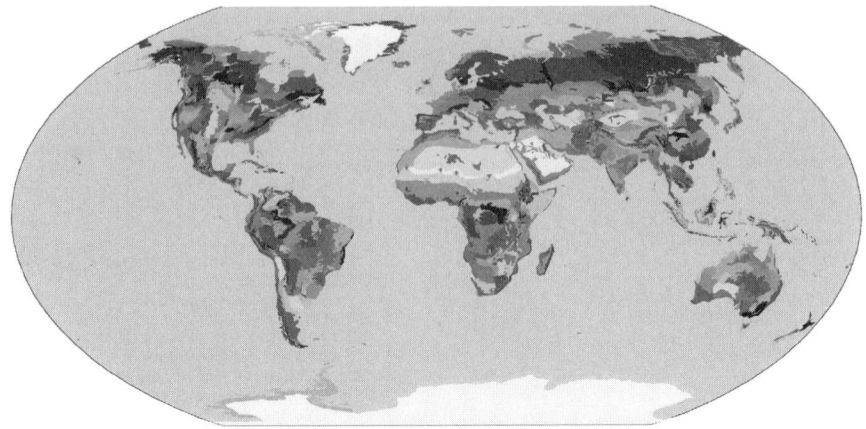

FIGURE 4-7. WWF ecoregions of Earth. (*Source*: According to Olson et al., 2001.)

boundaries in adjacent systems. This system was modified slightly for boreal and polar regions, where biogeographic characteristics are relatively homogenous across very large areas (Olson et al., 2001).

The Olson WWF delineation of ecoregions recognizes 867 ecoregional units across the globe (Figure 4-7). Ecological engineers designing within the U.S. and Canada should use Omernik's delineation, and sites outside the U.S. and Canada should comply with Olson's WWF ecoregion delineation system.

OTHER LAND CLASSIFICATION SYSTEMS

There are other land area classification systems that ecological engineers should be familiar with. The Natural Resources Conservation Service (NRCS) uses a slightly different version of the ecoregion concept. The NRCS delineates major land resource areas (MLRAs). MLRAs are delineated based on physiography, geology, climate, water source and use, soils, biological resources, and land use. The MLRA reflects the actual use of the land, rather than the natural biotic community. MLRAs are used by federal, state, and local agencies in developing conservation and commodity programs (NRCS, 2006).

The USDA plant hardiness zones can also be a useful way of visualizing land areas. The coldest temperature that a region may experience defines plant hardiness zones. North America has 11 zones. Each zone is 10 degrees warmer than the previous zone. Nurseries often describe a plant's ability to survive by the hardiness zone in which it is expected to thrive. A downside of the plant hardiness concept is that it considers only temperature. This might be okay if unlimited irrigation water is available. But if plants are expected to survive with minimum maintenance, then the precipitation pattern must also be considered. Plant hardiness maps for the U.S. and most other countries are available from the National Gardening Association (www.garden.org).

Climate Change and Ecoregions

Because biomes and ecoregions are defined by climate, changes in climate will ultimately result in changes in ecoregions. This change is already being experienced. In 2003, the USDA's plant hardiness zones were revised to reflect milder winters in much of the U.S. The southern Rocky Mountain pinyon pine forests are in dramatic decline, and may disappear in a decade (Dale et al., 2001; Adams et al., 2009). The cumulative impact of altered frequency and intensity of ground fire, drought, and the added pressure from pathogens are reducing the areal extent of pinyon pines (*Pinus edulis*) across the U.S. Southwest. Human impact on ecoregions may also force a reconsideration of the boundaries. Everything is connected. As land use changes, the biological community also changes (see Chapter 8). As the community changes, the boundaries of at least the finer-scale ecoregions will change with them.

An emerging challenge for ecological engineers will be to design ecosystem services in response to these global patterns of change. The methods for designing landscapes for changing climate are not yet developed. This will be a major challenge for ecological engineers for the first half of the twenty-first century. The utility of ecoregions becomes clear in this context. If an ecoregion is projected to become more arid (or more humid), the ecological engineer can anticipate the shift in community structure by evaluating the ecoregion that most approximates the anticipated shift. The ecological engineer should evaluate the similar Omernik Level III ecoregions within a Level II category for one with similar characteristics. If no similar system exists at Level II, the ecological engineer can move up to Level I to find ecoregions of similar characteristics, but with the anticipated range of temperature and rainfall. This process gives the ecotechnology design team a palette of potential ecoregion characteristics to anticipate over time. This approach also provides a mechanism for the Ecosystem Services Advisory Committee (ESAC) to understand and communicate shifting ecosystem dynamics.

Land Use Change and Ecoregions

Land use change is the single largest ecosystem pressure exerted by human beings on the landscape. The rate and extent of land use change on Earth is hard to fathom; hectares of rainforest converted to agriculture per day or per second do not adequately capture the global impact of our terraforming activities. Foley et al. (2005) demonstrated that in the past 200 years, human beings have redefined biomes across the planet, making agriculture (row crops, meadows, and grazing lands) the largest biome on the planet (Chapter 2).

The ecological engineer must explicitly quantify land use change within an ecoregion in order to understand the impact of human activities on a site. Information for the assessment of land use change within an ecoregion is becoming more accessible as land use geographic information system (GIS) data layers become more common. An ecotechnology team should be able to map land use change within an ecoregion over the past 20 years (from 1990) in most of the U.S. and Europe. These changes should focus on increases in percent impervious cover, habitat fragmentation with roads and other linear elements, fragmentation of habitat patches, and loss of forest and riparian land use.

It can be helpful to characterize land use change in adjacent ecoregions as well. This can provide ecoregion-level comparisons of impacts of aggregate human activities. Analysis of land use change impacts in similar ecoregions that have already undergone more extensive land use change can provide valuable insights into the potential impacts of land use decisions within an ecoregion that has not yet been altered. The land use change impacts can be associated with loss of key species, degradation of water resources, decreases in residence quality of life satisfaction metrics, and other key indicators. When expressed as a percent of the total land area of an ecoregion, land use change metrics can be powerful metrics for framing and prioritizing ecosystem services for restoration and conservation.

Olson et al. (2001) offer several cautions for using ecoregions. Delineating biomes and ecoregions is somewhat arbitrary, as most landforms can be seen as continuums at some scale. Ecoregion boundaries were selected by best professional judgment of ecologists experienced in the delineation methods described in this chapter, using data available at the time, but as indicated, global biogeography is changing at increasing rates. The boundaries between ecoregions must be considered wide gray lines, rather than sharp delineations. Of special interest is the area near the boundaries of ecoregions. These regions of transition exhibit characteristics of each of the neighboring ecoregions. As a result, these transitional regions are areas of unusually high biological diversity, because they offer high density of ecotones and landscape mosaics.

The ecoregion defines place in an ecological sense. From a geochemical perspective, water flow is the defining characteristic of place. The watershed or catchment provides a valuable context for ecological management, and is the subject of Chapter 5.

> The land ethic simply enlarges the boundaries of the community to include the soil, waters, plants, and animals, or collectively: the land.
>
> —Aldo Leopold

References

Adams, Henry D., Maite Guardiola-Claramonte, Greg A. Barron-Gafford, Juan Camilo Villegas, David D. Breshears, Chris B. Zou, Peter A. Troch, and Travis E. Huxman, Temperature sensitivity of drought-induced tree mortality portends increased regional die-off under global change-type drought, *Proceedings of the National Academy of Sciences USA*, 106: E107, 2009.

Bailey, Robert G., Description of the Ecoregions of the United States, Rocky Mountain Research Station, United States Forest Service, Ft. Collins, CO, 1995, www.fs.fed.us/land/ecosysmgmt/index.html (accessed September 24, 2009).

Bailey, R.G., *Ecoregions: The Ecosystem Geography of Oceans and Continents*, Springer-Verlag, New York, NY, 1998.

Dale, Virginia H., Linda A. Joyce, Steve Mcnulty, Ronald P. Neilson, Matthew P. Ayres, Michael D. Flannigan, Paul J. Hanson, Lloyd C. Irland, Ariel E. Lugo, Chris J. Peterson, Daniel Simberloff, Frederick J. Swanson, Brian J. Stocks, and B. Michael Wotton, Climate change and forest disturbances, *BioSciences*, 51 (9): 723–734, 2001.

Dinerstein, E., D.M. Olson, D.J. Graham, A.L. Webster, S.A. Primm, M.P. Bookbinder, and G. Ledec, *A Conservation Assessment of the Terrestrial Ecoregions of Latin America and the Caribbean*, World Bank, Washington, DC, 1995.

Foley, J.A., R. DeFries, G.P. Asner, C. Barford, G. Bonan, S.R. Carpenter, F.S. Chapin, M.T. Coe, G.C. Daily, H.K. Gibbs, J.H. Helkowski, T. Holloway, E.A. Howard, C.J. Kucharik, C. Monfreda, J.A. Patz, I.C. Prentice, N. Ramankutty, and P.K. Snyder., Global consequences of land use, *Science*, 309: 570–574, 2005.

Holdridge, L.R., Determination of world plant formations from simple climatic data, *Science*, New Series, 105 (2727): 367–368, April 4, 1947.

Kottek, Markus, J. Grieser, C. Beck, B. Rudolf, and F. Rubel, World map of the Köppen-Geiger climate classification updated, *Meteorologische Zeitschrift*, 15 (3): 259–263, 2006.

Millennium Ecosystem Assessment, Current State and Trends Island Press, Washington, D.C. 2005, www.millenniumassessment.org/en/Condition.aspx.

NRCS, Major Land Resource Areas Database, USDA, 2006, http://soils.usda.gov/survey/geography/mlra/.

Olson, David M., Eric Dinerstein, Eric D. Wikramanayake, Neil D. Burgess, George V. N. Powell, Emma C. Underwood, Jennifer A. D'amico, Illanga Itoua, Holly E. Strand, John C. Morrison, Colby J. Loucks, Thomas F. Allnutt, Taylor H. Ricketts, Yumiko Kura, John F. Lamoreux, Wesley W. Wettengel, Prashant Hedao, and Kenneth R. Kassem, Terrestrial ecoregions of the world: A new map of life on Earth, *BioScience*, 51 (11): 933–938, 2001.

Omernik, J.M., Ecoregions: A spatial framework for environmental management, in W.S. Davis and T.P. Simon, eds., *Biological Assessment and Criteria: Tools for Water Resource Planning and Decision Making*, Lewis Publishers, Boca Raton, FL, pp. 49–62, 1995.

Pielou, E.C., *Biogeography*, John Wiley & Sons, Hoboken, NJ, 1979.

Ricketts, T.H., et al., *Terrestrial Ecoregions of North America: A Conservation Assessment*, Island Press, Washington, DC, 1999.

Ricklefs, Robert E., *The Economy of Nature*, W.H. Freeman and Company, New York, NY, 1996.

Udvardy, M.D.F., A classification of the biogeographical provinces of the world, IUCN Occasional Paper no. 18, International Union of Conservation of Nature and Natural Resources, Morges, Switzerland, 1975.

Wallace, Alfred Russel,The Geographical Distribution of Animals; With A Study of the Relations of Living and Extinct Faunas as Elucidating the Past Changes of the Earth's Surface, 2 volumes, Macmillan & Co., London, UK, May 1876, http://books.google.com/books?id=8lAPAAAAYAAJ&pg=PR3#v=onepageq=f=false (accessed September 24, 2009).

White, F., The Vegetation of Africa: A Descriptive Memoir to Accompany the UNESCO/AETFAT/UNSO Vegetation Map of Africa, UNESCO, Paris, 1983.

Whittaker, Robert H., *Communities and Ecosystems*, 2 edition, Macmillan Publishing Inc, New York, 1975.

World Wildlife Fund, About Our Earth, Major Habitat Types, Terrestrial Ecoregions, www.panda.org/about_our_earth/ecoregions/about/habitat_types/selecting_terrestrial_ecoregions/ (accessed September 25, 2009).

5
Defining Place: The Watershed

We forget that the water cycle and the life cycle are one.

—Jacques Yves Cousteau

INTRODUCTION

Water is essential to life on earth. Water is also one of the driving forces shaping Earth, and thus Earth's ecosystems. Hydrologic processes drive landscape processes, the fate and transport of pollutants, and the shape of the landscape. Understanding these processes leads to an understanding of the ecosystems and ecosystem services that arise from the landscape. Watersheds are the minimum unit of ecosystem management, and thus design. In the Omernik ecoregion framework, watersheds are nested within ecoregions at Level II, because Omernik used watersheds as delineating criteria. Ecotechnology teams designing a site should characterize the ecoregions and watersheds within which the site resides, and the adjoining watersheds and ecoregions.

A watershed, or catchment, is the area of land that drains to a particular point on a stream, lake, or estuary. For water resources, the watershed is the most logical unit of management. When management begins at the watershed, the water resource becomes the focal point; managers develop a more complete understanding of overall conditions in the area and the stressors affecting those conditions (EPA, 1996). Management from the watershed level allows the manager and ecological engineer to consider factors beyond chemical pollution in protecting water quality. These factors include habitat destruction, geomorphologic changes, and changes in land use.

This chapter develops fundamental concepts in watershed processes. Understanding watershed processes exemplifies the three axioms of ecological engineering: (1) everything is connected, (2) everything is changing, and (3) we are all in this together. When watersheds are the unit of ecosystem management, ecological engineering projects are automatically placed in the larger context of the local ecosystem.

Watershed Services

Ecosystem services include provisioning, regulating, cultural and supporting services (Millennium Ecosystem Assessment, 2005). These services are derived from ecoregion landforms. Watersheds provide services from all of these groups, including (Smith et al., 2006):

Provisioning services: Freshwater supply, food production, fish production, timber and building materials supply, medicines, and hydroelectric power
Regulating services: Buffer runoff, soil water infiltration, groundwater recharge, baseflow maintenance, flood prevention and reduction, landslide reduction, soil protection, erosion and sediment control, surface and groundwater quality protection
Cultural services: Aquatic recreation, landscape aesthetics, cultural heritage and identity, artistic and spiritual inspiration
Supporting services: Habitat, flow regime regulation

The various groups of ecosystems that form watersheds, including forests, grasslands, wetlands, croplands, urban areas, riparian zones, streams, and lakes, provide the infrastructure that supports watershed services (Smith et al., 2006). Understanding these ecosystems is key to ecological design within a watershed.

Watershed Characteristics: Physical Description

Much as with ecoregions, characteristics describing the watershed are the area, geology, topography, drainage network, soils, land use/land cover, precipitation, and hydrology. Geographic information system (GIS) software has greatly enhanced our ability to gather this information and rapidly characterize a watershed. However, the data may also be compiled manually, if necessary.

Area
The drainage area of a watershed, along with the regional climate, determines the amount of water flowing out of the watershed. The watershed can vary in size from a residential yard to a large portion of a continent. Manual delineation of a watershed is done on a topographic map (Figure 5-1) with the following procedure: Start from the point of drainage to the receiving waterbody (estuary, stream, or lake) and draw a line perpendicular to the contour lines upslope to the ridgeline (or divide) on each side of the stream. Once the ridgeline

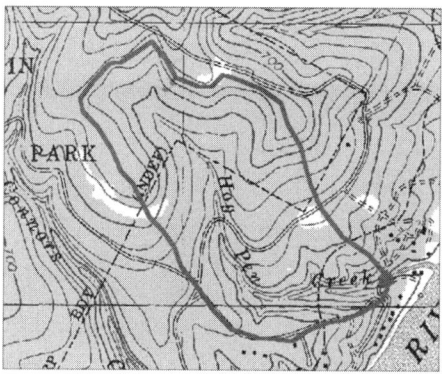

FIGURE 5-1. Watershed delineation. Watersheds can be delineated manually on topographic maps. Starting from the discharge point of the watershed, draw a line perpendicular to contours up to the ridgeline on each side of the valley. Then, trace along the ridgeline to delineate the watershed divide.

is reached, turn the line upstream and continue drawing along the ridgeline until you intersect the line going up the opposite side of the valley. The watershed area can then be determined as the area within the boundary diagrammed.

Watersheds are seldom delineated manually. Several GIS programs have watershed delineation routines built in. These are common to GIS and computer-aided design (CAD) programs, including ArcGIS (ESRI, Redlands, CA), MicroStation (Bentley Systems, Inc., Exton, PA), and AutoCAD (Autodesk, Inc., San Rafael, CA).

For programmatic proposes and uniformity, watersheds are often delimited by government agencies. In the United States, the Natural Resources Conservation Service and the United States Geological Survey (USGS) delineate watersheds in a hierarchical scheme with sub-watersheds nested inside watersheds inside larger drainage basins. These delineations are referred to as hydrologic units, and each is identified by a hydrologic unit code, or HUC. Data are available from many government databases describing HUC areas (www.nationalatlas.gov and various state repositories). In the U.S., there are six levels of delineation, with smaller HUCs nested within larger units (USGS, 2009). In order of descending area, the HUCs divide the country hydrologically into 21 regions, 222 subregions, 352 accounting units, and 2,150 cataloging units (Figure 5-2). Parts of the country are further delineated into watersheds and sub-watersheds.

A regions is defined by a two-digit HUC and may be the drainage basin of a major river or the combined drainage area of several rivers, such as

86   Ecological Engineering Design

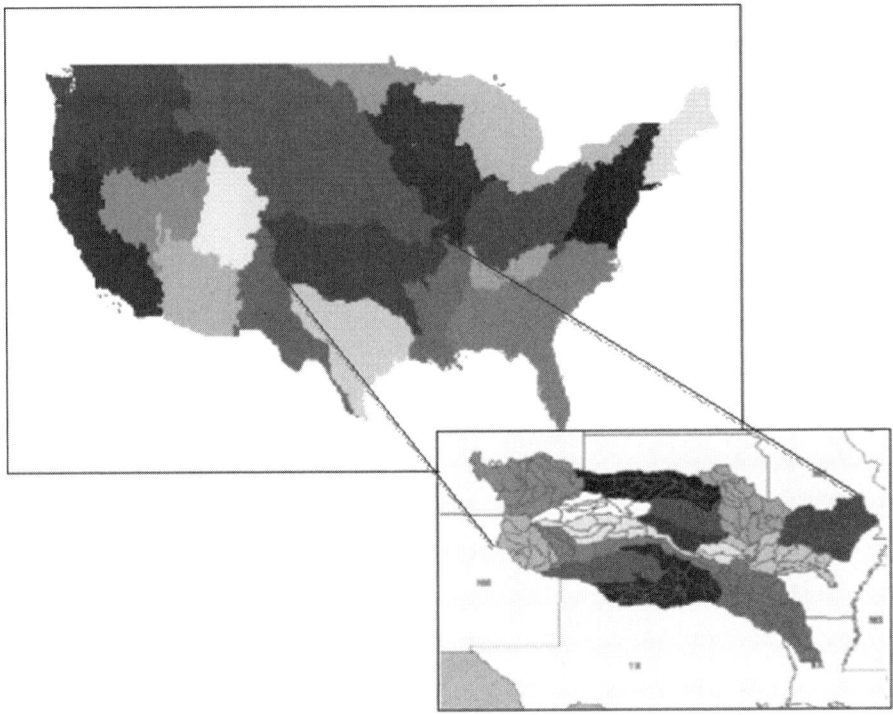

FIGURE 5-2. Hydrologic unit codes for the continental United States. The large map shows the two-digit HUCs referred to as regions. The smaller map draws out region 11, the Arkansas-White-Red region, and shows six-digit accounting units and eight-digit cataloguing units. Four-digit subregions are combinations of the accounting units. (*Source*: Data from the National Atlas.)

New England. Subregions divide regions and include the area drained by a river system, a section of a river and its tributaries in that reach, a closed basin or basins, or a group of streams forming a coastal drainage area (USGS, 2009). Subregions are defined by a four-digit HUC. The first two digits are the same as the larger regional HUC, and the last two define the subregion.

Accounting units, or basins, subdivide the subregions. They are used by the USGS for managing national water data (USGS, 2009). Accounting regions are defined by a six-digit HUC. The last two digits of the HUC are the accounting unit; the preceding digits describe the region and subregion. Cataloging units are smaller hydrologic unit areas, commonly referred to as sub-basins, and are defined by an eight-digit HUC. Most cataloging units are larger than 700 square miles in area (USGS, 2009). A cataloging unit is a distinct hydrologic feature.

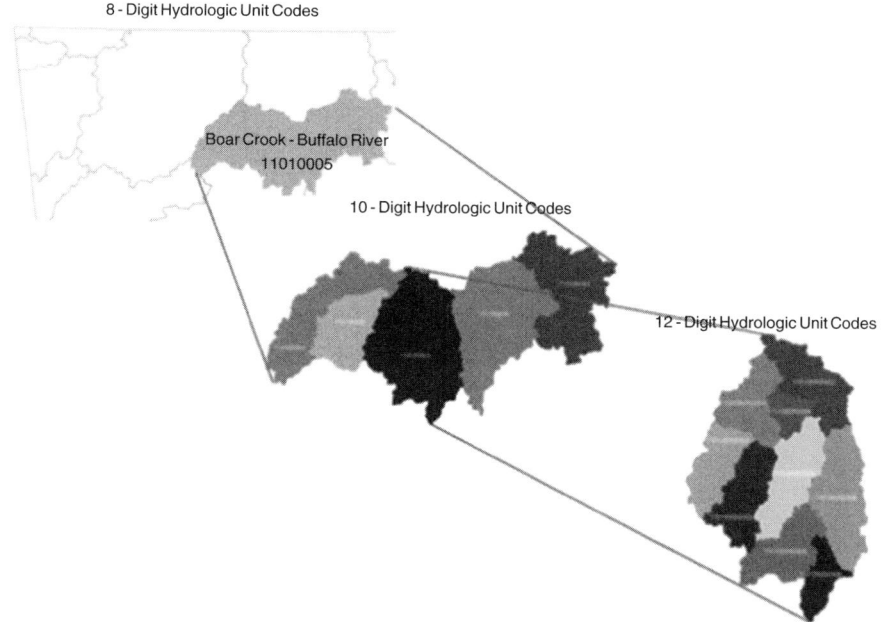

FIGURE 5-3. Eight, ten, and twelve-digit hydrologic unit codes (HUCs) in the Buffalo National River, Arkansas. The eight-digit HUC is the cataloguing unit, ten-digit HUCs are referred to as watersheds, and twelve-digit HUCs as sub-watersheds.

Maps of the hydrologic units are available from the USGS, as are GIS datasets containing the HUCs.

The delineation of hydrologic units is a continuing process. Complete sets of 10-digit and 12-digit HUCs are now available from the National Atlas (USGS, 2009). The unit defined by a 10-digit HUC is referred to as a watershed, and the unit defined by 12 digits is a sub-watershed (Figure 5-3). The National Hydrologic Dataset (USGS, 2009) provides data on hydrologic unit areas down to the 12-digit level. Many attributes of the delineated areas are available in the dataset.

Watershed Shape and Topography

On a map, watersheds may take many shapes, including fan-shaped and elongated (Figure 5-4). For two watersheds of a given area, the fan-shaped watershed will have a higher peak discharge of stormwater than the elongated watershed, because water from all parts of the watershed will tend to arrive at the outlet at close to the same time. In the elongated watershed, on the other hand, much of the runoff water from the downstream portion of the watershed will have already

88   Ecological Engineering Design

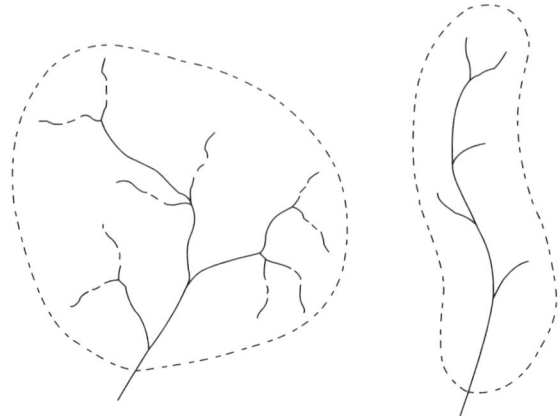

FIGURE 5-4. A watershed's shape affects peak discharge in watershed streams. Fan-shaped watersheds collect water more quickly, resulting in higher per hectare discharge.

passed the watershed outlet when water from more upstream parts of the watershed arrives at the outlet (Ward and Elliot, 1995).

Understanding the shape of the watershed can give the ecotechnology team insights into hydrologic processes. The Gravelius index describes a watershed's shape as the ratio of the perimeter of the watershed to the perimeter of a circular watershed having the same area:

$$Kg = P/2(\pi A)1/2$$

where
  Kg = Gravelius index
  P = watershed perimeter
  A = watershed area

The more different the Gravelius index is from 1.0, the more elongated the watershed. Fractal geometry has shown that a watershed's perimeter is sensitive to the scale used for measurement. Therefore, comparing the index from two watersheds is pointless unless the measurement scale is known. The Gravelius index should be used with some caution. However, for comparing watersheds delineated at the same scale, it has utility.

Topography
Topography is the description of an area's surface, including both natural and man-made features. Watershed topography may influence

the local climate through the watershed's aspect or direction of slope, elevation with respect to surrounding terrain, and aspects of runoff hydrology. Description of the watershed should consider the general elevation, slope, relief (difference between high and low elevation), watershed orientation, stream density, and man-made features such as roads, reservoirs, levees, and the like. The characteristics of topography affect how much precipitation infiltrates or runs off, water storage, runoff speed, and soil water content.

Stream density refers to the miles of stream or river per unit of watershed area. Drainage density provides information about the runoff potential of a watershed. Watersheds in arid regions and with high permeability will have low drainage density, whereas a watershed with high precipitation and clay soils may have much higher density of streams (Pidwirny, 2006).

Geology
The local geology—the physical shape, structure, and composition of the area—is the backbone on which the watershed is built. The underlying bedrock is the parent material of the soil of the watershed. That underlying material also largely determines the natural mineral composition of water in the watershed and the permeability of the soils. Geologic formations define many of the physical characteristics of watersheds such as the location of ridgelines that form watershed divides and the gradient of streams.

A special consideration is areas of Karst terrain. Karst, named after the city in Poland where the terrain was first described, refers to an area with limestone geology and multiple solution channels (Figure 5-5). Caves, springs, and sinkholes characterize Karst terrain. In Karst areas, there is a direct connection between surface water and groundwater. Water may inflow into the ground through sinkholes and outflow at springs and seeps. In Karst areas, the water table may rise and fall rapidly, and surface water pollution problems are transferred to groundwater. The Edwards Aquifer near Austin, Texas, is an example of Karst terrain.

Soils
Ecosystem functions performed by the soil include: acting as a medium for plant growth, providing water storage and purification, recycling organic materials and nutrients, providing habitat for soil organisms, and serving as an engineering medium. Understanding the soil and how it provides these functions is fundamental to understanding place.

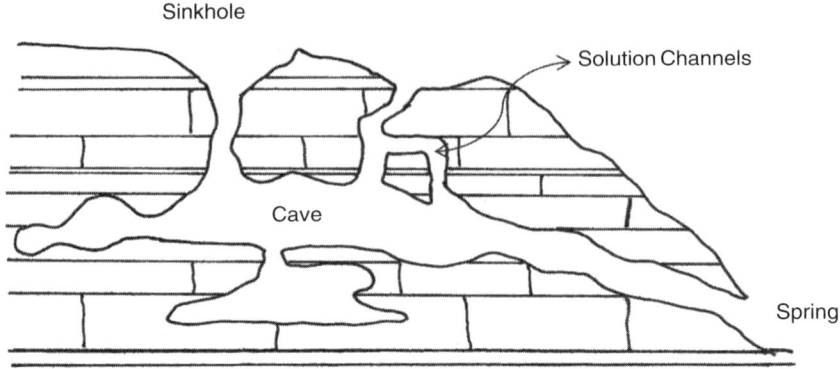

FIGURE 5-5. Diagram of Karst geology showing limestone layers, solution channels, sinkholes, and springs.

A full description of a watershed includes the soils and how they support those five functions. In the United States, data on soils and their properties are available from the Natural Resources Conservation Service, through their soil surveys. The State Soil Geographic (STATSGO) Database provides information on general soil associations in an area. The Soil Survey Geographic (SSURGO) Database provides more detailed data regarding the location and properties of specific soils; these databases are available at: http://websoilsurvey.nrcs.usda.gov/app/HomePage.htm.

Most watershed services in one way or another depend on the watershed's soils. Other than precipitation that falls directly on a lake or river, all of the water in a waterbody flows through or over soil. The soil's infiltration capacity determines water storage and runoff volume. Stored water that moves through the soil and underlying bedrock provides baseflow to streams. At the same time, microorganisms in the soil consume pollutants and transform them into less problematic materials. The soil determines the organic and mineral content of the water.

Land Use, Land Cover
Land use is the human employment of the land including settlement, cultivation, pasture, rangeland, recreation, and so forth (Meyer and Turner, 1994). Land cover is the physical state of the land. Land cover is the quantity and type of surface vegetation, water, and earth material. Changes in land cover cause changes in ecosystem function and environmental processes at local, regional, and global scales. Changes

include climate, biodiversity, and pollution of water, soils, and air (Ellis, 2007).

Watershed description should include discussion of the native land cover of the area, the current land cover, and the direction of change. United States Government Land Office (GLO) maps are a source for information on native land cover in the U.S. These maps show the location of forests, savannas, prairies, wetlands, streams, creeks, and other natural objects with remarkable precision, considering the surveying tools available. GLO maps are available from many regional libraries. Some states have converted the paper maps to GIS databases. Current land cover is generally determined from publicly available GIS databases. Land cover may also be delineated manually from aerial photographs. Land cover change can be determined by comparing databases from different years. Metadata (description of the data) must be checked when comparing databases from different time periods, because delineation criteria may not be the same for the different surveys.

Landscape structure in a watershed includes different plant community types (Figure 5-6). Boundaries between the communities are often distinct. The combination of communities can be described as

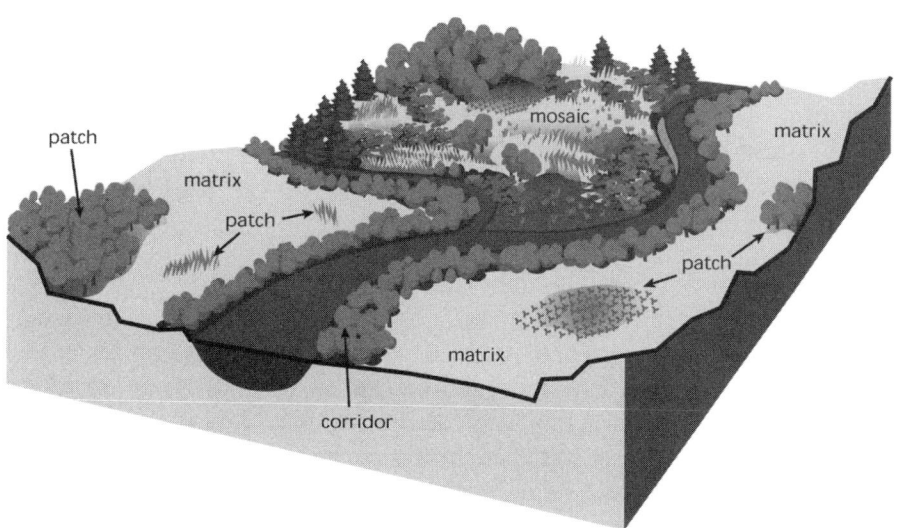

FIGURE 5-6. Landscape structure of watershed, including the matrix or dominant landform, patches, and corridors. (*Source*: From *Stream Corridor Restoration Principles and Practices*. 1998. Federal Interagency Stream Restoration Working Group [15 federal agencies of the U.S.].)

matrix, patches, and corridors. The matrix of the landscape is the land cover that is dominant in, and interconnected over, the majority of the landscape—that is, hardwood forest, row crop agriculture, grassland, and so forth. Patches are land cover types that differ from the dominant form and are located within the matrix. Corridors are linear patches that connect different patches, such as stream channels or riparian corridors.

Of the characteristics listed in this section, land use and land cover are the most apt to be modified in land development. Land cover also influences most watershed services. Through manipulation of land cover, we can control many aspects of watershed hydrology and water quality, as well as biodiversity and air quality.

Watershed Hydrologic Characteristics

Water on earth cycles continuously between the seas and oceans, the atmosphere, and the land, in a process referred to as the hydrologic cycle (Figure 5-7). The cycle is driven by solar energy. The sun heats water molecules—mostly in the oceans, but also in any other body of water—and causes a phase transition from liquid to gaseous (evaporation) or from solid to gaseous (sublimation). The gaseous form of water, or water vapor, then becomes part of the atmospheric circulation. When the mass of air becomes supersaturated with water vapor and there are condensation nuclei present, the water vapor will condense into water droplets or ice crystals. Typically, condensation happens when the air mass shifts from a warm condition to a cooler condition. When the drops or crystals become large enough, they fall as rain or snow. For watershed hydrology, the concern is that portion of the hydrologic cycle that produces precipitation over a land surface and the resulting runoff.

Precipitation
There are three types of precipitation. Frontal precipitation occurs when a cold air mass pushes in under a warm air mass, and the warm air is lifted. As the warm air rises, the air cools and expands. As air cools, its water-holding capacity decreases. The rising air becomes supersaturated, and precipitation may occur. Frontal rain is typically widespread along the boundary between the warm and cold air masses. Orographic precipitation occurs when an air mass encounters a mountain and is forced to rise. As the air rises, it cools and its water-holding capacity decreases. If there is enough moisture in the air mass and condensation nuclei are present, then precipitation will occur. Conversely, on the leeward side of a mountain, the air mass descends and warms.

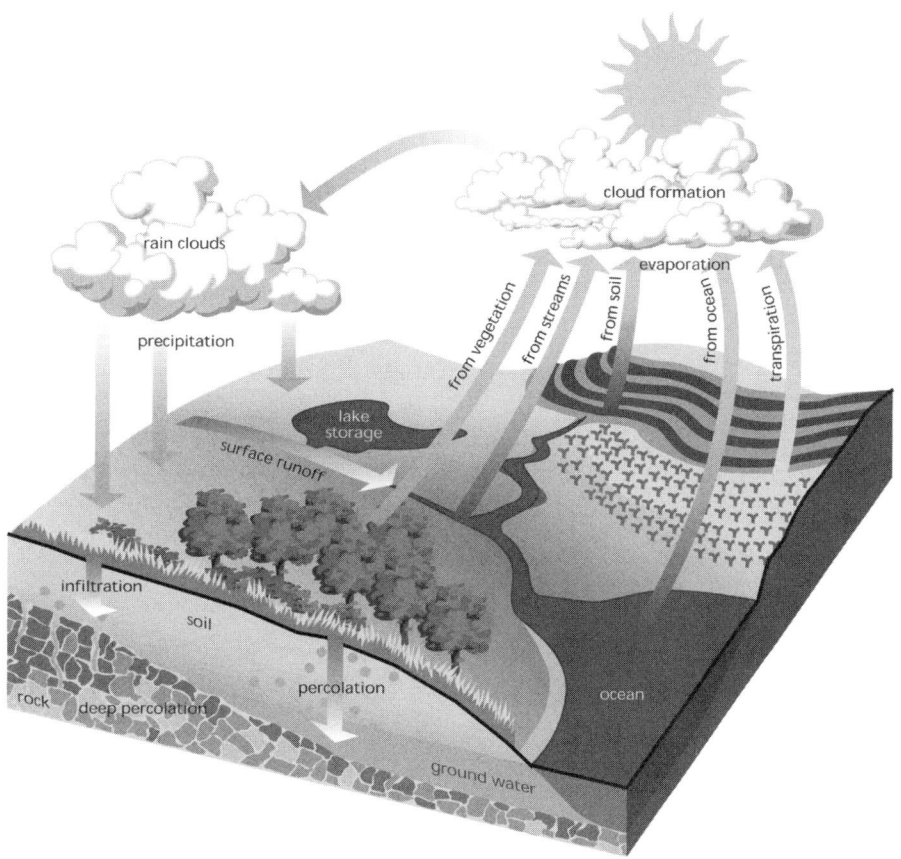

FIGURE 5-7. The hydrologic cycle is a continuous movement of water from surface reservoirs to the atmosphere and back to the reservoirs. Note that not all precipitation becomes runoff. Abstractions from precipitation include interception by plants, infiltration and percolation, and surface storage. (*Source*: From *Stream Corridor Restoration Principles and Practices*. 1998. Federal Interagency Stream Restoration Working Group [15 federal agencies of the U.S.].)

Warming air increases its water-holding capacity, and precipitation decreases. Thus, a rain shadow, an area of low rainfall, frequently exists on the leeward slopes of mountain ranges. Finally, convective precipitation occurs when air is forced to rise by heating of the land surface below the mass. As the land warms, the overlying air rises and cools. Once again, as the air cools, it may become supersaturated, so that water vapor condenses, and precipitation may occur. Thunderstorms are the typical form of convective precipitation.

Precipitation and temperature are the defining characteristics of the various biomes of the earth. Precipitation varies widely from location to location. However, annual mean precipitation does not tell the complete story. Annual and seasonal variability also must be considered. Fayetteville, Arkansas, for instance, has a normal annual precipitation of roughly 117 cm (46 in). But Fayetteville's annual precipitation varies from a low of 55 to a high of over 175 cm (21 to 69 in). The precipitation in Fayetteville also varies seasonally, with heavy rainfall during the spring and late fall when frontal systems move through the area, and relatively less rainfall during the summer and winter. Seasonality of rainfall is region-specific (Figure 5-8). Some areas, such as the Southwest United States and Southeast Asia, experience a pronounced seasonal or monsoonal rainfall. Phoenix, Arizona, averages slightly less than 20 cm (8 in) of precipitation per year (range 7.6 cm (3 in) to 50 cm (20) (National Weather Service), but the average rainfall for the months of June through September is 15.4 cm (6 in). Almost all of the precipitation received in Phoenix occurs during the summer monsoon months.

Precipitation Frequency Analysis

Extreme precipitation events are described by the frequency with which the event may occur and the duration of the event. The frequency of a precipitation event is normally described by the event's

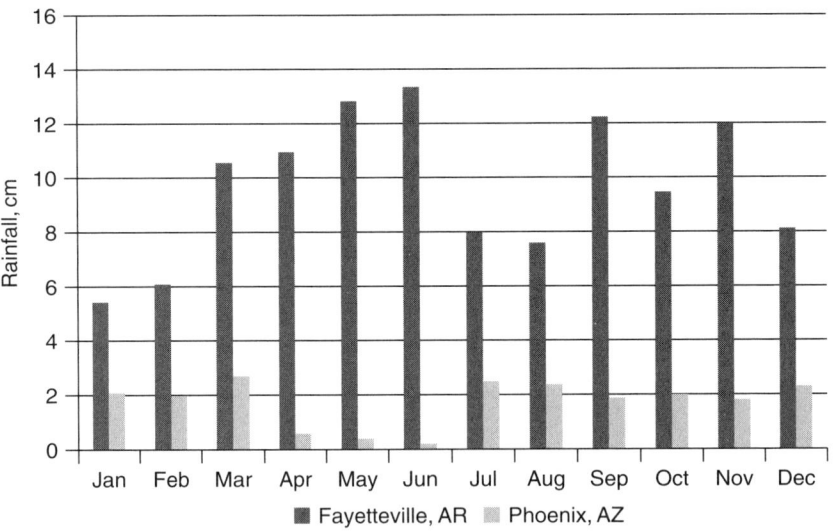

FIGURE 5-8. Seasonality of rainfall in Fayetteville, AR, and Phoenix, AZ, United States. The strong seasonality of rainfall in Phoenix is caused by monsoonal flow of moisture over the area during the summer. In contrast, the rainfall pattern in Fayetteville is influenced by frontal systems moving through the region in the spring and fall.

return frequency. An event with a return frequency of five years is one that on average will be equaled or exceeded once every five years. If T is the return frequency of a storm, then the probability that that storm will occur in any given year is $1/T$. So our five-year storm has $1/5 = 0.20$ or 20 percent chance of occurring in any particular year, even if a storm of similar magnitude occurred the previous year. Duration is simply the time from the start to the end of precipitation. Thus a five-year, 24-hour storm is a storm lasting 24 hours that has precipitation depth that is equaled or exceeded on average once every five years.

Rainfall intensity varies with location. In the United States, the National Weather Service's Technical Paper no. 40 (Hershfield, 1961) provides isopluvial maps (maps showing lines of equal rainfall intensity) giving rainfall depths for 30-minute to 24-hour storms and for return frequencies from 1 to 100 years. These data provide an indication of how hard it will rain in a given watershed.

Abstractions from Precipitation and Runoff

Precipitation is the ultimate source of all water flowing in streams and rivers. However, not all precipitation becomes streamflow. When precipitation occurs, some amount of water is intercepted and held, or abstracted, by watershed vegetation. The amount of interception depends on the type, density, and growth stage of the vegetation, on rainfall intensity, and on wind speed. Interception is likely not significant in an intense storm, but may account for a fair portion of annual rainfall. Another portion of the precipitation will be evaporated from surface water sources or transpired through vegetation, in a process referred to as evapotranspiration. Evapotranspiration is also not significant in an individual storm event, but may account for 40 to 70 percent of annual precipitation in a watershed. Water that falls to the surface initially fills small depressions on the surface. This portion of the precipitation is referred to as surface storage or detention. Surface storage may not be significant in natural systems, but it is a variable that can be manipulated through design. Next, water seeps into the soil, in a process referred to as infiltration. The rate of infiltration depends on the physical properties of the soil. Infiltration generally is a large abstraction from rainfall, both during individual storms and annually. Finally, when all surface storage is filled, and either the rainfall intensity exceeds the infiltration rate or the ground is saturated, water will begin to flow over the surface of the ground. This flow is called stormwater runoff or simply runoff.

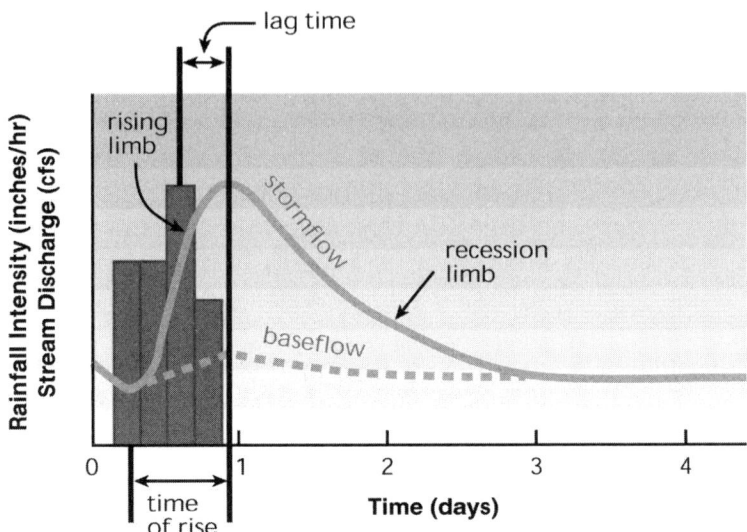

FIGURE 5-9. Stylized stream discharge hydrograph. The solid blocks represent incremental rainfall. Lag is the time from the centroid of rainfall to the peak of the hydrograph. The area under the stormflow line and the x-axis represents the volume of streamflow. (*Source*: From *Stream Corridor Restoration Principles and Practices*. 1998. Federal Interagency Stream Restoration Working Group [15 federal agencies of the U.S.].)

Mathematically, runoff is the difference of total precipitation and the sum of all abstractions. Runoff may also be referred to as effective rainfall. Runoff eventually flows downslope to a receiving stream (Haan, Barfield, and Hayes, 1994).

Of the water that infiltrates into the soil, some percolates to deep groundwater and recharges local aquifers. Some of the infiltration is held by the soil as soil moisture, and the rest flows through the soil in a process called interflow. Interflow may return to the surface and contribute to runoff. That process is called return flow.

A hydrograph shows a stream's discharge over time (Figure 5-9). The graph directly shows the discharge (q, cubic meters per second, or cms) of a stream at any particular time. Volume (V, cubic meters, or cm) of water moving past a point in a stream is the discharge rate multiplied by the amount of time (t, sec):

$$V = qt$$

The area under the hydrograph between any two points (a and b) in time is equivalent to the volume of water discharged in that time interval.

$$V_{a,b} = \int_a^b qt_{dt}$$

Hydrographs also show how quickly stormwater collects, and the length of storm events.

Hydrographs may be divided into two parts: baseflow and stormflow (Figure 5-9). Baseflow is the water that flows in streams between storms. Baseflow is precipitation that percolates into the groundwater and moves slowly through substrate before reaching the channel. Baseflow sustains streamflow during periods of little or no precipitation (FISRWG, 1998). Stormflow is that portion of the hydrograph that reaches the stream channel over a short timeframe, through overland or underground routes (FISRWG, 1998). Both elements are important to watershed services provided by the stream, and the stream channel must be adapted to both conditions.

Stream Characteristics

A stream corridor consists of the channel, its floodplain, and a transitional upland fringe (Figure 5-10). The channel has water flowing at least part of the year. Channels may be ephemeral, flowing only during or immediately after storms; intermittent, flowing only during certain times of the year; or perennial, flowing continuously during both wet and dry times. Geomorphologists divide a stream longitudinally into three segments: the headwaters, transfer, and depositional zones. In the headwaters zone, water flows relatively quickly downslope. Streams merge and flow down gentler slopes with some meanders, in the transfer zone. Finally, in the depositional zone, streams wander around the floodplain in wide meanders. From a process point of view, headwaters are the source of sediments; the transfer zone, as the name implies, moves sediment through the watershed; and the depositional zone is where sediment settles.

Stream order is a convenient way to describe the location of a stream in a watershed (Figure 5-11). In Strahler's stream-ordering system, a stream with no tributaries is a first-order stream. When two first-order streams merge, they become a second-order stream. The confluence of two second-order streams forms a third-order stream, and so forth. Low-order streams are generally headwaters, midrange-order streams are in the transfer zone, and high-order streams are depositional.

The biological community of a stream changes in a somewhat predictable pattern, going from low- to high-order streams. Vannote's (1980) river continuum concept visualizes low-order streams (first to

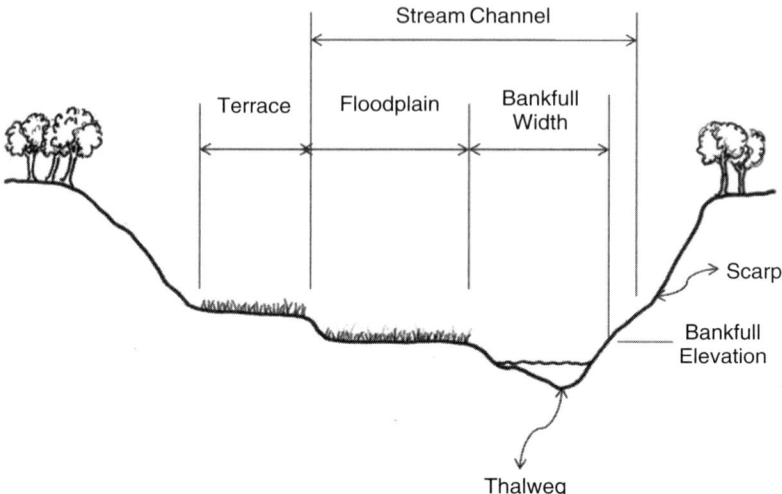

FIGURE 5-10. Cross section of a stream valley.

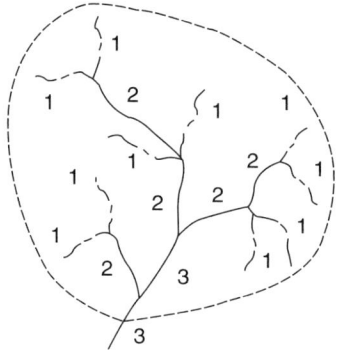

FIGURE 5-11. Strathler's stream-ordering system. Headwater streams are first order. At the confluence of two first-order streams, the stream becomes second order. When two second-order streams combine, the result is a third-order stream.

third order) as being shaded by riparian vegetation. The shading prevents sunlight from reaching the stream and limits photosynthesis. The food source for aquatic organisms is material that falls into the stream from the surrounding landscape (allochthonous materials). Mid-order streams (fourth to sixth order) become wider, and as a result the riparian vegetation does not cover the entire stream width. With the introduction of sunlight, photosynthesis dominates biological production. In these mid-order rivers, most of the food source is produced within the stream (autochthonous materials), and little comes from the landscape. Finally, in higher-order streams, the water becomes more turbid, reducing sunlight penetration and photosynthesis, causing the stream to revert to allochthonous production. In higher-order streams, the allochthonous material comes from the drifting downstream of particulate matter produced upstream. Stream biotic communities reflect their location within the continuum.

The floodplain is the relatively flat portion of the stream valley, slightly higher than the stream channel. Floodplains are occasionally inundated during storm events. Floodplains provide temporary storage for floodwater and sediment produced by the watershed. Floodplains may also provide refugia for aquatic species during floods. High-order streams have a predictable annual flood cycle. The flood-pulse concept (Junk et al. 1989) describes the movement of nutrients from the floodplain to the stream, as well as the distribution of fish with respect to the flooding. The transitional upland fringe serves as a transition zone between the floodplain and the surrounding landscape. It forms the boundary of the stream corridor (FISRWG, 1998).

Watershed Water Quality Characteristics

Characteristics of water quality in a stream or reservoir include physical, chemical, and biological aspects. A healthy stream is one in which these three aspects all support the function of the stream. Physical characteristics include water temperature, turbidity, color, and suspended solids, as well as the condition of the stream channel, banks, and flow. Chemical characteristics are normally measured with laboratory equipment and include pH, alkalinity, hardness, dissolved oxygen, nutrients (nitrogen and phosphorus), metals, oil and grease, and organic compounds. Biological characteristics are the microbial content, including pathogenic organisms, the benthic animals, and aquatic communities of insects and fish.

Water Quality Standards
The designated uses of a waterbody are those uses that society has specified as necessary or desirable for that waterbody. In the United States, designated uses are specified by state water quality agencies delegated by the EPA for implementation of the Clean Water Act. Designated uses may include recreation, propagation and growth of aquatic life, production of edible and marketable natural resources, domestic water supply, industrial and agricultural water supply, and aesthetic values. The states then develop water quality standards expressed as

pollutant constituent concentrations, levels, or narrative statements representing a quality of water that supports a particular use. Since water quality is affected by both point source (specific discharges) and nonpoint source (diffuse) pollution, the designated uses and water quality standards form a foundation on which watershed development activities may build.

Watershed Management Planning/TMDLs

A TMDL, or Total Maximum Daily Load, is theoretically the total amount of a particular pollutant that a stream can assimilate per day without impairment of any of the stream's designated uses. The equation for a TMDL is:

$$TMDL = WLA + LA + BL + MOS$$

where
 WLA = is the waste load allocation, or the mass of a pollutant allowed to be discharged from point sources
 LA = is the load allocation for nonpoint sources of the pollutant
 BL = is the background load from natural sources
 MOS = is a margin of safety

In the United States, TMDLs are legal documents developed by the state agency delegated to enforce provisions of the Clean Water Act. The TMDL will give the allowable load of a pollutant to a stream and will set targets for load reduction from the various point and nonpoint sources. A TMDL implementation plan describes how the state proposes to meet the various load reduction requirements.

In watersheds, as in all ecosystems, everything is connected. A change in land use or land cover will result in some change in geomorphology and hydrology, causing changes in stream hydraulics which, in turn, change stream function and ultimately can result in changes in the population of the aquatic community. Change is natural, and in fact necessary, for biodiversity. Although stream systems are resilient, accelerated change caused by human use of a watershed may adversely impact the ecosystem.

Watershed Human Impacts

Human use of a watershed includes both direct modifications to the stream and stream corridor and modifications of land use and land cover in the watershed. Many of these modifications involve trade-offs in watershed services.

Common hydrologic modifications of streams include dams and impoundments, channelization, and diversions (FISRWG, 1998). Dams may be built to enhance domestic, agricultural, and industrial water supply; for hydroelectric power production; to enhance transportation; or for aesthetic and recreational purposes. Channelization is typically done to enhance the capacity of a stream to move water, or to enhance the usability of a parcel for development. Stream modification through impoundment or channelization changes a stream's balance between discharge and sediment transport. Either erosional or depositional problems may develop. Both impoundment and channelization expose the water to more sunshine and increase productivity. Diversions, either for irrigation or water supply, have the obvious impact of removing water from the system.

Land use changes also have impact on the hydrology and water quality in a watershed. Watershed modifications include conversion to agriculture, livestock grazing, forest removal, mining, urbanization, and recreation. The primary issue with land use modification is that it changes the ecosystem, with cascading effects downstream. Everything is connected. In some areas, most of the native landscape matrix has been replaced by agriculture, leaving native vegetation only in patches. Tillage and soil compaction interfere with the soil's capacity to regulate the flow of water in the landscape. Surface runoff may increase, and the soil's water-holding capacity may be decreased. Increased erosion and sediment transport are common. Other aspects of agriculture, such as irrigation and field drainage, also alter the natural hydrology of the watershed. Finally, applied chemicals such as fertilizer, pesticides, and herbicides may leach into the groundwater or wash off in surface runoff (FISRWG, 1998). Best management practices (BMPs) are available to ameliorate the impact of agriculture on soil and water quality. Consult the Natural Resources Conservation Service for BMPs relevant to agriculture in the local watershed.

Impacts from grazing of livestock are the loss of vegetative cover due to its consumption or trampling, and streambank erosion from the presence of livestock. Cattle may also deposit manure and urine directly into the water or near surface water, where it may leach or run off into streams, ponds, lakes, and the like. However, grazing animals are a natural and integral part of most grassland ecosystems. Properly managed grazing lands may have positive impacts as well. Key issues are the effects of cattle on soils, water quality, riparian vegetation, and biodiversity (CAST, 2002).

Three general activities of forestry have potential to impact water quality and other aspects of watershed ecology. These activities include tree removal, materials transport, and site preparation. If poorly planned, these activities will have negative impacts on forest resources and on the services of forests, including conservation of biodiversity, and carbon and water cycling (FAO, 2005). BMPs for forestry activities are available from the U.S. Environmental Protection Agency.

Mining impacts the local environment through vegetative clearing, soil disturbance, altered hydrology, and contaminants (FISRWG, 1998). Today's mining practices are much less damaging than those used previously. However, many abandoned mines remain that still impact the environment, and today's less extreme methods are not totally without impact. Describing the watershed should include discussion of any mining that has occurred or is occurring. Include in the discussion any known acid mine drainage problems.

Urbanization has many impacts on the environment, including loss of vegetation, reduction in infiltration and evapotranspiration, increased surface runoff, hydrologic modification, and nonpoint source pollution. In an urbanized area, pavement and other impervious surfaces replace much of the native vegetation. With increases in impervious area, stormwater runs off more quickly and has little chance to come into contact with the soil. As a result, the water purification service provided by the soil is lost, and the loss of infiltration increases runoff. Downstream effects include increased flood peaks and reduced baseflow in the stream. Increased runoff may cause increased channel erosion, resulting in loss of habitat and other valuable stream ecological services. As little as 10 percent impervious cover in a watershed can result in stream degradation (FISRWG, 1998). Key characteristics to evaluate are population density, percent imperviousness, and the trend of each of these.

Recreational activities may also impact the environment. All-terrain vehicles, for instance, can cause increased erosion and habitat reduction. Hiking or equestrian activities also may result in soil compaction and increased erosion. In areas where boating can occur, propeller wash and wave action from wakes can increase bank erosion and re-suspend bottom sediments (FISRWG, 1998). BMPs are available to minimize the impact of recreational activities. These BMPs may be behavioral, vegetative, or structural.

Removal of natural riparian zone vegetation has several impacts on stream systems. In low-order streams, material falling into the stream from riparian vegetation is the main source of carbon feeding the system. The shade provided by riparian vegetation also helps to moderate temperature in the stream and reduces growth of algae. Riparian vegetation also removes pollutants, provides refugia during flood events, and armors the bank against excessive erosion.

Not all human use is negative. In many areas, society has recognized the value of natural landscapes for their ecological services. Much land has been set aside as parks, preserves, or working landscapes. The watershed description should note the extent of such preservation in the local watershed.

Human characteristics of the watershed can to a large extent be described by noting the percentage of the watershed in each land use, the population density and trend, the degree of imperviousness, and any lands under conservation protection measures.

Summary of Watershed Characteristics

The watershed is the smallest unit for effective ecosystem management. A description of a watershed includes:

- Biome and ecoregion setting
- Watershed area
- Watershed shape
- Topographic features
- Underlying geology
- Soils
- Land use/land cover
- Precipitation pattern: annual, seasonal, extreme events
- Hydrology
- Stream corridor characteristics
- Hydrologic modifications
- Population: total, density, trend
- Imperviousness
- Agricultural uses and trends
- Type and intensity of recreational uses
- Protected lands
- Local water quality standards and regulations
- Existing management plans

> In the field of watershed conservation and flood-control there is a definite issue between the engineers who tend to rely on engineering works alone, and the foresters and biologists who insist that vegetative cover and soil fertility are indispensable adjuncts to dams and dykes. The latter viewpoint is the naturalistic one. It is strongly supported by current research findings.
>
> —Aldo Leopold, *Game Management,* 1932

Further Readings
Leopold, Luna, *A View of the River*, Harvard University Press, 290 p. 1994.

References
Brady, Nyle C., and Ray R. Weil, *The Nature and Properties of Soil*, 13th ed., Prentice Hall, Upper Saddle River, NJ, 2002.
CAST, Environmental Impacts of Livestock on U.S. Grazing Lands, William C. Krueger and Matt A. Sanderson, co-chairs, Council for Agricultural Science and Technology, 2002, www.cast-science.org (accessed August 7, 2009).
Chang, Mingteh, *Forest Hydrology: An Introduction to Water and Forests*, 2nd ed., CRC Press, New York, NY, 2006.
Ellis, Earl, Land-use and land-cover changes, The Encyclopedia of Earth, 2006, www.eoearth.org/articles/land-use_land-cover_changes (accessed August 2, 2009).
EPA, Why Watersheds?, United States Environmental Protection Agency, Office of Water, Oceans and Watersheds, Washington, DC, 1996, www.epa.gov/owow/watershed/why.html (accessed July 23, 2009).
FAO, Environmental impacts of forest utilization and mitigation practices, Food and Agricultural Organization of the United Nations, Rome, Italy, 2005, www.fao.org/forestry/11787/en/ (accessed August 7, 2009).
FISRWG, Stream Corridor Restoration: Principles, Processes, and Practices, Federal Interagency Stream Restoration Working Group, 1994. GPO Item no. 0120-A, SuDocs no. A 57.6/2:EN 3/PT.653, October, Haan, C.T., B.J. Barfield, and J.C. Hayes, Design Hydrology and Sedimentology for Small Catchments, Academic Press, Boston, MA, 1998.
Harter, Thomas, *Watersheds, Groundwater and Drinking Water*, The Regents of the University of California, Division of Agriculture and Natural Resources, Oakland, CA, pp. 9–13, 2008.
Hershfield, David M., Rainfall Frequency Atlas of the United States for Durations from 30 Minutes to 24 Hours and Return Periods from 1 to 100 Years, U.S. Department of Commerce, Weather Bureau, Washington, DC, 1961.
Institute of Water Research, Watershed Approach: Land Use Effects on Water Quality and Quantity, Institute of Water Research, Michigan State University, East Lansing, MI, 1997.
Junk, W.J., P.B. Bayley, and R.E. Sparks, The flood pulse concept in river-floodplain systems. In *Proceedings of the International Large River Symposium*, edited by D.P. Dodge, pp. 110–127, Canadian Special Publication of Fisheries and Aquatic Sciences, Ottawa, Ca., 1989.
Meyer, William B., and Billie Lee Turner, *Changes in Land Use and Land Cover*, Cambridge University Press, Cambridge, UK, 1994.
Millennium Ecosystem Assessment, Ecosystems and Human Well-Being, World Resources Institute, Washington, DC, 2003.
National Weather Service, www.wrh.noaa.gov/twc/monsoon/monsoon.php.

O'Keefe, Thomas C., S.R. Elliot, R.J. Naiman, and D.J. Norton, Introduction to watershed ecology, U.S. EPA Watershed Academy Web, www.epa.gov/watertrain (accessed July 24, 2009).

Pidwirny, M., Stream Morphometry, in Fundamentals of Physical Geography, 2nd ed, www.physicalgeography.net/fundamentals/10ab.html (accessed July 27, 2009).

Smith, M., D. de Groot, D. Perrot-Malte, and G. Bergkamp, Pay–Establishing payments for watershed services, International Union for Conservation of Nature, Gland, Switzerland, 2006; IUCN Reprint, 2008.

Strahler, A.N., Quantitative analysis of watershed geomorphology, *American Geophysical Union Transactions*, 38: 913–920, 1957.

USGS, The National Atlas, Hydrologic Units, www-atlas.usgs.gov/articles/water/a_hydrologic.html (accessed July 23, 2009).

Vanote, R.L. G., W. Minshall, K.W. Cummins, J. R. Sedell, and C.E. Cushing. The river continuum concept. *Canadian Journal of Fisheries and Aquatic Science*, 27: 130–137, 1980.

Ward, Andrew D., and William J. Elliot, *Environmental Hydrology*, Lewis Publishers, Boca Raton, FL, 1995.

6
Defining Place: The Site

> Whether we and our politicians know it or not, Nature is party to all our deals and decisions, and she has more votes, a longer memory, and a sterner sense of justice than we do.
>
> —Wendell Berry

INTRODUCTION

Wendell Berry's work demonstrates that a farmer must have an intimate relationship with his land and its secrets to sustainably reap its harvests. The same is true in ecological design. The ecosystem is the palette on which design is based. A project site occupies a specific place in the ecosystem. The site may be characterized as belonging to a biome, ecoregion, and watershed, but the site itself also has defining characteristics relative to its contribution to the ecosystem. This chapter explores those characteristics as they relate to how the site can be used for human purposes.

Each site contributes to the ecosystem services of a place. The goal in ecological design is no net loss of ecosystem services. To simplify, only those services that are supportive of the ecosystem are measured. Characterization of the site's physical, hydrological, biological, and climatic conditions sets the baseline for determination of these services.

Physical Characterization

Important physical characteristics for defining the site are its area, geology, topography, slope, relief, and aspect.

Area is simply the space covered by the site. If there is a site survey, it will give the area. If there is not a current survey, one should be completed prior to any development. Most geographic information system (GIS) and computer-aided design (CAD) programs have a function for determining the area of a specified site. On a site with regular sides, the area can be determined by dividing the site into polygons and computing area using geometric relationships.

The geology of a site is the underlying strata of the site and the processes shaping those strata. Geology and climate largely determine the soil that is present on a site.

Topography refers to surface features of the site, including landforms, streams and other waterbodies, rock outcrops, and human-produced features.

Slope refers to the angle of a site from horizontal. Slope is measured as the vertical rise in surface elevation divided by the horizontal distance over which that rise takes place ($S = $ Rise/Run). Slope may be measured along a particular hillside for a specific runoff calculation, or it may be measured as the average slope of the site.

Relief refers to the total change in surface elevation over the site. Relief is simply the elevation of the highest point on the site minus the elevation of the lowest point.

Aspect is the horizontal angle of a slope from north. Aspect is expressed in degrees and provides information relative to the site's exposure to the sun or prevailing weather patterns. For instance, a hill slope facing southeast may have an aspect of $135°$.

Hydrological Characterization

The objective of the hydrological characterization is to determine predevelopment conditions regarding water moving onto, across, and through the site. In most areas, regulations set the maximum change allowed from the predevelopment condition as a result of the development. The developer also has liabilities regarding damage caused by changing the hydrology of a site. From an ecological standpoint, the goal is to match predevelopment hydrology when development is completed.

Start a hydrologic characterization by looking downstream. How close is the nearest watercourse? What type of waterbody receives drainage from the site? Is it a ditch, stream, lake or reservoir, or wetland? Are there structures downstream that must be protected? Are there other sensitive areas?

If the stream runs through the site, is there a flood plain? What is the width of the floodplain? What is the condition of existing riparian vegetation? Are there any wetland areas on the site that have to be preserved or mitigated?

After looking downstream, turn around and look upstream. What is coming onto the site that will have to be managed? Is the upstream run-on liable to change as the upstream area develops?

Precipitation

Consider both the average annual precipitation at a site and the variability of that precipitation. Generally, these data are available locally from government agencies, chambers of commerce, or TV stations. In mountainous areas, however, care should be taken to ensure that the data are relevant to the site. In those areas, annual precipitation may be sensitive to elevation and the aspect of the slope relative to prevailing winds.

On an annual scale, precipitation onto a site is balanced by losses from evapotranspiration, infiltration, and runoff. For a specific site, infiltration and runoff from a specific storm are the critical characteristic. In most areas of the United States, local regulations dictate consideration of the impact of a storm of a specified return interval or probability of occurrence. The impact of 2-, 10-, 25-, 50-, and 100-year storms should be evaluated, at a minimum. Characteristics that must be considered include the total amount of precipitation in the storm, the duration of the storm, and the intensity of rain.

Precipitation volume is normally measured as depth, centimeters or inches. The depth is the depth that would accumulate over an area if there were no infiltration, evapotranspiration, or runoff. Depth of precipitation times the area over which it falls is the volume of precipitation.

Example: 1 cm of rain falls on a site with area of 1 hectare. What is the total volume of rainfall?

$$1 \text{ cm} \times \frac{1 \text{ m}}{100 \text{ cm}} \times 1 \text{ hectare} \times \frac{10{,}000 \text{ sq. m}}{\text{hectare}} = 100 \text{ cu. m}$$

$$= 100{,}000 \text{ liters}$$

Duration of precipitation is measured as time, in units of minutes, hours, or days. Intensity is the total precipitation divided by the duration. Typically, intensity has an inverse relation with duration (Figure 6-1).

Natural rainstorms do not follow a specified pattern. For design purposes, it is necessary to develop a synthetic design storm for analysis. Statistical analysis is used to generate probability distributions for rainfall depth for different durations; Haan, Barfield, and Hayes (1994) give a good description of precipitation analysis.

In the United States, depth of the design storm is typically taken from The National Weather Service's Technical Paper 40 (TP 40) (Hershfield, 1961) or a similar document. TP 40 contains depth,

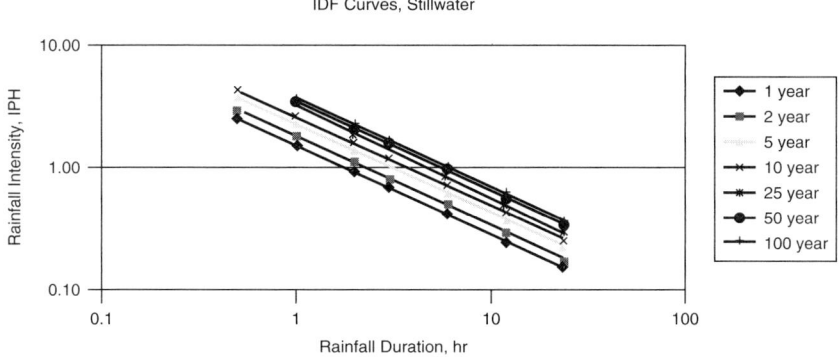

FIGURE 6-1. Time-intensity-duration curve.

duration, and frequency data for storm durations of 30 minutes to 24 hours and frequencies of 1 year to 100 years. For storm durations of less than 60 minutes, HYDRO-35 (Frederick et al., 1977) supersedes TP 40. TP 40 is useful for areas east of the Rocky Mountains. In Western states, the National Oceanic and Atmospheric Administration's Atlas 2, (NOAA, 1973) replaces TP 40.

Rainfall intensity varies throughout a storm. TP 40 and other documents give the total rainfall for specified durations. The Natural Resources Conservation Service (NRCS), formerly the Soil Conservation Service (SCS), developed a method for generating synthetic storm patterns, or "type curves," based on the 24-hour rainfall for the desired frequency. In the SCS method, dimensionless rainfall temporal patterns (Figure 6-2) were developed for different regions of the United States (Figure 6-3). The x-axis of the type curve gives the time since the start of the synthetic storm, in hours. The y-axis gives the cumulative portion of total precipitation as a decimal fraction.

To use the SCS type curves:

1. Find the 24-hour rainfall for the desired location and frequency from TP 40 or other appropriate document.
2. Find the type curve that should be used for the site from Figure 6-3.
3. Find the most intense portion of the curve (steepest segment) that includes the desired duration.
4. Find the portion of the 24-hour storm that occurs during the desired duration.
5. Find depth by multiplying the portion occurring by the total rainfall depth.

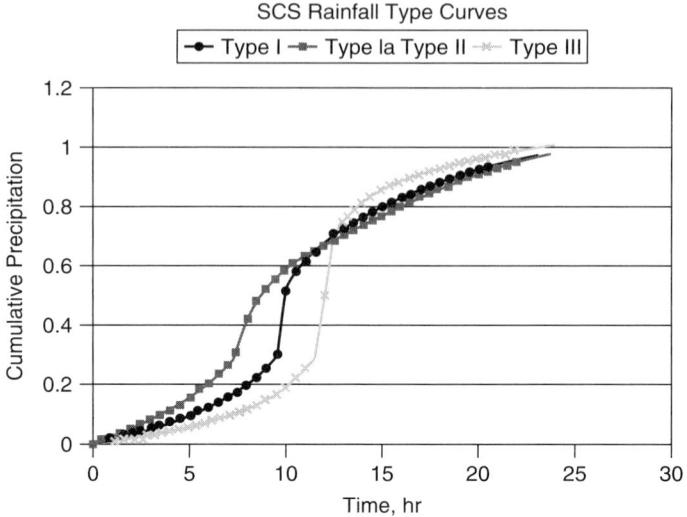

FIGURE 6-2. SCS type curves for dimensionless 24-hour rainfall.

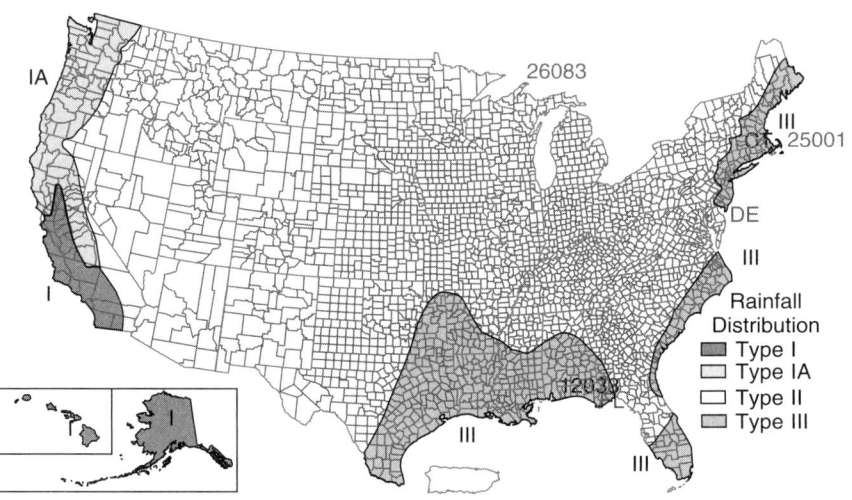

FIGURE 6-3. Applicable regions of the U.S. for SCS type curves. (*Source*: Natural Resources Conservation Service.)

Infiltration and Runoff

Infiltration is the process of water moving into and through the soil. Infiltration is the major abstraction from precipitation during a major storm event. The rate of infiltration is determined by soil properties, vegetation, and antecedent moisture.

Infiltration is limited to the infiltration capacity of the soil or the rainfall rate, whichever is less. The infiltration rate is measured in length per unit of time (i.e., cm/min or in/min). The rate is measured with a double-ring infiltrometer.

In dry soil, the initial infiltration rate will be high. As the soil becomes saturated, the infiltration rate reduces. The ultimate infiltration rate in saturated soil is the saturated hydraulic conductivity of the soil. The most restrictive layer of soil present limits the infiltration rate. Therefore, the infiltration rate is measured at the most restrictive layer in the soil profile.

Horton (1940) and Holton (1961) developed empirical formulas to estimate the infiltration rate of a soil over time. The equations require estimation of constants that are specific to soil hydrologic group and vegetation. Green and Ampt (1911) developed an equation for infiltration based on Darcy's law. Infiltration in Green and Ampt's equation is a function of the hydraulic conductivity of the soil and the pressure differential above and below the wetted front as water moves downward. In computer modeling of watershed hydrology, the Green and Ampt infiltration equation is frequently used because it is more theoretically correct than other methods. For a complete discussion of infiltration equations see Haan, Barfield, and Hayes (1994).

Characterizing the hydrology of a site requires knowledge of the volume of runoff, the peak discharge rate, and the relationship of runoff rate to time (the stormwater hydrograph). At the watershed scale, continuous simulation models now commonly develop these attributes. These models do not perform well at the site scale. Simplified approaches to estimating hydrological parameters follow.

Effective Rainfall and Runoff
If rainfall intensity exceeds the infiltration rate, and all other abstractions are sated, runoff will occur. The amount of rainfall that becomes runoff during a storm is called the effective rainfall. Abstractions other than infiltration and surface storage can be ignored in a storm event. So, effective rainfall equals precipitation less infiltration and surface storage.

Effective rainfall, or runoff, is expressed by units of length. One cm (inch) of runoff is the runoff produced by one cm (inch) of effective rainfall over a specified site. Effective rainfall is computed directly from Holton's, Horton's, or Green and Ampt's equation in hydrological models. For small sites, the computations are tedious. The curve number approach to estimating runoff developed by the SCS (1985) provides a simplified approach to estimating runoff directly.

The curve number (CN) is a characteristic of soil type and cover. The SCS divided soils into four hydrologic soil groups (HSGs) based on their infiltration capacity. Group A soils have the most rapid infiltration, and group D have the slowest. Generally, A soils are sandy, and D soils are heavy clays. The CN represents the potential for a soil to produce runoff from a given rainstorm.

The NRCS provides HSG classification for a large number of soils on its web soil survey (http://websoilsurvey.nrcs.usda.gov/app/HomePage.htm). Soils survey data are given for undisturbed soil. For disturbed soils, Table 6-1 gives estimated HSG by soil texture.

Curve numbers are tabulated for many land uses. Table 6-2 provides CNs for several general land uses. More detailed tables can be found in many hydrology textbooks or local drainage manuals. The NRCS recommends development of curve numbers locally. Practically, development of curve numbers is a research project that should be undertaken by local agencies.

In the CN method, effective rainfall or runoff is estimated by the relationship:

$$Q = \frac{(P - 0.2S)^2}{P + 0.8S}, \quad P > 0.2S \quad (6.1)$$

Q is the effective rainfall or runoff in inches or mm, P is precipitation, and S is a parameter given by:

$$\begin{aligned} S &= \frac{1{,}000}{CN} - 10, \quad (Q, P, \text{ and } S, \text{ in.}) \\ S &= \frac{25{,}400}{CN} - 254, \quad (Q, P, \text{ and } S, \text{ mm}) \end{aligned} \quad (6.2)$$

CN is the composite curve number for the entire area over which effective rainfall is estimated. To compute the composite CN, divide the area into units with similar soils and land use, select CNs for each unit from an appropriate table, and then compute an area-weighted CN for the entire area. Units have to be consistent.

TABLE 6-1 Hydrologic Soil Group Classification by Soil Texture

HSG	Soil Texture
A	Sand, Loamy Sand, or Sandy Loam
B	Silt Loam or Loam
C	Sandy Clay Loam
D	Clay Loam, Silty Clay Loam, Sandy Clay, Silty Clay or Clay

Source: Brakensiek and Rawls, 1983, as reported by Haan et al., 1994

TABLE 6-2 Curve Numbers (CN) for Use with the National Resource Conservation Service's Stormwater Runoff Calculation Method

Land Use	Condition	Hydrologic Soil Group			
		A	B	C	D
Open Space	Poor	68	79	86	89
	Fair	49	69	79	84
	Good	39	61	74	80
Paved areas		98	98	98	98
Commercial		89	92	94	95
Industrial		81	88	91	93
Residential	8 units per acre	77	85	90	92
	4 units per acre	61	75	83	87
	2 units per acre	54	70	80	85
	1 unit per acre	51	68	79	84
Pasture	Poor	68	79	86	89
	Fair	49	69	79	84
	Good	39	61	71	78
Woods	Poor	45	66	77	83
	Fair	36	60	73	79
	Good	30	55	70	77

Source: Taken from the Texas Department of Transportation's Hydraulic Design Manual, 2009.

Effective rainfall is also dependent on the moisture state of the soil at the time the rainfall starts, or the antecedent condition. The NRCS defined three antecedent conditions, I, II, and III. Antecedent condition is defined by the NRCS as the amount of rain that has fallen during a five-day antecedent period. Antecedent condition I is less than 0.5 inches during the dormant season and less than 1.4 inches during the growing season. For antecedent condition II, the five-day rainfall is between 0.5 and 1.1 inches during the dormant season and between 1.4 and 2.1 inches during the growing season. Antecedent condition III occurs when precipitation has exceeded either of the aforementioned. The CN method described here gives effective rainfall or runoff for antecedent condition II. For conditions I or III, adjust the CN by the following relationships (Chow et al., 1988):

$$CN(I) = \frac{4.2CN(II)}{10 - 0.058CN(II)} \quad (6.3)$$

$$CN(III) = \frac{23CN(II)}{10 - 0.13CN(II)} \quad (6.4)$$

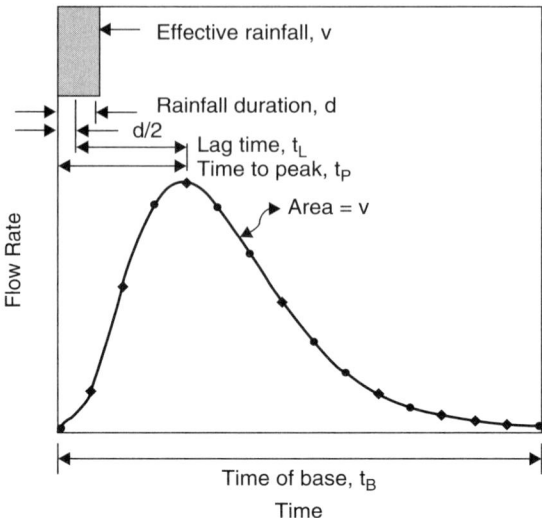

FIGURE 6-4. Example discharge hydrograph showing elements.

The CN method described here, or effective rainfall computed by Horton's, Holton's, or Green and Ampt's equations, provides the volume of effective rainfall. To fully describe the hydrology of a site, a runoff hydrograph is needed. A runoff hydrograph displays the relationship of discharge (volume per unit time) over time (Figure 6-4).

Peak Runoff Rate

Methods for computing the peak runoff rate or peak discharge from a site include the "rational method," the NRCS graphical peak discharge method, and unit hydrograph approaches. The unit hydrograph is discussed separately in a following section.

The rational method is widely used to compute peak discharge from small sites. The rational equation is:

$$q = kCiA \qquad (6.5)$$

where q is the peak discharge (cms or cfs), k is a conversion factor (0.00278 in metric units or 1.00833 in English units), C is an empirical coefficient, i is the intensity of rainfall (mm/hr or in./hr), and A is the area of the site (hectares or acres). When working in English units, the conversion factor k is often ignored.

The coefficient C is frequently tabulated for different land uses and soil types. Values for C can be found in any standard hydrology text. Area (A) can be calculated by standard geometric or trigonometric methods. Rainfall intensity is assumed to vary inversely with storm duration. Rainfall intensity is dependent on the duration of the storm event. The rational method assumes that the most intense rainfall occurs over the time of concentration (t_c) of the watershed. Time of concentration is the time that it takes for rain falling on the most remote location in the watershed to reach the outlet. Once t_c is found, then rainfall intensity can be found from a local rainfall intensity-duration-frequency curve (Figure 6-1) (Hershfield, 1961).

Several methods are available for estimating t_c. The methods provide widely varying results. Check to assure yourself that the assumptions of the method chosen apply to your situation. Time of concentration (t_c) can be found by summing the time of travel for various flow segments as water travels from the most remote point to the outlet. These segments include overland flow, shallow concentrated flow, and larger channels. Once the segments have been defined, t_c is calculated with:

$$t_c = \sum_{i=1}^{n} \frac{L_i}{v_i} \tag{6.6}$$

where L_i is the length of segment i, and v_i is the velocity in segment i. For overland flow or shallow concentrated flow, v_i can be calculated by the relationship:

$$v = aS^{1/2} \tag{6.7}$$

where v is velocity, a is a coefficient, and S is the slope of the surface or channel. Consult Haan, Barfield, and Hayes for values of a. The NRCS Technical Release No. 55 (TR55) provides graphs relating velocity for overland and shallow concentrated flow to slope and land cover. Overland flow is limited to less than 300 feet or 100 meters. Overland and shallow concentrated travel time can also be estimated with a relationship developed by Overton and Meadows (1976) based on Manning's equation. The relationship is:

$$t_t = \frac{0.007(nL)^{0.8}}{P_2^{0.5} S^{0.4}} \tag{6.8}$$

where t_t is time of travel in hours, n is Manning's n (Table 6-3), L is the segment length in feet, P_2 is the 2-year, 24-hour rainfall in inches, and S is slope. This relationship assumes steady rainfall, uniform flow, and minor effects from infiltration. Manning's n for overland and shallow concentrated flow is provided in TR55. A condensed version is

TABLE 6-3 Values for Manning's n for Overland and Shallow Concentrated Flow

Ground Cover	Manning's n
Smooth pavement	0.011
Cultivated soil, <20% residue cover	0.06
Cultivated soil, >20% residue cover	0.17
Short grass prairie	0.15
Light turf	0.20
Dense turf	0.35
Short grass	0.03–0.35
Lawns, tight soils	0.15–0.33
Range	0.13
Pasture	0.03–0.05
Row crops	0.07–0.20
Woods, light underbrush	0.40
Woods, dense underbrush	0.80

Source: Modified from TR55 and Water Quality: Prevention, Identification, and Management of Diffuse Pollution, Novotny and Olem, 1994.

provided in Table 6-3. For channel flow, use Manning's equation:

$$v = \frac{k}{n} R^{2/3} S^{1/2} \tag{6.9}$$

where k is 1 in the metric system and 1.49 for U.S. customary units; R is the hydraulic radius (area divided by wetted perimeter); and S is channel slope.

Kirpich (1940) developed an empirical equation for time of concentration based on length of the watershed and the difference in elevation between the most remote point and the outlet. Kirpich is frequently used with the rational method. The equation is:

$$t_c = 0.0078 L^{0.77} (L/H)^{0.385} \tag{6.10}$$

where t_c is in hours, L is the maximum length of flow in feet, and H is the difference in elevation. Kirpich's equation does not consider the flow resistance, nor channel roughness.

The NRCS also estimates time of concentration with a method based on watershed lag time. Lag time is the time between the centroid of rainfall and peak runoff. The relationship is:

$$t_l = 0.6 t_c \tag{6.11}$$

where t_l is the lag time. To compute lag time, use the equation:

$$t_l = \frac{L^{0.8}(S+1)^{0.7}}{1900 Y^{0.5}} \tag{6.12}$$

where L is the hydraulic length of the watershed in feet, S is related to curve number as in the SCS runoff equation (Equation 6.2), and Y is the average land slope in percent. The equation is good for CNs between 50 and 95.

Another method for computing peak discharge is the NRCS TR55 method. TR55 is available as a computer program from the NRCS Hydrology and Hydraulics website: www.wsi.nrcs.usda.gov/products/w2q/H&H/Tools_Models/tool_mod.html. TR55 uses a unit peak discharge approach. The unit peak discharge (q_u) is the discharge in cfs per inch of runoff per square mile. Peak discharge is related to unit peak discharge by:

$$q_p = q_u A Q F_p \tag{6.13}$$

where q_p is the peak discharge, A is area in square miles, Q is runoff in inches, and F_p is a factor adjusting for detention in ponds and swamps. TR55 provides tables for F_p. TR55 also provides a graphical solution for q_u. The equation for computing q_u is:

$$\log(q_u) = C_o + C_1 \log t_c + C_2 (\log t_c)^2 \tag{6.14}$$

C_o, C_1, and C_2 are coefficients related to the ratio of I_a/P and can be found in any standard text on hydrology. P is the precipitation in inches, and $I_a = 0.2S$. S is from the NRCS runoff equation (Equation 6.2).

The Discharge Hydrograph

The last piece of data to go along with the runoff volume and peak discharge is the shape of the runoff hydrograph (see Chapter 5). One approach to developing a runoff hydrograph is the unit hydrograph. A unit hydrograph is a graph of the runoff that occurs from a rainfall excess of one inch. Assumptions are that rainfall is uniform over the study area, that there is a uniform rainfall excess rate, and that runoff rate is proportional to runoff volume for a rainfall excess of a given duration. To develop a unit hydrograph, first find the time to peak discharge, then find the peak discharge, then select an appropriate time of base and shape. Unit hydrographs can be used to graph complex storms if rainfall excess data are available for each portion of the storm, but that is beyond the scope of this discussion. The concern here is to describe a condition that can be used as a target for post-development conditions.

Time to peak (t_p) for a unit hydrograph is the lag time (t_l) plus one-half of the runoff excess duration (D):

$$t_p = t_l + D/2. \tag{6.15}$$

For an NRCS unit hydrograph (SCS, 1972), peak flow (q_p) can be estimated by:

$$q_p = \frac{484\,A}{t_p} \tag{6.16}$$

where A is the watershed area in square miles. The actual discharge from the watershed (Q_p) is found by multiplying q_p by the ratio of actual rainfall excess to the unit rainfall excess. The NRCS uses a single triangle unit hydrograph or a curvilinear unit hydrograph. In a single triangle unit hydrograph, the duration of runoff or time of base (t_b) is equal to 2.67 times the time to peak. For curvilinear hydrographs, the NRCS has a dimensionless unit hydrograph (Figure 6-5). They also provide tables of q/q_p vs. t/t_p so a unit hydrograph can be developed in a computer program.

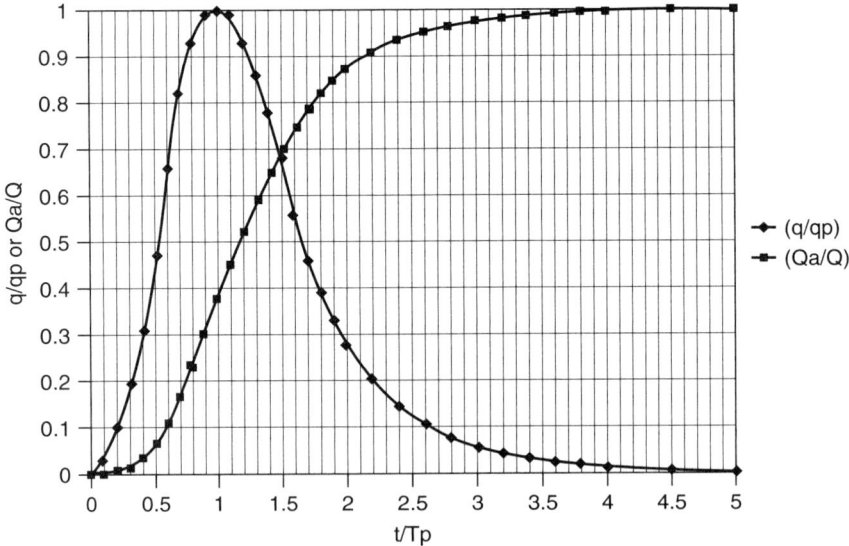

FIGURE 6-5. NRCS dimensionless unit hydrograph.

The goal of hydrologic characterization of the site is not to reproduce actual storm hydrographs. Storms will seldom repeat themselves. The goal is to generate a synthetic hydrograph that resembles a design storm. The hydrology of a site is influenced by the physical characteristics of the site and also influences the physical characteristics of the site. Biology is also influenced by and influences hydrology. The next section considers site biological characteristics.

Biological Characterization

Biological characteristics of a site include both the vegetation and soils existing on the site. Vegetation on a site supports multiple ecosystem services, including both local and global climate regulation, air and water cleansing, water supply and regulation, erosion and sediment control, hazard mitigation, pollination, habitat, waste decomposition and treatment, human health and well-being, food and nonfood renewable products, and cultural benefits (ASLA, 2009). Soil scientists consider the soil to be a matrix of biotic and abiotic materials. Because of the living nature of soil, it is included in this section on biological characteristics. Soils provide services of global climate regulation, air and water cleansing, water supply and regulation, erosion and sediment control, habitat, waste decomposition and treatment, and food and nonfood renewable products (ASLA, 2009).

The Land Matrix
Four basic terms can be used to describe the land cover or vegetation in an area: matrix, patch, corridor, and mosaic (see Chapter 5). The land matrix is the dominant land cover of the area (FISRWG, 1998). The matrix is dependent on scale. For instance, at the biome scale, the matrix may be "eastern hardwood forest" or "tall grass prairie." At the local scale, where site analysis is done, the matrix may be described in more specific terms, but it is still the dominant land use. A site may be in a low-density residential matrix, a cool season grass matrix, or an oak-hickory forest matrix, and so on. A patch is a nonlinear area or polygon that is less abundant than and different from the matrix (FISRWG, 1998). A patch may be a hickory forest within a cool season grass matrix, for example. A corridor is a special type of patch that links other patches (FISRWG, 1998). Typically, corridors are linear or elongated, such as streamside zones or buffers. Finally, a mosaic is a collection of patches none of which is dominant enough to be interconnected throughout the landscape (FISRWG, 1998).

In the United States, the native land matrix of an area can be found in the original Government Land Office (GLO) surveys and field notes. These maps, developed by surveyors in the 1800s, divided the country into townships and sections, and also identified the dominant land cover and significant patches. These GLO maps are valuable for developing restoration goals. The existing land use matrix of an area can be determined through examination of aerial photographs or land use/land cover (LULC) GIS datasets. In most areas, LULC datasets are available free of charge from the state GIS coordinator or from the National Atlas (USGS).

One measure of the amount of vegetation on a site is the biomass density index (BDI) (ASLA, 2008). The BDI is the density of plant layers covering the ground. The BDI is a quantitative measure of the vegetation on a site. It contains no information about the quality of that vegetation. However, it does provide a baseline against which to measure the degree of preservation of vegetative cover during site development. Data for calculating the BDI can be taken during the initial site assessment. The protocol developed by the Sustainable Sites Initiative is as follows: Determine the percent of the site area in each cover class (Table 6-4), multiply the percentage by the weighting factor, and find the sum of the products.

Example: A site contains 15 acres. Five (5) acres are forest with understory. Nine (9) acres are in tall grass meadow. Total impervious area including a house and drives is 0.1 acres. The remaining 0.9 acres are turfgrass. The BDI is:

$$\text{Forest: 5 acres/15 acres} \times 100 \times 5 = 167$$

$$\text{Meadow: 9 acres/15 acres} \times 100 \times 2 = 120$$

$$\text{Impervious: 0.1 acres/15 acres} \times 100 \times 0 = 0$$

$$\text{Turfgrass: 0.9 acres /15 acres} \times 100 \times 1 = 6$$

$$BDI = 167 + 120 + 0 + 6 = 293$$

Soil

In his book *Out of the Earth: Civilization and the Life of the Soil*, Daniel Hillel describes soil as "the fragmented outer layer of the earth's terrestrial surface in which the living roots of plants can obtain anchorage and sustenance, alongside a thriving biotic community of microscopic and macroscopic organisms." That is an oversimplified definition. Soil is the foundation of terrestrial life on earth. It provides footing for the vegetation as well as serving as the source of nutrients

TABLE 6-4 Values for Different Vegetation Classes Used in Computing the Biomass Density Index

Description of vegetation class	Value
Trees with understory	5
Trees, no understory (less than 10% herbaceous/shrub cover)	4
Shrubs	2
Herbaceous annuals and perennials and/or succulents	2
Tall grasslands	2
Turfgrass	1
Green roofs	0.5
Impervious cover or bare ground	0

Source: Modified from The Sustainable Sites Initiative, ASLA 2008.

and water. Soil is also the primary component of the earth for cleansing and recycling materials. Understanding the soil is necessary for ecological design.

Soils are described more completely in Chapter 7, "Defining Place: Soils as a Living Organism." The paragraphs that follow are a brief description of characteristics that should be specifically considered in site design.

Soil Textural Class

Soil textural class refers to the percentage of sand, silt, and clay in the soil. Textural class determines properties such as soil drainage, plant available water, runoff potential, water transportability, and erodibility. Textural class can be roughly determined in the field by feeling the soil. Sand is loose and granular; individual particles or grains can be seen and felt. Silt has smaller particles than sand, but still has a gritty feel. When silt is squeezed between the thumb and the finger, it will form a ribbon, but the ribbon will have a broken appearance. Clay is fine-textured. Moist clay will form a long flexible ribbon when squeezed between the thumb and finger (Spangler, 1966). An exact determination of textural classification requires a sieve analysis. Generally, soils are grouped in 12 textural classes, including, from the coarsest to finest: sands, loamy sands, sandy loams, loam, silt loam, silt, sandy clay loam, clay loam, silty clay loam, sandy clay, silty clay, and clay. With a sieve analysis of a soil including the percent of sand, silt, and clay, the soil textural class is determined from the textural triangle (Figure 6-6).

Soil Engineering Characteristics

Engineers are typically concerned with the ability of a soil to support earthworks, transportation facilities, and structures. Of critical

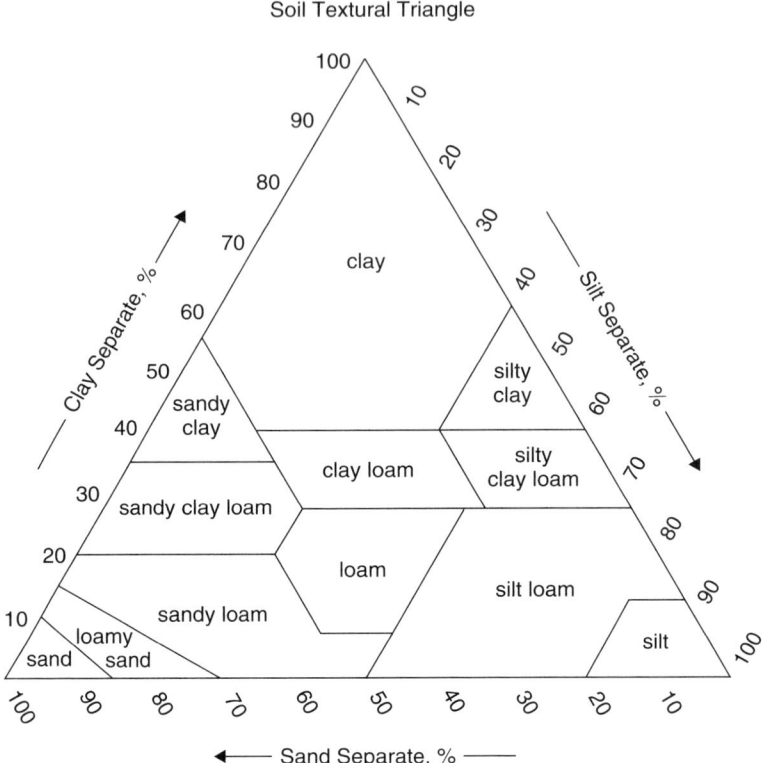

FIGURE 6-6. Soil textural triangle.

importance are the soil's resistance to shear failure, its ability to change volume under load, and its tendency to shrink or swell under varying moisture conditions. The following tests should be run if structures or pavements are to be considered (Brady and Weil, 2002):

- Unconfined compression: Good for cohesive soils, those with more than about 15 percent clay content.
- Angle of repose: For non-cohesive soils, the angle of repose represents the steepest slope to which a soil can be piled without slumping.
- Proctor density test: The maximum density that a soil can be compacted to under optimum moisture conditions.
- Consolidation: The degree to which a soil can be compressed.
- Atterberg limits: Measures of the shrink-swell potential of the soil. The plastic limit is the moisture content at which a clay soil will begin to behave as a plastic malleable plastic mass. The plastic

limit is found at the point where the clay will roll into a 3-mm-diameter rope. The liquid limit is the moisture content (percentage) at which the clay begins to flow as a viscous liquid when jarred. The difference between the plastic limit and the liquid limit is the plasticity index. Soils with plasticity indices of greater than about 25 are expansive clays and are generally not good structural soils.

Soil Hydrologic Characteristics

The hydrological characteristics of a soil will be used to determine the water balance of a site. Characteristics of importance are:

- Soil hydrologic group: See hydrologic characterization, page 112 in this chapter.
- Infiltration capacity: Design of drainage features and construction of stormwater hydrographs require knowledge of the infiltration capacity of the soil. Infiltration capacity is typically tested with a single- or double-ring infiltrometer test (ASTM D3385). The test is conducted in the field, is relatively simple, and provides reliable results. Each different soil in the project area should be tested, and the test should be conducted on the most confining layer of the soil (Brady and Weil, 2002). Consult a soils textbook for instructions on conducting the double-ring infiltrometer test.

Soil organic matter is responsible for, or a factor in, many soil properties, including its cation exchange and water-holding capacity, nutrient retention and release, and formation and stabilization of soil aggregates (Brady and Weil, 2002). Since all organic matter contains carbon, organic matter also relates to the amount of carbon stored in the soil. Soil contains approximately three times as much carbon as the atmosphere. Soil organic matter is commonly measured with a modified Walkley-Black method or a loss-on-ignition method (Walkley and Black, 1934).

Chemical Characteristics

The key chemical characteristics of soils are the soil pH, nutrient and metal concentration, and cation exchange capacity. Cation exchange capacity is the sum total of exchangeable cations that a soil can absorb (Brady and Weil, 2002).

Preliminary data can be found in soil surveys developed by the NRCS, available from the soil survey geographic database. More precise data should be compiled prior to final design. Geotechnical engineering firms provide testing of soils for engineering applications.

Biological and chemical tests are provided in most of the U.S. from the Cooperative Extension Service or environmental testing laboratories.

Biological, hydrological, and physical characteristics are in many ways a reflection of the climate of the area. At continental scale, the biome is primarily determined by climate. Local climate conditions may vary from large-scale conditions and must be considered as well.

Climatological Characterization

Conditions that must be considered to characterize climate are the prevailing wind speed and direction, temperature, available solar energy, and the amount, timing, and kind of precipitation. Precipitation has been discussed previously.

Wind Direction and Speed
A wind rose (Figure 6-7) is a graph that shows the distribution of wind direction and wind speed. The value of a wind rose depends on the

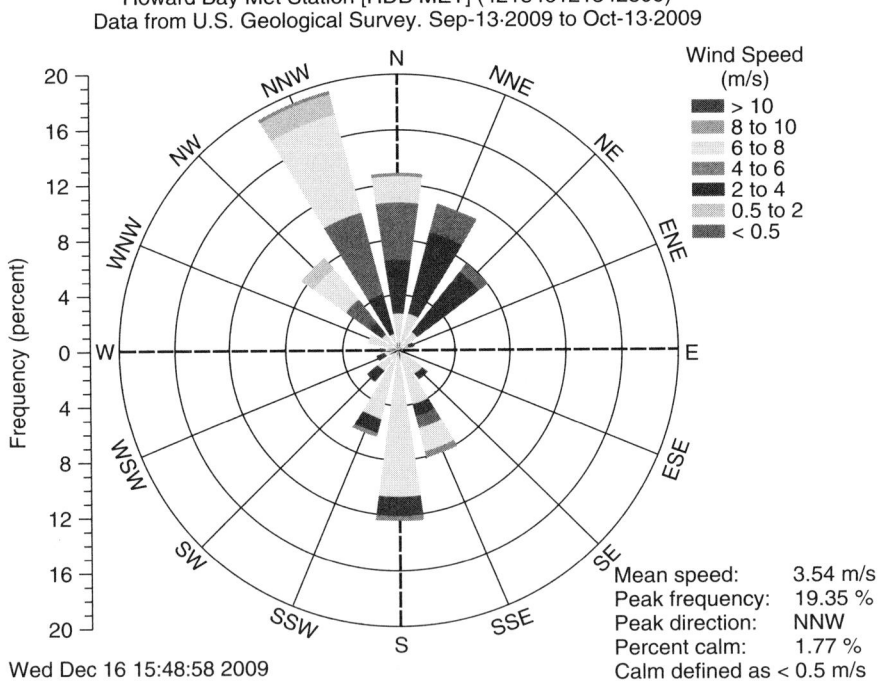

FIGURE 6-7. Wind rose. Figure generated with the USGS Wind Rose Data Grapher. Available at: http://or.water.usgs.gov/cgi-bin/grapher/graph_windrose_setup.pl. Accessed May 2, 2010.

volume of data that goes into producing it. Since wind characteristics vary by season, seasonal wind roses are needed. Probably the best source of wind data for a site is nearby airports.

Temperature
Necessary temperature data are the mean, minimum, and maximum temperature and the duration of seasonally cool or warm periods. Mean, minimum, and maximum temperatures for the United States can be obtained from the National Weather Service's Regional Climate Centers. Local media may also have historical records of temperature.

A degree heating or cooling day is a surrogate measure of the energy required to heat or cool a structure. Heating or cooling days are determined by the difference between the average temperature for a day and a base temperature. In the United States, 65°F is generally used as the base temperature. Heating and cooling days can be computed from daily temperature data. These data have been compiled by local heating, ventilation, and air conditioning companies.

Solar Energy
Days of sunshine and the angle of the sun to the horizon are useful data for estimating available solar energy. Days of sunshine are available from nearby weather stations. The solar angle can be easily computed from the site's latitude and aspect.

Plant Hardiness Zone
Plant hardiness zones are a guide to which plant species will survive in a particular region (see Chapter 4). North America is divided by the USDA into 11 hardiness zones based on increments of 10°F in minimum low temperature that is expected. Zone 6, for instance, is an area where the minimum temperature will be between 0 and 10°F.

Summary

The following checklist is a guide for site characterization:

General Information:

☐ Ecoregion characterization (Omernik, 1987, or Bailey and Hogg, 1986).

☐ Watershed: What is the USGS hydrologic unit code? Is there a Total Maximum Daily Load for this stream? Is there an active watershed management plan? If so, what special conditions need to be considered?

☐ Site Location: Latitude and longitude, layman's description of the site location, and legal description.
☐ Area (hectares or acres).
☐ Geology (provide description from available geological maps).
☐ Significant topographic features: Streams or rivers, riparian vegetation, floodplains (regulatory and otherwise), wetlands or bogs, landforms, rock outcrops, structures, utilities, average slope, relief, aspect of major slopes.

Hydrology:
☐ Downstream conditions: What type of waterbody receives runoff from this site (lake, stream, wetland, etc.)? How far is it to the nearest stream? What is the capacity of downstream conveyance? Are there habitable structures, transportation facilities, or other improvements that may be endangered by runoff from this site?
☐ Upstream contribution to site: Where will water run onto the site? What is the volume of and peak rate of run-on? Is there potential for the runoff to change because of future conditions?
☐ Precipitation: What is the annual mean and variability? What are the monthly means and their variability? What are the 2-, 10-, 25-, 50-, and 100-year, 24-hr rainfall amounts? Is a depth-duration-frequency curve available?
☐ Runoff: Where will runoff discharge from the site? What is the area tributary to each site discharge? What is the composite curve number for each tributary area? What runoff volume will each site discharge during 2-, 5-, 25-, 50-, and 100-year events? What will the peak flow be at each location where runoff discharges from the site, for 2-, 5-, 25-, 50-, and 100-year events? What is the shape of runoff hydrographs at each site discharge point?

Biology:
☐ Land cover: Under predevelopment conditions, what was the dominant land matrix? Under current conditions, what is the dominant land matrix? What significant patches and corridors exist on the site? What is the current land use? What is the site's biomass density index?
☐ Soils: Provide the following: soil type and textural class, unconfined compression strength (if structures are to be designed), angle of repose, proctor density, consolidation, Atterberg limits, hydrologic soil group, infiltration capacity, soil organic matter (%), pH, nutrient and metal concentration, and cation exchange capacity.

Climate:
☐ What is the dominant wind direction and speed from the local wind rose? What is the mean, minimum, and maximum temperature by season? On average, how many days of sunshine occur annually? What is the average number of degree heating and cooling days? What is the angle from the horizon to the sun on the winter and summer solstices?

Defining the biome, ecoregion, watershed, and local conditions of a site provides context for which design can proceed. Preservation or restoration of ecological services provided by a site requires an understanding of ecological processes. The next section of this book provides information on those ecological processes.

> Land, then, is not merely soil; it is a fountain of energy flowing through a circuit of soils, plants, and animals.
>
> —Aldo Leopold, *A Sand County Almanac*, 1949

Further Readings
Hillel, Daniel, *Out of the Earth: Civilization and the Life of the Soil*, University of California Press, Berkley, CA, 1991.

References
ASLA, Sustainable Sites Initiative. The Sustainable Sites Initiative: Guidelines and Performance Benchmarks. Available at www.sustainablesites.org/report, 2009.
Bailey, R.G., and H.C. Hogg, A world ecoregions map for resource reporting, *Environmental Conservation*, 13(3): 195–202, 1986.
Brady, Nyle C., and Ray R. Weil, *The Nature and Properties of Soils*, 13th ed., Prentice Hall, Upper Saddle River, NJ, 2002.
Brakensiek, D.L., and Rawls, W.J., Green-Ampt infiltration model parameters for hydrologic classification of soils, in J. Borrelli, V.R. Hasfurther, and R.D. Burman, eds., *Advances in Irrigation and Drainage Surviving External Pressures, Proceedings*, American Society of Civil Engineers Specialty Conference, New York, 1983.
Chow, V.T., D.R. Maidment, and L.W. Mays, *Applied Hydrology*, McGraw-Hill Series in Water Resources and Environmental Engineering, 1988.
FISRWG, Stream Corridor Restoration: Principles, Processes, and Practices, Federal Interagency Stream Restoration Working Group, GPO Item no. 0120-A, SuDocs no. A 57.6/2:EN 3/PT.653, October, 1998.
Frederick R.H., V.A. Meyers, and E.P. Auciello, Five to 60 minute precipitation frequency for the eastern and central United States, National Oceanic and Atmospheric Administration Technical Memorandum NWS HYDRO-35, U.S. Department of Commerce, Washington, DC, 1977.

Green, W.H., and G. Ampt, Studies of soil physics, part I –The flow of air and water through soils, *Journal of Agricultural Science*, 4: 1–24.
Haan, Charles T., B.J. Barfield, and J.C. Hayes, *Design Hydrology and Sedimentology for Small Catchments*, Academic Press, New York, NY, 1994.
Hershfield, D.M., Rainfall Frequency Atlas of the United States, Technical Paper no. 40, U.S. Department of Commerce, Weather Bureau, Washington, DC, 1961.
Hillel, Daniel, *Out of the Earth: Civilization and the Life of the Soil*, University of California Press, Berkeley, CA, 1991.
Holton, H.N., A concept for infiltration estimates in watershed engineering, U.S. Department of Agriculture ARS 41–51, 1961.
Horton, R.E., Approach toward a physical interpretation of infiltration capacity, *Proceedings of the Soil Science Society of America*, 5: 339–417, 1940.
Kirpich, P.Z., Time of concentration of small agricultural watersheds, *Civil Engineering*, 10(6), 1940.
NOAA, Precipitation Frequency Atlas of the Western U.S., NOAA Atlas II, National Oceanic and Atmospheric Administration, Superintendent of Documents, U.S. Government Printing Office, Washington, DC, 1973.
Novotny, Vladimir, and Harvey Olem, *Water Quality: Prevention, Identification, and Management of Diffuse Pollution*, Van Nostrand Reinhold, New York, NY, 1994.
Omernik, J.M., Ecoregions of the conterminous United States, Map (scale 1:7,500,000), *Annals of the Association of American Geographers*, 77(1): 118–125, 1987.
Overton, D.E., and M.E. Meadows, *Storm Water Modeling*, Academic Press, New York, NY, 1976.
Scott, Don, *Soil Physics*, Iowa State University Press, Ames, Iowa, 2000.
SCS, *National Engineering Handbook*, Section 4, Hydrology, Soil Conservation Service, Chester, PA, 1972, 1985.
SCS, Hydrology Technical Note no. N4, Soil Conservation Service, Chester, PA, 1986.
SCS, *Urban Hydrology for Small Watersheds*, 2nd ed., Technical Release no. 55, U.S. Department of Agriculture, Washington, DC, 1986.
Spangler, Merlin Grant, *Soil Engineering*, International Textbook Company, Scranton, PA, 1966.
Texas Department of Transportation, *Hydraulic Design Manual*, Texas Department of Transportation, Austin, TX, 2009.
Walkley, A., and I.A. Black, An examination of the Degtjareff method for determining soil organic matter and a proposed modification of the chromic acid titration method, *Soil Science*, 37: 29–37, 1934.
Ward, Andy D., and William J. Elliot, *Environmental Hydrology*, Lewis Publishers, New York, NY, 1995.

7
Defining Place: Soils as a Living Organism

> Man, despite his artistic pretensions and his many accomplishments, owes his existence to a six-inch layer of topsoil and the fact that it rains.
> —Author unknown

INTRODUCTION

The Soil Science Society of America defines soils as "a complex mixture of minerals, water, air, and organic matter forming the surface of the land." It is hard to overstate the importance of soil to ecological design. Simply put, soil is the basis of terrestrial life on Earth. Soil is life, and soil is alive.

In 1997, the National Resources Conservation Service placed a value of $19 per ton (O.907 MG) on soil for its contribution to air quality, water quality, nutrients and plant yield, and water regulation. A 6-inch layer of soil over an acre of land weighs approximately (2,000,000 pounds) 1,000 tons. That makes the value of soil over $19,000 per acre ($47,000 per hectare). But air and water quality are only two of the functions of soil.

Soil functions in our environment can be put into six general roles, including:

1. Providing a medium for the growth of plants
2. Atmospheric regulation through emitting and sequestering gases and dust
3. Providing habitat for soil organisms
4. Water regulation (see Chapter 6).
5. Nutrient and carbon cycling (see Chapter 9)
6. Acting as an engineering material

Beyond these six general roles, soils are related in some way to every one of the ecosystem services identified in the ecosystem assessment (Table 7-1). Because of its role in so many ecosystem functions, soil is truly priceless.

With so many roles related to ecological services, soil is one tool with which the ecological engineer must be familiar. Chapters 5 and 6 reviewed the importance of soils to watershed and site assessment. This chapter considers soil characteristics, including morphology, the form and arrangement of soil features; physics, the physical properties and processes of the soil; fertility, the soil's support of life; and ecology, the interaction of abiotic and biotic characteristics of soil. This chapter will provide an introduction to soils from an engineering perspective. Extensive study of soils is recommended as a component of ecological engineering education.

Morphology

Soil morphology is the study of the form and arrangement of soil features. Five factors control formation of soil: the parent material, weathering, organisms, topography, and time (NASA, 2005). Parent material is the material from which the soil formed. The parent material may be weathered bedrock, organic matter, older soils, or possibly a deposition layer from water, wind, volcanic activity, or mass wasting. Weathering is breaking down of the parent material by forces of heat, rain, ice, snow, sunshine, and the like. Organisms living in and on the soil affect the way the soil processes minerals and nutrients. Soil organic matter is dead organisms decomposed by soil bacteria. Local topography affects characteristics of the soil. Soils on hillsides directly facing the sun will be drier than soils on the opposite side of the hill. Time allows the other four factors to assert themselves.

Texture determines much of a soil's physical characteristics, including its water-holding capacity and its ability to transmit or transport water. Texture refers to the relative percentage of clay, silt, and sand-sized particles (see Chapter 6). Figure 6-6 is a tool for determining textural class.

Porosity is the open space between soil particles. High-porosity soils hold lots of water, low-porosity soils hold less. Permeability is the degree of connectivity between soil pores. Fine-grained soils such as clays or silts have numerous pores between the small soil particles and, as a result, high porosity. Coarse-grained soils such as sand have fewer pores, but those pores are more effectively connected to each other than those in fine-grained soils. The connectivity of pores gives coarse-grained soils more permeability than fine-grained soils. Many soil chemical properties are also determined by soil texture. Fine-grained

TABLE 7-1 The Relationship of Soils to Provisioning of Ecosystem Services

Category of Ecosystem Services	Subcategory		Relationship to Soil
Provisioning	Food	Crops	Soils provide media for plant growth, nutrients, and water supply.
		Livestock	Soils provide for growth of grasses necessary for livestock production. Livestock waste material is largely returned to and processed by the soil.
		Capture Fisheries	Soils are essential to maintaining water quality, reducing flooding, and sustaining baseflow in streams and rivers.
		Aquaculture	The engineering properties of soils provide for construction of dykes and ponds used for fish production.
		Wild Foods	Soil is the medium for growth of wild plants and is at the base of the food chain for edible fauna.
	Fiber	Timber	Soils provide the medium for growth of trees.
		Cotton, Hemp, Silk	Healthy soil provides for economical production of fiber plants.
		Wood Fuel	Soils are required for timber growth.
	Genetic resources		Even in highly disturbed ecosystems, soils may retain the native seed bank for decades.
	Biochemical and pharmaceuticals		The soil contains bacteria and fungi that frequently provide the base material for pharmaceutical products.
	Fresh water		Soils absorb, hold, release, alter, and purify water.
Regulating	Air quality		Soils emit and absorb gases and dust.
	Climate	Global	Soil carbon storage and sequestration help regulate atmospheric CO_2 levels.
		Regional	Healthy soils with vegetative cover provide a cooling effect locally.
	Water flow		Healthy soils and the vegetation they support absorb precipitation and slowly release it to maintain baseflow in rivers and streams.
	Erosion		Healthy soils support deep rooting and resist erosion.
	Water treatment		Soil bacteria consume organic matter in wastes and convert pathogenic bacteria to more useful bacteria.
	Disease		Soil bacteria consume pathogenic bacteria.

(*continues*)

TABLE 7-1 (*continued*)

Category of Ecosystem Services	Subcategory		Relationship to Soil
	Pest		Healthy soils provide habitat for both predator and prey, keeping pests in check naturally.
	Pollination		Native vegetation supported by soil provides habitat for insects responsible for pollinating plants.
	Natural hazard		Soils help ameliorate flooding and maintain vegetation during droughts. Soils support vegetation that absorbs the energy of violent storms.
Supporting	Soil formation	These services are not directly used by human beings.	Healthy soil provides the environment for conversion of minerals and organic material into humus, thus building the soil.
	Photosynthesis		Soils are the medium for plant growth, providing for photosynthesis
	Primary production		Soils harbor autotrophic bacteria and provide media supporting plant growth.
	Nutrient cycling		Soils provide storage of nutrients and assist in mineralization of organic nutrients.
	Water cycling		Soils provide for infiltration of precipitation, reducing runoff and maintaining baseflow.
Cultural	Spiritual and religious values		Most religions recognize humankind's responsibility to be stewards of the land.
	Aesthetic values		Vegetation, which requires healthy soil, is a key ingredient of "scenic" landscapes.
	Recreation and ecotourism		Gardening, which requires soil, is one of the top forms of recreation in the U.S.
	Education		Learning about soil gives a good foundation for all ecological studies.

soils have more surface area per unit volume than coarse-grained soils. Because of the surface area, fine-grained soils such as clays have more ability to adsorb chemical elements than sands.

For engineering purposes, it is necessary to also know the type of clay present in a soil. Montmorillonite, beidellite, and saponite clays have high shrink and swell potential. These expansive clays change volume in response to moisture content. Forces exerted by the swelling of the clays may crack foundations of buildings and buckle roads.

Other clays do not exhibit shrink and swell and are more easily worked for structural support.

A fundamental characteristic of soils is the distinction between topsoil and subsoil. Topsoil is the outermost layer of soil, usually 0–6 in. thick. The topsoil has the highest concentration of organic matter and organisms. Most soil biological activity takes place in the topsoil. Subsoil is the layer beneath the topsoil. Subsoil is much less fertile than topsoil.

Expanding on the idea of topsoil and subsoil, soil scientists describe a series of visually distinct soil horizons or layers (Figure 7-1). Soil horizons can be identified in a soil profile such as a roadside cut or an open trench. Starting from the surface, soil horizons are labeled as O,

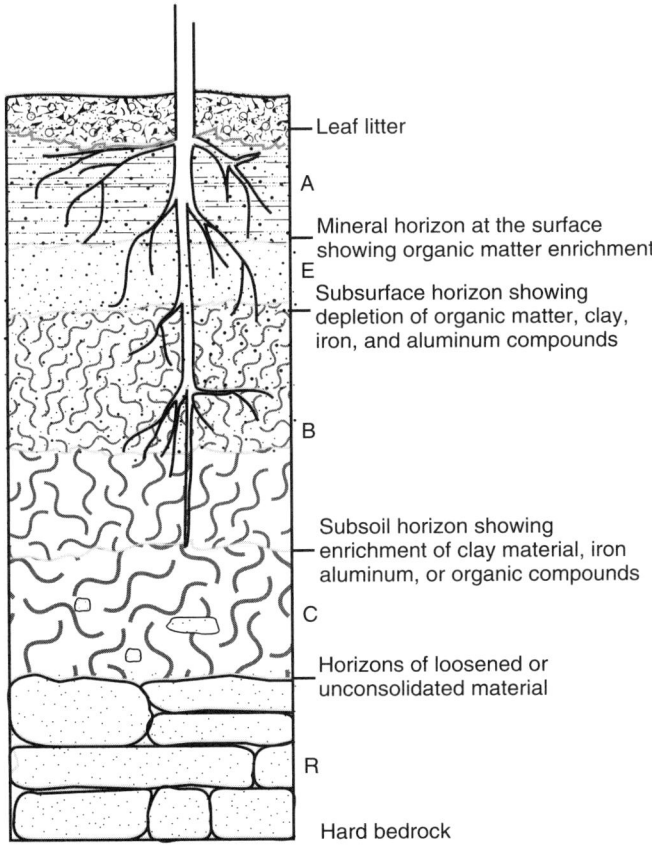

FIGURE 7-1. A soil profile typically exhibits, from surface downward, O, A, E, B, C, and R horizons. (*Source*: Figure from Arizona Cooperative Extension, Department of Agriculture, University of Arizona.)

A, B, and C. Sometimes an E horizon exists between the A and the B horizon, and some scientists define an R horizon below the C.

The O horizon is mostly plant litter and detritus. This horizon contains most of the soil biota. Plant matter is decomposing and adding humus and organic matter to the lower horizons. Below the O horizon is the A horizon, or the topsoil. The A horizon is the first true soil layer. In the A horizon, organic matter mixes with minerals. Organic matter in the A horizon makes this horizon darker than lower horizons. Eluviation, leaching of material from the A horizon by the downward movement of soil water, moves clays, minerals, and organic matter out of the A horizon into lower horizons. Plant roots tend to be densest in the A horizon. An E may exist just below the A horizon, especially in forested soils. The E horizon is a light-colored layer of sand and silt.

The B horizon lies below the A and E horizons. The B horizon is a zone of illuviation, or accumulation. There is relatively less organic material in the B horizon than in higher horizons, but mineral materials are accumulated. The materials in the B horizon either leach from higher layers or have weathered from the parent material of the soil. The B horizon stores water that is used by plants between precipitation events.

The C horizon is the deepest layer in most soils. This horizon consists of the unconsolidated parent material of the soil. The physical characteristics of the C horizon largely determine the utility of the soil for structural engineering functions. Below the C horizon is typically bedrock, sometimes referred to as the regolith or R horizon.

Color provides understanding of soil characteristics. Surface soil that is dark brown or black is high in organic matter. Soils in lower horizons tend to become lighter as organic matter becomes depleted. Color is also affected by soil moisture and oxidation state. Moist soils are darker than dry soils. Well-drained soils have oxidized forms of iron and manganese and tend to be yellow to brown. Frequently saturated soils go anoxic and will have reduced states of minerals. These soils will be gray to blue. Gray or blue mottling in the soil profile is an indication of a seasonal water table.

Soil structure is the combination of primary soil particles into secondary particles, units, or peds. These secondary units may be, but usually are not, arranged in the profile in a way that gives a distinctive characteristic pattern. Size, shape, and degree of distinctness of secondary particles are the basis for sorting soils into classes, types, and grades, respectively (Brady and Weil, 2002). Primary structure may be granular, platy, blocky, or prismatic (Whiting et al., 2009). Granular structure is found in topsoil with high organic matter that has been worked by earthworms. Platy structure consists of thin horizontal layers, or plates. These layers may be the result of parent materials laid down during sedimentary processes, or they may be from compaction by machinery. Blocky structure consists of cubes roughly 5 to 50 mm across. Blocky structure in the B horizon is an indication of good drainage in the soil. Prismatic structure consists of vertically oriented pillars up to 150 mm across. Prismatic structure is common in arid soils and also humid soils with poor drainage. Expansive clays are associated with prismatic soil structure (Brady and Weil, 2002).

The combination of properties of soil material that determines its resistance to crushing and its ability to be molded or changed in shape is the soil's consistence. Such terms as loose, friable, soft, plastic, and sticky describe soil consistence (Brady and Weil, 2002).

Soil organic matter is living organisms, carbonaceous remains of once living organisms, and organic compounds produced by soil metabolism (Brady and Weil, 2002). Organic matter makes up 1 to 6 percent of topsoil and helps to maintain a friable structure, making nutrients available to plants and retaining soil moisture.

The density of soil is the soil's weight per unit volume. Density may be measured either as the bulk density or as particle density (Brady and Weil, 2002). Soil consists of soil particles, air, and water. Typically, just over half of a soil is particles, and the remaining void space is filled with air or water. Bulk density (ρ_b) refers to the mass of noncompacted dry soil per unit of volume (v_t) (Brady and Weil, 2002). Particle density (ρ_s) is the mass of soil particles per unit volume of the solid phase of the soil (v_s) (Brady and Weil, 2002). Particle density is normally in the range of 2.6 to 2.7 g/cc. A value of 2.65 g/cc is normally assumed. Soil porosity (p_t) is the total volume of space in the soil that is filled with air or water (Brady and Weil, 2002). Bulk density, particle density, and total porosity are related by the equation (Brady and Weil, 2002):

$$\rho_b = (1 - p_t)(\rho_s)$$

Since particle density is normally assumed to be 2.65 g/cc, knowing either bulk density or total porosity allows computation of the other property.

A tremendous amount of information is acquired through observation of the soil profile. That information, coupled with understanding of how soil interacts with other parts of the environment, makes a powerful tool for ecological design. Physics provides the relationship between physical forces and soil properties.

Soil Physics

Soil physics deals with the physical properties and processes of soils. Of special importance is the movement of water and gases into and out of the soil and the forces that drive that movement.

Temperature

The rate at which chemical and biological processes occur in the soil is exponentially related to the temperature of the soil. In other words, as the temperature of the soil increases, the processes proceed more rapidly. This characteristic makes temperature one of the most important factors affecting soil.

Temperature is a measure of the thermal state of matter considered in reference to its ability to transfer heat. The heat content of a soil is the kinetic energy of the particles of which the soil is composed. The heat capacity of a soil is the amount of temperature change in the soil in response to the absorption or release of heat. Heat capacity may be either gravimetric or volumetric. Gravimetric heat capacity (C_g) is the heat required to raise the temperature of 1 kg of soil by 1 degree Kelvin. Volumetric heat capacity (C_v) is the amount of heat required to raise the temperature of 1 cubic meter of soil by 1 degree Kelvin. Multiplying the gravimetric heat capacity by the bulk density of the soil gives the volumetric heat capacity (Scott, 2000):

$$C_g \times \rho_b = C_v$$

Because soil is a matrix of solid, liquid, and gaseous materials, all three phases of the matrix must be considered in computing the heat capacity. The capacities are additive, so the total heat capacity of the soil is simply the sum of the capacities of each phase times the fraction of the soil in that phase (Scott, 2000):

$$C_v = SC_s + \theta_v C_w + f_a C$$

where S, θ_v, and f_a are the volume fractions of soil solids, water, and air (m^3/m^3); and C_s, C_w, and C_a is the volumetric heat capacity of the soil solids, soil water, and soil air respectively (J/m^{3}K). From this, it is seen that the heat capacity of soil is dependent on the mineral composition and on its water and air content (Scott, 2000).

The amount of energy reaching the soil is related to the amount of solar radiation, radiation from the sky, conduction from the atmosphere, condensation, and rainfall or irrigation (Scott, 2000). Energy is lost from the soil through radiation from the earth to the sky, conduction to the atmosphere, evaporation, and rainfall or irrigation (Scott, 2000). Vegetative cover regulates soil temperature through insulating the soil from environmental temperature factors and by transpiration of water from plant surfaces. Soil color also will affect temperature. Light soils reflect more energy than dark soils so they will be somewhat cooler.

Heat may be transferred through soil by convection or conduction. However, conduction is the most important mechanism (Scott, 2000). Conduction of heat through the soil follows Fourier's law:

$$H = -k\delta T/\delta z$$

where H is the heat flux density (J/m2s or W/m2), k is the thermal conductivity (J/msK or W/mK), and $\delta T/\delta z$ is the thermal gradient (K/m).

Aeration
Soil aeration is the process by which gases consumed or produced in the soil are exchanged for gases in the atmosphere. The aeration status of the soil can be characterized by measuring aeration porosity, composition of soil atmosphere, gaseous diffusion rates, oxidation-reduction potential, and the respiration quotient (Scott, 2000).

The aeration porosity (f_a) is the air-filled portion of the soil matrix. Since air and water compete for the same void space, aeration porosity is indirectly related to volumetric soil water content. Void space in the soil matrix depends on soil texture and structure. Void space typically increases as grain size decreases. Structure that is granular or crumbly will be more porous than other structures (Scott, 2000).

The composition of the soil atmosphere can be used to indicate the state of aeration of the soil. Soil oxygen sensors are available for measuring the concentration of O_2 in the soil atmosphere. If soil O_2 concentration falls significantly below the atmospheric O_2 concentration, then restricted gas exchange across the soil-air boundary is indicated.

Oxygen diffusion rate (ODR) is the flux density of oxygen across a soil surface. ODR can be measured with an oxygen diffusion meter (www.Eijkelkamp.com). Lower ODR rates mean poorer soil aeration.

Oxidation-reduction potential, or redox, is a measure of the electron availability potential in a chemical or biological system. Redox is the tendency of a chemical species to gain electrons and hence be reduced. Redox is closely related to the presence or absence of oxygen. The absence of oxygen is related to the presence of a soil water table that restricts soil aeration (Scott, 2000).

The respiratory quotient (RQ) is the ratio of the volume of CO_2 released by the soil to the volume of O_2 consumed. Under aerobic condition, RQ is 1. For anaerobic respiration, RQ will be increased (Scott, 2000).

Soil Water Movement

The abiotic components of soil include solids, water, and air. As with air, water is continually moving into, out of, and through the soil. The movement of water into the soil is the process of infiltration. Infiltration was covered in detail in Chapter 6 and is not repeated here.

Water moves through soil in accordance with Darcy's law, "the volume discharge per unit time from a one-dimensional column of soil is proportional to the column cross-sectional area and the total hydraulic head loss":

$$Q/At = K(\Delta H/L)$$

where
Q is the volumetric discharge (m^3).
A is the cross-sectional area (m^2).
t is time (s).
$\Delta H/L$ is the hydraulic gradient (m/m).
K is hydraulic conductivity (m/s).

Under saturated conditions, k is 10^{-5} m/s for sandy soil, and from 10^{-6} to 10^{-9} in clayey soils (Scott, 2000). Conductivity is related to pore size. A is the total cross-sectional area of the soil. The amount of area taken up by the solid phase of the soil is accounted for in k.

The energy in water is typically expressed in terms of "head" or meters above a datum. The hydraulic gradient is equivalent to the energy gradient and is simply the difference between the values of the head of the water across the area of interest. For flow of soil water under saturated conditions, the hydraulic gradient is the sum of gravitational head (H_g) and pressure head (H_p) (Scott, 2000). At any point, H_g is the height of the point surface over an arbitrarily placed reference level. H_p is the height of the water column above that point.

For unsaturated conditions, Darcy's law is:

$$q = -K(\Theta_v)(\delta H/\delta z)$$

where
q is the soil water flux density (m/s).
H is the total soil water potential (m).

z is the spatial coordinate (m).
$K(\Theta_v)$ is the hydraulic conductivity.

The total hydraulic potential (H) is the sum of the gravitational potential (H_g) and the matric potential (H_m):

$$H = H_g + H_m$$

Matric potential is the force exerted on water by the attraction of soil particles to water molecules (Scott, 2000). Matric potential is the force that causes water to rise up capillaries. Matric potential is negative.

When soils are unsaturated, hydraulic conductivity is not constant. As the soil dries, the attraction of soil particles becomes much greater. At low percentage of saturation, K becomes much smaller, perhaps by orders of magnitude. Under these conditions, flow of water will occur very slowly.

Plants extract water from soils by exerting osmotic and pressure potential against the soil water. As soil dries, K may become so small that plants are not able to create enough potential to obtain water from the soil. When this occurs, the soil is at the wilting point (Scott, 2000). At the other end of the water availability spectrum is the field capacity. Field capacity is the amount of water held by the soil after the draining of excess water (Scott, 2000). The difference between field capacity and the wilting point is the plant available water (Scott, 2000).

Civil engineers are more interested in using soil for its structural properties than for its ecological or water transport properties. The critical characteristics are the soil's bearing capacity, lateral earth pressure, and slope stability. A soil's bearing capacity is the stress that may be placed on a soil by a foundation that results in shear failure. Lateral earth pressure is the force exerted on a wall by a column of soil. Slope stability is the ability of soils on an embankment to resist sloughing off or slumping. These characteristics are covered extensively in civil engineering texts.

Soil Fertility

Soil fertility refers to the characteristics of soil that support life. A fertile soil contains adequate amounts of essential nutrients and organic matter, has a near neutral pH, has good structure, and is well drained. On the other hand, the level of production in a soil can be no greater than that allowed by the most limiting of any of the essential growth factors. This idea is referred to as the principle of limiting factors.

Living organisms require at least 16 elemental nutrients. The atmosphere provides nonmineral elements, including oxygen, hydrogen, and carbon; the remainder come from the soil (Hodges, 2010). The primary nutrients are nitrogen, phosphorus, and potassium (N, P, and K). These nutrients are needed in relatively large amounts compared to other minerals. N, P, and K are present in the soil; however, they are also frequently added to soil in chemical or organic fertilizers. Secondary nutrients include calcium, magnesium, and sulfur (Hodges, 2010). Plants and animals need smaller amounts of the secondary nutrients than N, P, and K. The secondary nutrients also are present in the soil. In management programs, calcium and magnesium are frequently added to the soil during liming operations, and sulfur is mixed with the chemical fertilizer (Hodges, 2010). Micronutrients include boron, copper, chlorine, iron, manganese, molybdenum, zinc, and sometimes cobalt and nickel (Hodges, 2010). Micronutrients are critical to healthy systems, but they are needed in much smaller quantities than the primary and secondary nutrients. Micronutrients are present in soils and plant material (Hodges, 2010).

Soil is a mixture of biotic and abiotic material. The biotic component—including living organisms, detritus from past living organisms, and organic compounds—is the organic matter of the soil. Soils typically contain from 1 to 6 percent organic matter (Brady and Weil, 2002). This organic matter is mostly present in the O and A horizons. Dark crumbly topsoil is an indication of soil with high organic matter content.

Tilth is the term used to describe the health of a soil. Tilth is technically defined as "the physical condition of soil as related to its ease of tillage, fitness of seedbed, and impedance to seedling emergence and root penetration" (Whiting et al., 2009). Soil with good tilth has loose structure that allows air and water movement, and holds sufficient water and nutrients for plant growth. Organic matter is an essential element of good tilth. Organic matter alters the structure of the soil, providing better drainage, and making moisture more available to plants. Organic matter also provides many of the essential nutrients for plant growth, and assists in cycling of those nutrients back to plant available forms (see Chapter 9).

A soil's cation exchange capacity (CEC) is the degree to which it can absorb and exchange cations (WSU, 2004). Cations are positively charged particles (NH_4^+, K^+, Ca^{2+}, etc.). Soil particles have negatively charged surfaces. Cations adsorb on these negative surface charges. Once adsorbed, cations are not easily dislodged by leaching with water. However, cations can be exchanged by other cations. The sorbed cations provide a reserve of nutrients from which plants can draw. Thus, CEC can increase soil fertility (WSU, 2004).

CEC is highly related to soil texture. Soils with more clay and organic matter have more available surfaces for exchange and have higher CECs (WSU, 2004). Clays have a very high surface area per unit volume and hence many available surfaces. CEC also increases with pH (WSU, 2004).

Anion exchange capacity is the degree to which a soil can adsorb anions (negatively charged particles). Anion exchange increases as soil pH decreases. In the range of normal soil pH, Anion exchange is not a significant process (WSU, 2004).

Soil Ecology

Daniel Hillel (1991) describes soil as a self-regulating biological factory. The inputs to the factory are the soil's—its own material, water, and energy from the sun. Work done at the factory cleanses our environment, produces food and fiber, and supports terrestrial life. The study of how this factory works is the field of soil ecology. The interaction between the biotic and abiotic components of the soil and between the soil and external elements makes our world livable.

Abiotic components of the soil include sand, silt, and clay particles; water; and air. These components have been discussed extensively in the preceding sections of this chapter. The biotic components of soil include plant roots, bacteria, fungi, actinomycetes, nematodes, protozoa, arthropods, earthworms, and burrowing animals (Michigan State University Extension, 2004). These organisms interact with each other and with the abiotic components of the soil to maintain nutrient and carbon cycles (Chapter 9) and to increase primary productivity or photosynthesis in the ecosystem. Table 7-2 provides a very brief description of the biotic components of soil and their function in the ecosystem.

Soil quality is the measure of a soil's function, specifically its ability to: accept, hold, and release nutrients and other chemical constituents; accept, hold, and release soil water to plants, streams, and groundwater; promote and sustain root growth; maintain suitable soil biotic habitat; respond to management; and resist degradation. Characteristics important to high-quality soil include: organic matter, water-holding capacity, microbial biomass carbon and nitrogen, good structure, texture, bulk density, electrical conductivity, nutrient availability and release, pH, and biotic diversity (Michigan State University Extension, 2004).

TABLE 7-2 Soil Organisms and Their Function in the Soil Ecosystem

Component	Description	Function
Plant roots	Residues of plant material and living roots	Provide the source of carbon for soil organisms
Bacteria	Unicellular microorganisms a few micrometers in length and coming in a variety of shapes, including spheres, rods, and spirals	Along with fungi the most important decomposers Help bind soil particles into aggregates Involved in nitrogen cycle
Fungi	Eukaryotic organisms (cells contain a nucleus), including molds, yeasts, and mushrooms	The most important decomposer of resistant compounds Help to bind soil particles into aggregates Establish symbiotic relationships with plant roots and improve nutrient and water uptake
Actinomycetes	Soil bacteria	Function similar to fungi Give soils their distinctive earthy aroma Valuable to the pharmaceutical industry
Nematodes	Slender wormlike animals typically less than 2.5 mm long	Accelerate decomposition by grazing on bacteria, fungi, and plant residues
Protozoa		Help to accelerate decomposition by grazing on bacteria, fungi, and plant residues
Arthropods	An invertebrate animal with an exoskeleton, a segmented body, and jointed appendages	Also help accelerate decomposition
Earthworms	Big worms	Mix soil Create macropores that increase infiltration and soil drainage Increase aeration
Burrowing animals	Small mammals	Mix soil Create macropores

Source: Adapted from Michigan State University Extension Soil Biology website, www.safs.msu.edu/soilecology/soilecology.htm.

The soil is one of the largest reservoirs of carbon on Earth (Brady and Weil, 2002). Soil contains more carbon than the atmosphere and plant material combined. This fact makes soil a critical component of the global carbon cycle and a regulator of the planet's climate. Photosynthesis uses carbon dioxide from the atmosphere and converts it to organic carbon. Metabolism of organisms in the soil continually uses soil carbon and converts it back to carbon dioxide (CO_2). Under favorable conditions, atmospheric carbon is sequestered in the plant material and eventually becomes part of the soil's organic matter. Natural ecosystems maintain the soil's carbon content (Brady and Weil, 2002). Continuous no-till cropping can also maintain carbon storage

(Sundermeier et al., 2005). Poorly managed soil may become a source of atmospheric carbon.

In agricultural and other non-natural systems, managing soil ecology is key to sustainable use of the resource. On the other hand, lack of an ecosystem approach to soil management will result in degradation of ecosystem services provided by this resource. Ultimately, if we use the soil resource unsustainably, we will deplete its function, and our society will not be able to maintain itself.

Summary

Ecological engineering is at its core utilization of soils, water, and vegetation in a sustainable system to support human needs. Good soil provides the template for effective use of vegetation. Good soil and vegetation also provide remediation of contaminated water, maintaining our streams and lakes. In return, plants and water maintain fertile soil through the cycling of nutrients and water and through addition of organic matter.

Civilizations live and die by their management of their soils. When they recognize the value of the soil and develop sustainable food and fiber production systems, the civilizations thrive. Those that deplete their soils eventually collapse under bureaucracy attempting to support a society with no resources. Ecological design recognizes the relationship between resource management and societal success. To be successful, ecological designers must have an understanding of soil as a living organism and of the way that organism interacts with the cycles and processes that support our economy.

This section of this book provided definition of place, from place in the largest sense (the biosphere, biome, and ecoregion) to watershed, site, and down to the soil. The next section of this book reviews basic ecological processes and functions. The connection of those processes with the idea of "place" sets the stage for competent ecological design.

> When the land does well for its owner, and the owner does well by his land—when both end up better by reason of their partnership—then we have conservation.
>
> —Aldo Leopold

Further Readings

Hillel, Daniel, *Out of the Earth: Civilization and the Life of the Soil*, University of California Press, Berkeley, CA, 1991.

References

Brady, Nyle C., and Ray R. Weil, *The Nature and Properties of Soils*, 13th ed., Prentice Hall, Upper Saddle River, New Jersey, 2002.

Hillel, Daniel, *Out of the Earth: Civilization and the Life of the Soil*, University of California Press, Berkeley, CA, 1991.

Hodges, Steven G., *Soil Fertility Basics*, North Carolina State University, Soil Science Extension, n.d., www.soil.ncsu.edu/programs/nmp/Nutrient%20 Management%20for%20CCA.pdf (accessed January 16, 2010).

Michigan State University Extension, *Soil Ecology and Management: A Sustainable Agriculture and Food Systems Cluster, Department of Crop and Soil Sciences*, Michigan State University, East Lansing, MI, 2004, www.safs.msu.edu/soilecology/soilecology.htm (accessed January 17, 2010).

NASA, Soil Science Education, National Aeronautics and Space Administration, April 20, 2005, http://soil.gsfc.nasa.gov/, (accessed January 10, 2010).

Ritter, Michael E., *The Physical Environment: An Introduction to Physical Geography*, 2006, www.uwsp.edu/geo/faculty/ritter/geog101/textbook/title_page.html (accessed January 11, 2010).

Scott, Don C. Soil Physics, *Agricultural and Environmental Applications*. Iowa State University Press. Ames Iowa. 2000.

Sundermeier, Alan, Randall Reeder, and Rattan Lal, *Soil Carbon Sequestration—Fundamentals*, The Ohio State University Extension, Food, Agricultural and Biological Engineering, Columbus, OH, 2005.

Whiting, David, Adrian Card, Carl Wilson, Catherine Moravec, and Jean Reeder, *Managing Soil Tilth: Texture, Structure and Pore Space*, CMG GardenNotes 213, Colorado Master Gardner's Program, Colorado State University Extension, 2009, www.cmg.colostate.edu/gardennotes/213.pdf.

WSU, Tree Fruit Soil and Nutrition, Cation-Exchange Capacity, Tree Fruit Research and Extension Center, Washington State University, 2004, http://soils.tfrec.wsu.edu/webnutritiongood/soilprops/04CEC.htm.

8
Fundamental Principles of Ecology for Design

> There is sufficiency in the world for man's need but not for man's greed.
> —Mohandas K. Gandhi

INTRODUCTION

The only constant in ecosystems is change. The study of ecology is about understanding the impact of changes in the continuum of life on Earth. Living things (biotic) are inseparable from their nonliving (abiotic) environment (Odum, 1993). All living things are interconnected with the place in which they reside and with the other living things around them. The ecosystem is the sum of all the biotic and abiotic elements within a boundary, including the flow of mass and energy through the boundary (Figure 8-1).

The scale of ecosystems is defined by the observer for particular purposes; an ecosystem can be as large as the global biosphere, and as small as the gut of an organism. The boundaries of ecosystems are determined by the purpose of analysis. As such, those boundaries must be explicitly defined prior to the process of designing them. Ecosystems are often classified based on structure or function. Ecosystem structural macrofeatures commonly used to define ecosystem boundaries include biomes and ecoregions (see Chapter 4), but other characteristics, such as major plant species, or type of waterbody (river, lake, estuary), are frequently used. Generally, ecosystem boundaries are associated with landforms, geographic boundaries, geopolitical boundaries, or other prescribed systems (Table 8-1). These scales from an ecological perspective are relatively fixed (Ricklefs, 2008). Organisms are the unit of exchange of energy throughout the environment, and for survival, reproduction, and therefore adaptation (natural selection). Populations are the unit of evolution, whereby genes that enhance fitness emerge throughout a group of individuals within a place. Community is the unit of biodiversity, and therefore ecosystem function. A metacommunity is

146 Ecological Engineering Design

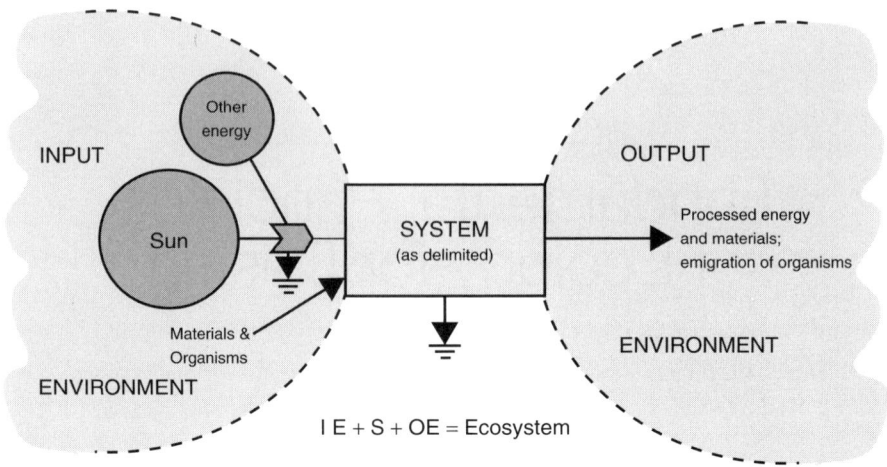

FIGURE 8-1. Conceptual model of ecosystems as open systems. (*Source*: Modified from Odum, 1993, *Ecology and Our Endangered Life Support System*).

TABLE 8-1 Structural Ecosystem Boundary Scales Compared to Geopolitical Boundary Scales and Ecological Design Scales

Ecological	Geopolitical	Ecological Design
Biosphere	World	Global
Biogeographic region	Nation	Continental
Biome	Region	Region
Ecoregion	State	Region
Watershed	County	Region
Biotic community	Town	Locale
Population	Neighborhood	Site
Organism	Individual	Element

a set of communities interconnected through dispersion corridors. The ecosystem is the arbitrarily defined unit of energy and nutrient cycling and flux. Global processes occur at the biosphere level.

As indicated in Chapter 3, the most common scale of design is site, with some local and very little regional design. However, as small geopolitical boundaries continue to fail to serve as effective entities for addressing increasingly complex ecosystem problems, regional planning and design has emerged. This is particularly evident in water resource management and transportation. The only truly global ecosystem design projects have been related to atmospheric processes such as ozone-depleting chemicals and greenhouse gas emissions.

TABLE 8-2 Ecosystem Classification Based on Energy Density

Ecosystem	Examples	Energy Flow(MJ/m² yr)
Unsubsidized	Forest, prairie, open ocean	5–50
Subsidized	Tidal estuaries, rain forests, riparian zones, wetlands	50–200
Human-Subsidized	Cropping systems, aquaculture, and pasture-grazed animal systems	50–200
Fuel-Power–Subsidized	Cities and industrial areas, wastewater treatment plants, electrical power generation facilities	500–15,000

Source: Modified from Odum, 1993.

Ecosystem function is less often used to define boundaries, and is generally associated with energy dynamics. Odum (1993) classified ecosystems based on sources and levels of energy (Table 8-2). The classification is based on the energy density of a system, defined as the amount of energy consumed per unit area per year (MJ/m2 yr). This notion of energy density has high utility for ecological engineers; it provides a framework for comparing function and structure of ecological designs with conventional designs. In general, ecological designs have much lower energy density than conventional designs (see Table 3-3). Unsubsidized ecosystems are still the dominant category across the globe, but agricultural crop and grazing lands now occupy over 40 percent of Earth's surface, making them the largest structural biome on the planet (Chapter 1).

The unit of design of ecosystem services is the ecosystem services element. This is a landform composed of a biotic community (plant, animal, microbes) and abiotic characteristics (soils, rocks, slope, drainage) and built structures (flow control devices, channels, topography). Each element within a site design combines with others to create a suite of ecosystem services (see Table 2-3). Multiple elements within a site or locale design would likely provide the same service (nutrient regulation, for example), creating a palette of redundant and reinforcing design elements for that service.

The most common energy density class in ecological design is a subsidized ecosystem, because of the high energy density associated with the system, and therefore high density of ecosystem services. One goal of ecological design is to replace high-energy, fuel-subsidized processes like wastewater treatment plants with human-subsidized versions of subsidized systems such as wetlands (Table 3-3). However, for a given level of ecosystem services (biochemical oxygen demand removal, disinfection, denitrification), a certain level of energy is required. That energy either comes from the sun (unsubsidized or subsidized) or is added (human-subsidized or fuel-power–subsidized). The obvious cost to achieve the same level of wastewater treatment from a wetland as from a wastewater treatment facility is geographic area (10–100 times as much); this is why the notion of energy density is so useful for the ecological engineer. The purpose of this chapter is to provide ecological theory for integrating energy density demands and ecosystem services provided by each element within a site, locale, and region.

Fundamental Principles of Ecology

Ecosystem functions are complex systems, but the structural elements of ecosystems interact and function based upon a relatively small set of fundamental principles (Ricklefs and Miller, 2000; Dale and Haeuber, 2001). The fundamental principles of ecology include:

1. The laws of thermodynamics govern the function of ecological systems.
2. The productivity of ecological systems is limited by the physical environment, defined by place.
3. Population processes regulate the structure and function of ecological communities, and therefore ecosystem services.
4. Organisms adapt and evolve in response to changes in the environment.
5. Ecological processes function across multiple time scales simultaneously.
6. Ecosystem functions are governed by networks of species interactions.
7. Ecosystem functions and structures are defined by the type, intensity, and duration of disturbances.
8. Ecosystems are defined by the size, shape, and spatial relationships of land cover (patches).

Ecosystems are open thermodynamic systems; energy flows into and out of them (Figure 8-1). The primary source of all energy for biota on Earth is Sol, our sun. Electromagnetic radiation from this star is captured and stored as phytochemical energy by primary producers (plants and algae and some bacteria) through the process of photosynthesis. This star energy is metabolized by the populations within an ecosystem, which converts it by respiration to kinetic energy to support activities, biomass for growth and reproduction, stored energy, and excreted energy. One organism's biomass and excreted energy are another organism's input energy. The cycling of mass and energy through ecosystems will be described in Chapter 9.

The productivity of an ecosystem is a function of the amount of sunlight entering the system, the amount of liquid water available, the availability of critical nutrients, and space. Areas on Earth with lots of sunlight and water (tropical rainforests, for example) have very high energy density, while areas from the same latitudes but with less water tend to have lower energy density (tropical deserts). This illustrates the effect of resource limitation on growth of a species. Thus, a defining characteristic of community interactions within ecosystems is competition for resources, especially those that are most limiting to growth of the individual, and thus population. Individual organisms respond to changes in resource availability and competition through mutations and natural selection against less advantageous traits—or in very high competition conditions, through selection for advantageous traits, resulting in evolution of the population. These processes will be discussed further in Chapter 10.

Ecosystems exist in states of dynamic change or dynamic equilibrium. Dynamic equilibrium is the product of redundant ecosystem functions, feedback controls, and other population-level processes. The cogs and gears of this complex machine are the organisms that reside on the land. As indicated in Chapter 2, the value of biodiversity is the stability and resiliency provided by a functioning biota within an ecosystem. These processes will be explored in Chapter 11.

ORGANISMS AND PLACE

Organisms in a place compete for resources to survive and thrive. Place defines the resources that are most limiting, and thus most in demand in the economy of nature (Ricklefs, 2008). Earth has a highly variable biosphere, as described in Chapter 4. The way organisms in a place deal with limited resources in a place is referred to as the life strategy for that species. In general, plants compete for energy (light and heat), water, nutrients (predominantly nitrogen, phosphorus, potassium, sulfur, and micronutrients), and space. Gymnosperms (flowering plants) also compete for pollinators and seed distribution. Many plants compete for bacteria and fungi (rhizofauna) in the rhizosphere. Animals compete for energy (food and heat), water, space (refugia), and mates. The variation of climate, topography, and connectivity of ecosystems over the past several billion years has resulted in adaptation of individuals and evolution of species, yielding the diversity of life on Earth.

The abiotic design in ecological engineering is not significantly different from environmental engineering. The biotic component is dramatically different. Ecological engineers are fundamentally planters and "ploppers," from a biotic perspective. We use plants and microbes to create the infrastructure or elements for ecosystem services to function. We select plants from one place and put them in another, or enhance the growing conditions for the indigenous plants. We bring microbes from one system (wetland, soil, stream) and plop them unceremoniously into the system we are designing. They adapt as a community and serve as the engines of ecosystem services. Fauna, especially macrofauna, are recruited to the site through various enticements, including structure, ecotones, and corridors.

Adaptation Processes

Ecological engineers design ecosystems for long time spans, often longer than 100 years. Design elements such as riparian canopy may take 50 years to develop to design potential. It is imperative, therefore, that the ecological engineer understand the way living things adapt to their environment, or evolutionary ecology. Designing ecosystem services requires an understanding of the natural history of a site and its organisms. If a site has legacy pollutants such as heavy metals, the design process must include selection of organisms adapted to these conditions for revegetation. The ecological engineer must understand the process of adaptation that drives evolutionary ecology.

Adaptations are the product of genetic mutations in individual organisms. These mutations occur at varying frequencies, depending on the genetic complexity and reproductive rate of the species. In extreme conditions, environmental phenomena (ultraviolet light, chemical contaminants) may accelerate mutations. The vast majority of mutations are deleterious, and therefore reduce the competitive fitness of the individual and its offspring. Some mutations are benign, and so persist in the genotype of the species. On rare occasions, a mutation will occur that provides a competitive advantage to the individual. If this competitive advantage is heritable, the offspring of this individual will be more successful than their peers, and thus over many generations the gene will become common in the species.

The approach evolutionary ecologists use to understand form and function in the context of location is called the adaptationist program (Ricklefs and Miller, 2000). The underlying premise is that the combination of traits best suited to survival in a place will emerge over time through evolution, within adaptive boundaries and physical limits.

This approach investigates how phenotypic expressions of form and function are limited by physical constraints, and how to measure fitness within populations (Ricklefs and Miller, 2000). Fitness is defined as "the genetic contribution of an individual's descendants to future generations of a population" (Ricklefs and Miller, 2000). Fitness does not just mean fecundity, or reproductive success. Fitness means the potential to have one's genes distributed across a population through successful selection. The distinction is that an evolutionarily fit individual has offspring that have offspring that have offspring. It is about survival through generations.

Successful adaptation to environmental conditions within a population means that the species must undergo genetic substitution of the adaptive gene through selection over many generations. For populations where no selection, mutation, migration, or selective mating occurs, the frequency of genetic characteristics (alleles and genotypes) remains constant (Ricklefs and Miller, 2000). The quantitative expression of this relationship is a modification of the Hardy-Weinberg law:

$$\Delta q = \frac{-sq^2(1-q)}{1-sq^2} \qquad (8.1)$$

where
   Δq is the change in allele frequency in the descendant population.
   q is the relative frequency of the new allele.
   s is the proportional reduction in fitness of the allele.

If $s = 1$ it is lethal. This simple relationship allows evolutionary ecologists to predict rates of adaptation across populations based upon observations of variability within their gene pool. Simply stated, when selection s is very low (no disadvantage), the alternative allele remains in the population, but when selection is high (>0.1), the alternative allele decreases, even among heterozygous alleles.

While the ecological engineer generally does not explicitly quantify fitness or adaptability of a species when designing ecosystem services, a more nuanced understanding of the process of heritability will help inform design decisions. The phenotypic variance (V_P) of a trait is the sum of its genotypic variance (V_G) and environmental variance (V_E):

$$V_P = V_G + V_E \qquad (8.2)$$

Genotypic variance is the product of additive variance (V_A), the expression of homozygous alleles; dominance variance (V_D), the

TABLE 8-3 Examples of Heritability of Traits of Animals and Plants

Trait	Organism	h^2	Source
Feedlot gain	Cattle (Bos primigenius)	0.86	Knapp and Clark, 1950
Grain yield/plant	Wheat (Triticum sp.)	0.85	Waqar-ul-Haq et al., 2008
Plant height	Corn (Zea maize)	0.70	Ricklefs and Miller, 2000
Dryland yield	Rice (Oryza sativa)	0.55	Zhao et al., 2006
Milk yield	Cattle (Bos primigenius)	0.40	Ricklefs and Miller, 2000
Body weight	Pigs (Sus domestica)	0.30	Ricklefs and Miller, 2000
Seeds per pod	Peanuts (Arachis hypogaea)	0.26	Songsri et al., 2008
Litter size	Mice (Mus mus)	0.15	Ricklefs and Miller, 2000

Source: Modified from Ricklefs and Miller, 2000.

expression of heterozygous alleles; and interaction variance (V_I), the influence that expression of one allele has on another:

$$V_G = V_A + V_D + V_I \qquad (8.3)$$

The heritability of a trait is the proportion of phenotypic variance due to genetic factors (h^2):

$$h^2 = \frac{V_A}{V_P} \qquad (8.4)$$

Comparison of heritability of different traits for different organisms gives the ecological engineer a sense of the adaptability of a species to a particular change in environmental conditions (Table 8-3). The importance of understanding heritability cannot be overstated—much of the advancement in grain production efficiency in the past 100 years has come through exploiting heritable traits in agronomic plants. The take-home message for the ecological engineer is: When designing ecosystem services for dynamic systems (almost all of them), diversity matters. Diversity in this case means interpopulation, so variability within a given species of plant, for example, is key to providing that population with the tools necessary to adapt to changes over time.

Responses to Environmental Variation

Form and function of biota are adaptations to the abiotic components of place. The thresholds in the ability of organisms to adapt represents their inherent design limitations in an environment. It is the environmental extremes, not averages, that define suitability of sites for particular species. The interaction between organisms also defines the viability or fitness of a species in a place (see Chapter 11).

This interaction between organisms (and phenotypes) can be understood using an evolutionary ideal called the evolutionarily stable strategy (ESS) (Ricklefs and Miller, 2000). The ESS for a site is that combination of organisms (phenotypes) that makes it impossible for others to invade. All the pieces fit tightly together, creating redundant and interconnected processes that result in high resource use efficiency.

Competition for resources is at the heart of an organism's response to environmental variation. At the biosphere level, the defining environmental variables are temperature and moisture on the landscape (Chapter 4). The most difficult ecosystems to design are those with wide ranges of temperature and moisture, because most organisms are adapted to one extreme or another. In extreme environments, polymorphic genotypes within a population often emerge, whereby one subgroup is adapted to one set of extremes, and the other subset to the other extreme. The species persists because of this range of adaptability, with enough of the genotype persisting during environmental extremes to carry the species through.

The individual within a population responds to environmental conditions in nongenetic ways, in that the genome of the individual does not change over its lifespan. However, the ways an individual can respond to environmental stresses are controlled by its genetics, and thus subject to natural selection (Ricklefs and Miller, 2000). The ability of an organism to respond to environmental conditions is called its phenotypic plasticity. The most extreme form of phenotypic plasticity is exhibited by facultative anaerobic bacteria, which have the capacity to metabolize carbon aerobically or anaerobically, depending on availability of oxygen. Plants can grow faster or slower and flower earlier or later, depending on a variety of environmental signals. Similarly, animals can grow faster or slower, reproduce or not, forage or hibernate, depending on environmental signals.

The range of environmental conditions within which an organism can persist is termed its niche. The theoretical range, or fundamental niche, is often limited by predation, competitors, pathogens, and other stresses, resulting in a smaller zone of survival, termed the realized niche (Ricklefs, 2008). Organisms with very low phenotypic plasticity, and therefore small realized niches, are specialists, in that they cannot adapt to changing environmental conditions very well. These organisms are often selected as indicators of environmental condition because of their sensitivity to change. These indicator species are often the first to disappear from a locale after disturbance (land use change, water contamination, and global climate change, for example). They also can serve as very valuable indicators of successful restoration of ecosystem services, as their presence indicates that the system is functioning at some level of competence.

LANDFORMS AND ECOSYSTEM FUNCTION

Ecological engineers design ecosystem services through design elements on the landscape that provide those ecosystem services. Those landform elements are composed of biotic and abiotic components. The biotic components interact with the abiotic components at all scales. Plants compete with neighboring plants for sunlight, water, and nutrients. Animals compete with other animals within a site for water, food, shelter, and mates (within species). The structure of ecosystem service elements on the site creates a metafunction of ecosystems at the locale and region level. The study of these processes is called landscape ecology. The landforms that emerge from combinations of elements should be designed to create the desired level of complexity, interaction, and redundancy. The role of the landscape ecologist in ecotechnology teams should be to assess and quantify the impacts of design options on these ecosystem functions.

Populations are distributed across the biosphere based on their ability to respond to and survive environmental extremes. The geospatial extent of their distribution is called their geographic range. The density and demographic characteristics of a population within a site characterize its spatial structure. A forest of red oaks in the Ozark Plateau may be heterogeneous or homogeneous in age, depending on the disturbance history of the forest (fire, clear cutting, windfall, disease). The demographics of the species determine the site's susceptibility to disturbance through disease, invasion, and other stresses. The structure of the site can influence the response of species to these stresses.

Patches, Corridors, and Connectivity

One of the most significant impacts human activities have on the landscape is fragmentation of landforms (Figure 8-2). Clusters of landform elements create patches. These patches can be connected by adjoining edges or corridors. Patches are typically where things live. The size of patches and diversity of landforms within them determine the diversity of populations that occupy them. The size and density of organisms within patches determines the density and robustness of ecosystem services they provide.

From an ecological engineering perspective, patches are defined by land cover type, and thus vegetation. Patches are also where animals

FIGURE 8-2. Fragmentation of forested landforms by agriculture, with riparian corridors. (*Source*: Photo by Matlock, 2009.)

interact. The mechanisms of interaction are described in Chapter 10, but the most common interaction is predator-prey. The ability for a predator to survive is dependent upon successful encounters with prey species. The ability of a prey species to survive is dependent upon avoidance of encounters with a predator species. This tension is affected by the density and distribution of populations within a patch, as well as by the structures within the patch that allow for these interactions.

Connections and corridors are critical for movement of organisms between patches. The process of dispersal is central to persistent ecosystem function. Dispersal limitations due to physical barriers often prevent a population from inhabiting a site that has appropriate habitat. This is a major flaw in the often-practiced *Field of Dreams* approach to ecological design, whereby the ecological engineer assumes that "if you build it, they will come."

Over time, a population at a site will likely experience conditions beyond its realized niche, resulting in site mortality. If a population is removed from the site and no replacement organisms (propagules) are available, the population will become extinct from the site and will not return. Dispersal mechanisms are critical for populations at a site to persist over time.

The dispersal strategies of many species are so successful that they are termed cosmopolitan in their distribution. Many algal species and insects, species of rats, and other species with highly plastic phenotypes and successful dispersion strategies live almost everywhere on Earth. However, the challenge for the ecological engineers is often to restore habitats for indicator species that, by definition, are very susceptible to site mortality. Thus, the design must often include the development of habitats that provide not only refugia within the realized niche constraints of the species but also connections to other organisms within the species, for propagation, genetic diversity, and, when necessary, site recovery.

The most common form of corridor in human-dominated ecosystems, both urban and agricultural, is the stream riparian corridor. River corridors represent ideal corridors for dispersal of individuals and propagules (Naiman et al., 2005). Verry et al. (2004) defined the geomorphic basis for riparian ecotones as the bankfull width plus 30 meters on each side (Figure 8-3). Riparian corridors also illustrate the scalar nature of landscapes—at closer scales corridors become a series of interconnected patches, as diversity of landforms becomes resolved.

Ecotones and Edge Effects

As indicated, one major impact human beings have on the landscape is fragmentation of landforms. Most nonhuman-dominated communities are characterized by gradients of change, rather than abrupt shifts. The effect of human landscape fragmentation at some scale is the increase across the landscape of sharp gradients and zonations (Odum, 1993). The transition zones from landforms create edges. These edges in ecosystem structures are called ecotones. Ecotones are actually transitional ecosystems more than physical edges. These structures are very recognizable; they include forest-meadow transitions, shorelines, soil demarcation zones, and urban fringe (Figure 8-4a and b).

The diversity of populations occupying and utilizing edges within landscapes is generally very high relative to patches (Naiman and Décamps, 1990). Pasitschniak-Arts and Messier (1998) found the highest species richness of rodents along the edges of meadows rather than in the prairies themselves, probably due to enhanced nesting materials. High prey density means high predator density. The reason ecotones are so robust is that they are where the action is. Ecotones represent transitional places where environmental conditions are in high flux; these areas are always in some state of transition. Thus, the shoreline has waves and tidal flux creating high disruption zones. The forest-meadow ecotone has high light availability for understory and meadow plants that cannot compete with trees, but tend to do very well at the

Fundamental Principles of Ecology for Design 157

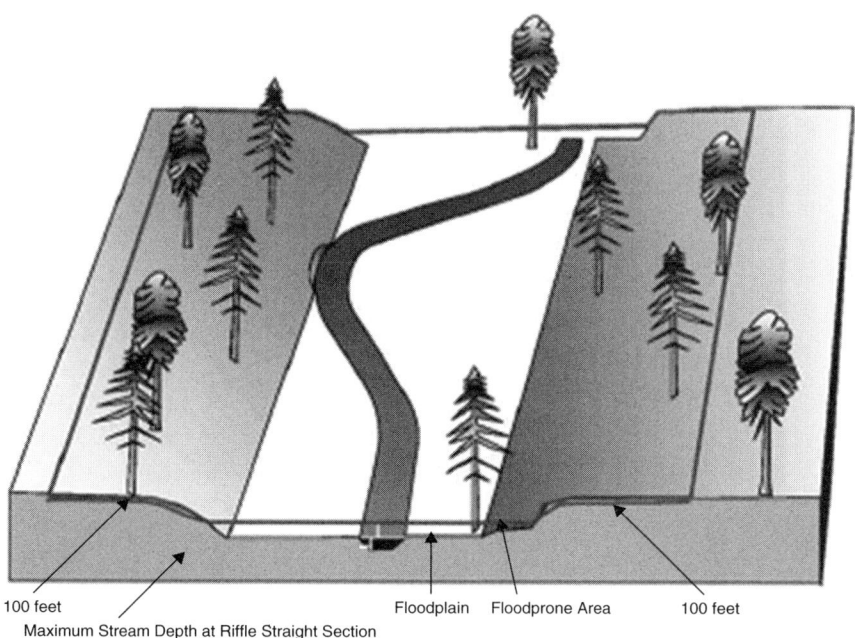

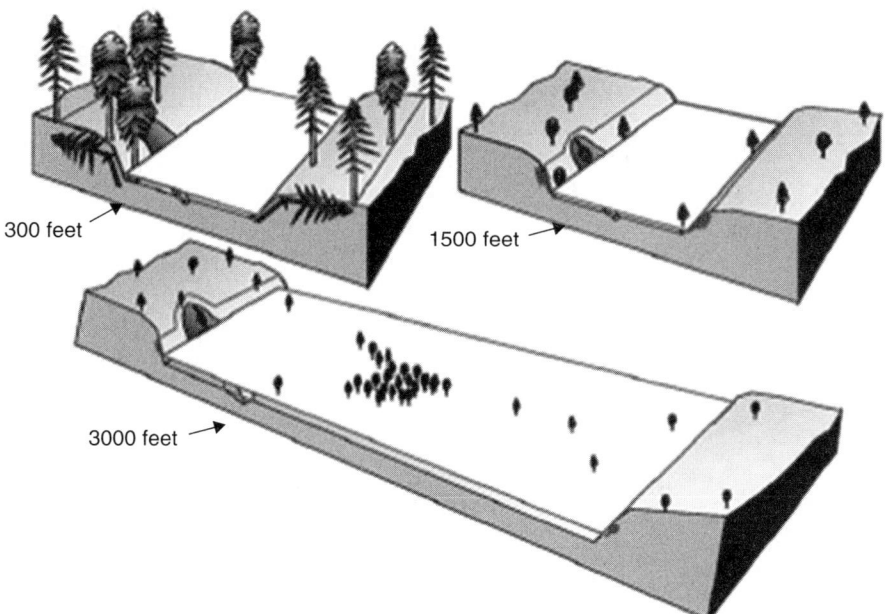

FIGURE 8-3. Geomorphic criteria for the riparian ecotone. (*Source*: Verry et al., 2004.)

158   Ecological Engineering Design

FIGURE 8-4. Ecotones across multiple landscapes. (*Source*: Photo by Matlock.)

edge. This complex transition zone creates opportunities for species that can compete in high-disturbance regimes. As a result, biodiversity tends to be highest at the edges of patches.

Landform Metrics

The ecological engineer rarely has the resources to restore large patches and thus smooth gradients. More often, the ecological engineer has small parcels of space within which to create ecosystem services. This spatial limitation drives the design to ecotones and patches. These landforms should be mapped at the site, locale, and region scale for analysis. The mapping process should include, explicitly, assessment of size and shape of patches, corridors, and dispersion barriers. This process can provide the initial assessment of the metacommunity within which the ecological design must persist (Holyoak et al, 2005).

Hof and Bevers (1998) proposed a geometric model for assessing patch and connectivity impacts on populations. They defined the expected number of organisms of population S to be a function of the size of the patch, the ideal density of individuals, and the probability that the patch is connected to other patches:

$$E(S) = \sum_{i=1}^{M} PR_i \, a_i \, A_i \tag{8.5}$$

where
 PR_i is the probability of patch i being connected.
 a_i is the expected density of organisms in connected patches (no distribution barriers).
 A_i is the size of patch i.

The geometric model provides a mechanism to optimize patches based on geometric shapes in an area for a population S based on ideal geometries (circle, rectangle).

The ability of a species to colonize a patch is generally dependent upon the resource availability of that patch to support the species. This bottom-up control process will dominate ecological design processes, as top-down–controlled ecosystems, where predators control prey populations, generally require time and stability to be established (Holyoak et al., 2005). Thus, for a species of a trophic rank j, the proportion of a patch occupied by that species (P_j) over time is a function of the fraction of the patch landscape (h) suitable for species j (h_j), the colonization rate (c_j), and the rate of extinction (e_j):

$$\frac{dP_j}{dt} = c_j P_j (h_j - P_j) - e_j P_j \tag{8.6}$$

For a top predator of species k to persist in a donor-controlled metacommunity with n levels, the fraction of the landscape suitable for that species (h_k) must be greater than the sum of the ratios of the extinction rates of the species upon which it depends (e_j) and the patch colonization rate for that species (c_j):

$$h_k > \sum_{j=1}^{n} \frac{e_j}{c_j} \qquad (8.7)$$

One of the clear consequences of habitat fragmentation is decline of the more rare species, as indicated in these relationships. The challenge for ecological engineers is to design ecosystem elements such that critical rare species can persist and flourish. Community dynamics will be discussed in more detail in Chapter 10

> The last word in ignorance is the man who says of an animal or plant: "What good is it?" If the land mechanism as a whole is good, then every part is good, whether we understand it or not.
>
> —Aldo Leopold

Further Readings
Odum, E., *Ecology and Our Endangered Life Support System*, Sinauer Associates Press, Sunderland, MA, 1993.
Ricklefs, R., *The Economy of Nature*, 6th ed., W.H. Freeman and Co., New York, NY, 2008.

References
Dale, V., and R. Haeuber, *Applying Ecological Principles to Land Management*, Springer Publishers, New York, NY, 2001.
Hof, J., and M. Bevers, *Spatial Optimization for Managed Ecosystems*, Columbia University Press, New York, NY, 1998.
Holyoak, M., M. Leibold, and R. Holt, *Metacommunities: Spatial Dynamics and Ecological Communities*, University of Chicago Press, Chicago, IL, 2005.
Naiman, R., and H. Décamps, *The Ecology and Management of Aquatic-Terrestrial Ecotones*, UNESCO, Paris, France, 1990.
Naiman, R., H. Décamps, and M. McLean, *Riparia*, Elsevier Press, Boston, MA, 2005.
Odum, E., *Ecology and Our Endangered Life Support System*, Sinauer Associates Press, Sunderland, MA, 1993.
Pasitschniak-Arts, M., and F. Messier, Effects of edges and habitats on small mammals in a prairie ecosystem, *Canadian Journal of Zoology*, 76(11): 2020–2025, 1998.
Ricklefs, R., *The Economy of Nature*, 6th ed., W.H. Freeman and Co., New York, NY, 2008.
Ricklefs, R., and Miller, G., *Ecology*, 4th ed., W.H. Freeman and Co., New York, NY, 2000.

Songsri, P., S. Jogloy, T. Kesmala, N. Vorasoot, C. Akkasaeng, A. Patanothai, and C. C. Holbrook. Heritability of Drought Resistance Traits and Correlation of Drought Resistance and Agronomic Traits in Peanut. *Crop Sci.* 48: 2245–2253, 2008.

Verry, E., C. Dolloff, and M. Manning, Riparian ecotone: A functional definition and delineation for resource assessment, *Water, Air, & Soil Pollution*, 4(1): 67–94, 2004.

Waqar-ul-haq, M.F. Malik., M. Rashid, M. Munir and Z. Akram. Evaluation and estimation of heritability and genetic advancement for yield related attributes in wheat lines. *Pak. J. Bot.*, 40(4): 1699–1702, 2008.

Zhao, D.L.; Atlin, G.N.; Bastiaans, L.; Spiertz, J.H.J. Cultivar weed-competitiveness in aerobic rice: heritability, correlated traits, and the potential for indirect selection in weed-free environments. *Crop Science* 46: 372–380, 2006.

9
Energy and Mass Flow through Ecosystems

> The Earth is bioregenerative: plants, animals and especially microorganisms regenerate, recycle, and control life's necessities.
> —Eugene Odum, *Ecology and Our Endangered Life Support System*

INTRODUCTION

Ecosystems are by definition systems bounded by observer-defined boundaries across which energy and mass flow. The biosphere, as Odum conceptualized it, is a life support system for all creatures on Earth, operating as a complex machine of interacting subsystems (Odum, 1993). Odum recognized the perils of oversimplification, but asserted that understanding and managing ecosystems required a systems approach to quantifying energy and mass flow through the component systems. Odum suggested applying systems modeling approaches to simulate ecosystem processes; he defined five components of an ecosystem element (Figure 9-1):

1. Properties (P; state variables)
2. Forces (E; forcing functions)
3. Flow Pathways (F)
4. Interactions (I; interaction functions)
5. Feedback loops (L)

Properties or state variables are internal components of an ecosystem process; they can be chemical compounds (phosphate or nitrate ions), species of organisms (population of foxes as $P1$ and voles as $P2$), or assemblages within ecosystems (primary producers as $P1$, primary consumers [grazers] as $P2$, etc.). Forces (E) represent the driving energy that moves energy and mass through the system, the thermodynamic forcing function for the system. At the biosphere level, the forcing function is solar energy; for some extremaphiles, it might be

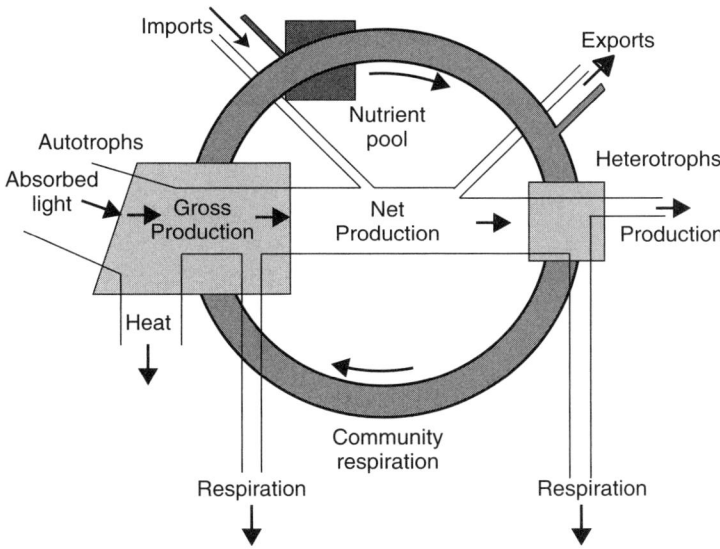

FIGURE 9-1. Systems diagram of ecosystems. (*Source*: Modified from Odum, 1993, *Ecology and Our Endangered Life Support System*.)

volcanic vents; for mammals, it would be primary production rates. The flow pathways (F) represent movement of energy or mass across boundaries, with some rate function (flux). Interactions (I) are defined by state variables (P), and can be chemical reactions, competition for sunlight between plant species, predator-prey interactions, or competition for mates when *P1* and *P2* are individuals within a population. The presence of feedback loops (L) gives the system complexity, resulting in nonlinear relationships between the forcing function and state variables, and can result in homeostatic or threshold responses to changes in E, depending on the system. The utility of this conceptual framework is that it is scalable, with nested processes within systems (Figure 9-2).

The goal of this chapter is to provide a conceptual framework for understanding energy and mass flow through ecosystems. The chapter is organized to provide quantitative tools for design, as well as concepts necessary to interpret and create relationships between components. This conceptualization of energy and mass flux across ecosystems and within components is subjective in that ecosystem boundaries (and components and subcomponents) are defined by the observer. This approach represents a conceptual model of ecosystem function. When modeling complex processes, one should always heed the sage admonition of Dr. George Box: "All models are wrong; some

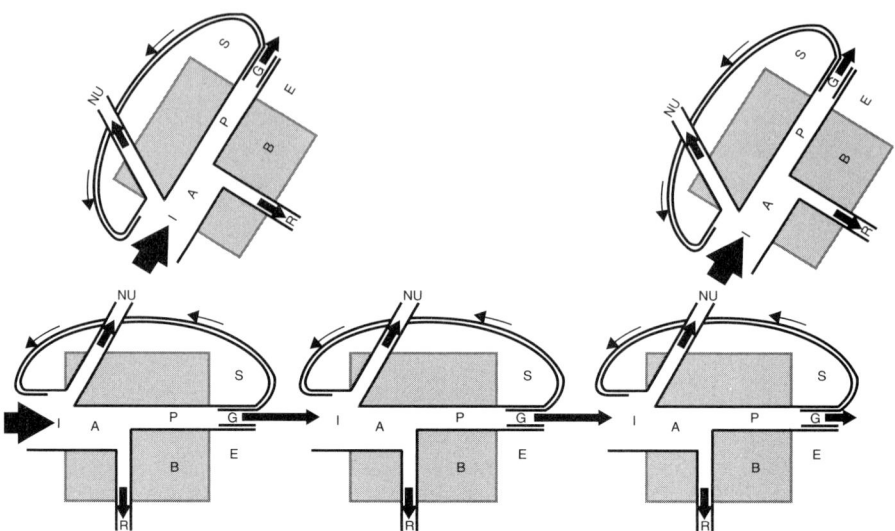

FIGURE 9-2. Nested systems represent ecosystems. (*Source*: Modified from Odum, 1993, *Ecology and Our Endangered Life Support System*.)

are useful." The key for the ecological engineer is to create useful constructs of ecosystem processes.

ENERGY FLOW THROUGH ECOSYSTEMS

The biosphere and all ecosystems contained within it are open thermodynamic systems. Energy in the form of electromagnetic waves flows into the biosphere from Sol, Earth's star, and is re-radiated out into space. Some of that energy is converted into biomass, which is then cycled through the biosphere, and ultimately re-radiated into space as heat. Energy from Sol captured by Earth's biosphere drives all biotic process, with minor exceptions for bacteria that harvest chemical and thermal energy from volcanic vents. Energy from Sol drives weather patterns, thus climate, and therefore controls biome distribution. Solar energy drives weathering of Earth's abiotic parent material into biotic soil. Ecosystem services are all dependent upon solar energy. The challenge for ecological engineers is to design systems that enhance utilization of this energy on the landscape.

Energy Balance in the Biosphere

Energy enters Earth from Sol as electromagnetic radiation, is transformed to heat and chemical energy, and cycles through the Earth's ecosystems. Ecosystems are thermodynamic systems; they are ultimately controlled by the laws of thermodynamics. The *First Law of Thermodynamics* is that energy cannot be created or destroyed, but rather changes from one form to another. As photons contact chlorophyll, electromagnetic energy is converted to chemical energy (carbohydrates), for example. Those carbohydrates are metabolized by organisms to produce work (locomotion) and convert biomass (growth). These are all forms of energy. The *Second Law of Thermodynamics* is that, in a closed system, energy at the end of a conversion process is less than at the beginning; no process is 100 percent efficient. At first glance this might seem to defy the first law, but there are two reasons it does not: (1) the energy that is lost is not destroyed; it is just converted to a form that is less dense or measurable (dissipated heat, for example), and (2) the system is closed by definition; that is, no new energy comes in. Earth is an open system; thus life, evolution, and biospheric processes are all open and entropic. Life is an energy cascade process, from high electromagnetic energy (sunlight) to abiotic elements (atoms). Within ecosystems, the second law describes why each transformation of energy results in less energy at the end, a critical concept in trophic dynamics. The *Third Law of Thermodynamics* is that at absolute 0° Kelvin, the entropy is zero—there is no more energy to give. The utility of the third law for ecological engineers is that it gives us a benchmark for how much energy is in a system (enthalpy).

A Newton (N) is the force of one kg accelerating at one m/s^2. Energy is the inherent force (potential or kinetic) that a system can exert, in Joules (J, equal to one newton meter). Energy is the potential to do work, or to be converted to power. Power is the rate of doing work, or the amount of energy flowing through a system per unit time, and is measured as J/s, or Watts (W).

On an average basis, Earth receives 5.42 E 24 J energy from Sol per year (Solomon et al., 2007). This energy is in balance in the biosphere; that is, all the energy that enters Earth's atmosphere leaves Earth (Figure 9-3). About 35 percent of this energy is reflected back to space from the atmosphere (1.9 E 24 J/a), and the remainder (3.52 E 24 J/a) is absorbed into the atmosphere or by Earth's surface (including photosynthesis).

As energy changes form throughout the biosphere, it creates gradients that result in weather patterns that are ultimately manifested as climate. If this energy balance were to shift as a result of atmospheric chemical changes, solar disruption, or other processes, the impact on Earth's climate would be dramatic. The forms of energy and their

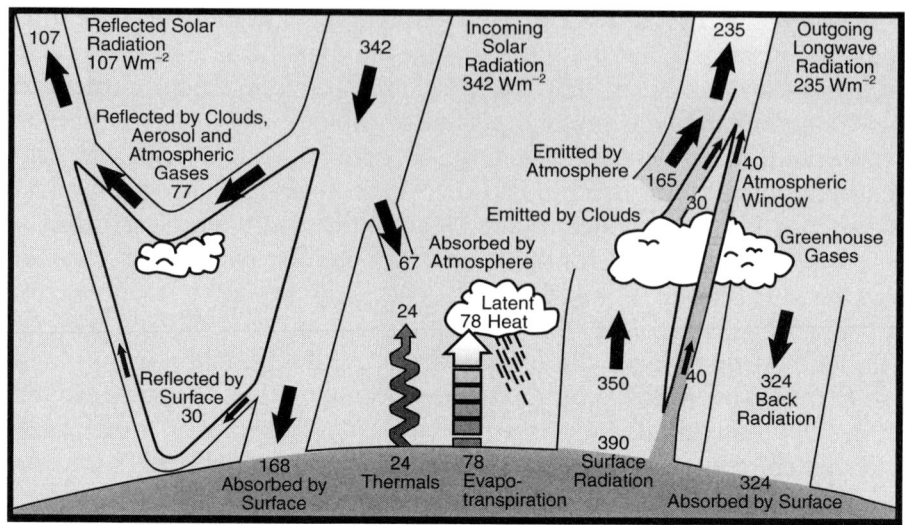

FIGURE 9-3. Energy flow through earth. (*Source*: Modified from Ricklefs and Schluter, 1993)

TABLE 9-1 Energy Expenditures in the Biosphere in Joules per Year (J/a), with Estimated Maximum Potential for Energy Conversion

Global Energy Form	Total Energy (J/a)	Potential Conversion (J/a) [percent of maximum]
Direct reflection	1.9 E 24	—
Atmospheric circulation (winds and waves)	3.0 E 23	1.0 E 18 [0.0003]
Atmospheric heating	1.1 E 24	—
Evaporation and precipitation	3.0 E 20	1.0 E 20 [33.33]
Surface heating	2.4 E 24	2.4 E 20 [0.010]
Photosynthesis	1.0 E 22	1.0 E 20 [1.00]

Source: Modified from Smith, 1981.

potential for use are summarized in Table 9-1. The potential of this energy to do work is remarkable; the challenge for the ecological engineer is to understand and anticipate these energy systems in ecosystem services design.

The unit of consumption of energy for power generation is typically the "metric tonne oil equivalent," or "mtoe." A toe is equal to 42 GJ energy. This is an incredible amount of energy; the density of energy in petrochemicals is the reason that the Industrial Revolution has been so successful, from a production perspective. The anticipated increase in demand for energy is staggering; the World Energy Outlook from the International Energy Agency (IEA) projects that world

Energy and Mass Flow through Ecosystems   167

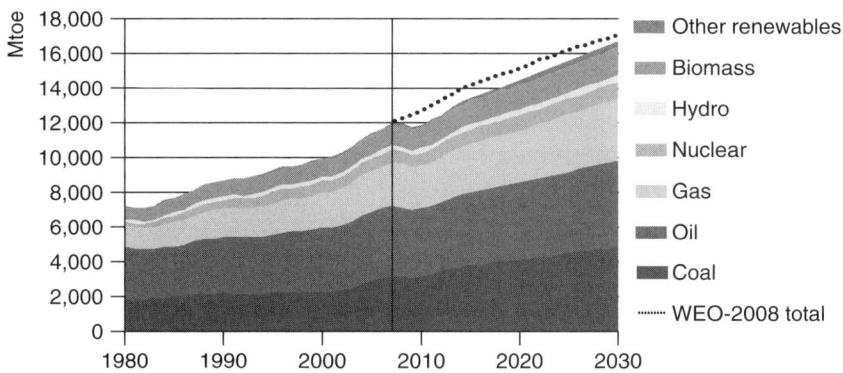

FIGURE 9-4. Projected energy consumption by source, 2030. (*Source*: Modified from Solomon et al., 2007.)

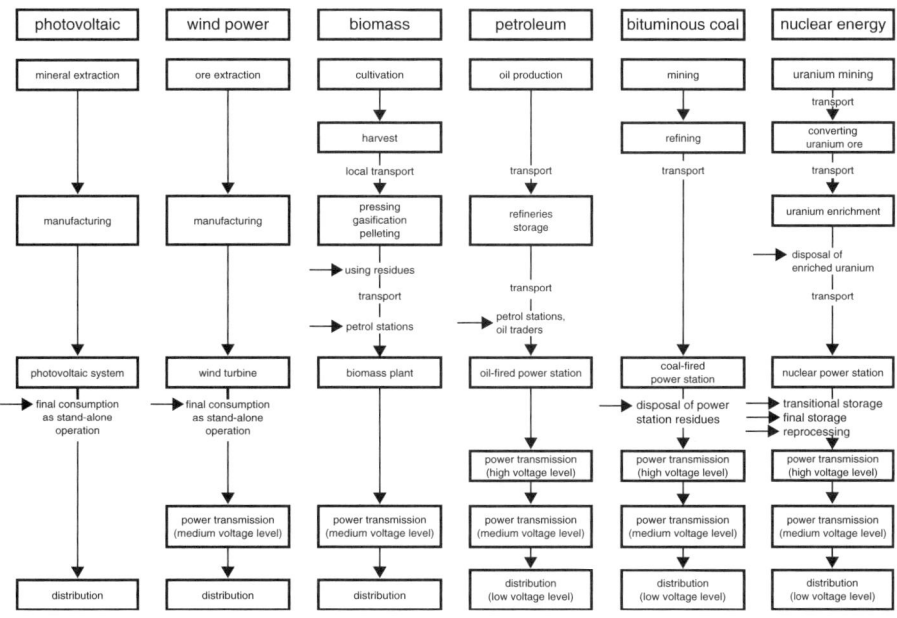

FIGURE 9-5. Energy source development pathways. (*Source*: Modified from Hegger et al, 2008.)

energy demands will increase by 45 percent by 2030 (an increase of 1.6 percent per year) (Figure 9-4).

The notion that there is enough energy in the biosphere to replace current petrochemical sources can be overly simplistic, if the elements required to create a diversified power grid are not considered. The primary sources of energy are photovoltaic, wind, biomass, petroleum, coal, hydropower, and nuclear. Others may become available at some point in the future, but this is our menu for now. Converting the energy of these sources to power in a distribution system presents a number of technical and logistical challenges. The impact of transitioning from carbon-heavy petrochemical and coal sources to less carbon-intensive sources can have unintended consequences. No energy supply is easy, cheap, or without impact (Figure 9-5). Ecological engineers must be able to evaluate and assess the impacts of designs on energy within a system, including human-generated energy. Mitch and Jorgensen (2004) described a goal of ecological engineering as low-energy inputs.

Emergy as a Unit of Analysis

Designing ecosystem services within some scalar framework requires inventorying and managing energy movement and storage through that system. Just tracking energy per unit time (J/a) does not adequately account for the complexity of a system to process and transform energy, the essence of ecosystem services. Mitch and Jorgensen (2004) explored Odum's Maximum Power Principle for ecological engineers. Odum (1993) described this principle as the competitive advantage of systems that convert more energy flow to useful work than other systems. Mitch and Jorgensen (2004) defined maximum power as the "optimum efficiency that results in highest power flow" through a system. There are a number of metrics for efficiencies within ecosystems; these will be described further in the next section, "Trophic Levels." The concept of *emergy* encompasses not only the energy moving through a system, but also the energy required to create the system itself (Odum, 1996). Thus, emergy is the energy memory or energy embodied within a system, or the amount of energy it takes to make something (Brown et al., 2000).

Ultimately, all energy in the biosphere is solar, so all systems could be represented as the amount of solar energy required to create them. Solar energy equivalents in emergy are *solar emjoules* (sej). The solar emergy of a product is thus the solar energy required to produce that product divided by the energy contained in that product (sej/J), or its transformity. The ratio of the solar energy required to produce a kg of corn is its emergy. The ratio of emergy to the energy in that kg of corn is its transformity.

Odum (1996) uses a forest production system example to illustrate these concepts. The emergy of spruce trees in a hectare (ha) of Swedish forest is approximately 3.0 E 14 sej/a, and the energy contained in the logs extracted from that ha of forest is 7.8 E 10 J/a. Thus, the transformity of the spruce logs is 3.85 E 4 sej/J (Odum, 1996):

$$Tr = \frac{Solar\ Emergy\ Flux}{Energy\ Flux} = \frac{3.0\ E\,14^{sej/a\,ha}}{7.8\ E\,10^{J/a\,ha}} = 3.85 E 3^{sej/j} \quad (9.1)$$

where transformity (Tr) is a measure of solar energy efficiency.

High transformities mean that the system or product takes a lot of solar energy to produce a unit of energy in the system. This concept is especially important as ecological engineers design systems to convert primary production into fungible forms of energy such as ethanol or other biofuels. Biofuels with high transformity will have other input requirements (nutrients, water, and other resources) associated with the inefficiency of conversion from sunlight.

TROPHIC LEVELS

As energy flows through the biosphere, it becomes partitioned into sectors based upon the complexity of the ecosystem. The most common sectors for partitioning within ecosystems are the trophic levels, or consumer-resource categories. Plants, algae, and some bacteria (autotrophic organisms) are the base of Earth's food chain, because they produce their own biochemical energy from sunlight (primary production). These organisms are consumed by primary consumers, also called herbivores. Herbivores are consumed by secondary consumers (carnivores) who may be consumed by tertiary consumers (also carnivores). All are ultimately consumed by detritivores, including fungi. Organisms with similar feeding strategies are grouped into guilds (seed eaters, for example, includes rodents, birds, and insects). The map of energy movement through an ecosystem's trophosphere is its food web. These can be organized by feeding relationships (connectedness), energy flow, or some integrated function (functional). All energy in these systems is influenced by the density of energy reaching the ecosystem.

Energy Density

The energy density of an ecosystem, as described by Odum (1993), can be divided into four categories (Table 9-2). Unsubsidized solar-powered ecosystems are those whose energy is entirely from the sun (upland forests, prairies, and open oceans, for example). While relatively low in energy density, these areas represent a large area of Earth's surface (approximately 40 percent). Naturally subsidized

TABLE 9-2 Ecosystem Energy Density Categories

Ecosystem Category	Total Energy Density (MJ/M^2)	Range of Energy Density (MJ/M^2)
Unsubsidized solar-powered ecosystems	10	5–50
Naturally subsidized solar-powered ecosystems	75	50–150
Human-subsidized solar-powered ecosystems	100	50–200
Fuel-powered urban-industrial ecosystems	7500	500–12,500

Source: Modified from Odum, 1993.

solar-powered ecosystems are those that receive energy in addition to solar energy, from nonhuman sources. These ecosystems are characterized by very high density mixes of ecosystem services; examples include estuaries, coral reefs, wetlands, and floodplain rainforests. The subsidies are energy (organic carbon), nutrients, and sediments introduced from outside the system boundary, usually through hydrodynamic processes.

Human-subsidized solar-powered ecosystems are the very familiar agricultural production systems. These systems are subsidized with exogenous fuel and fertilizer to produce food, feed, fiber, and fuel for human use. This category represents almost 45 percent of Earth's surface, if pasture and grazing lands are included. Fuel-powered urban-industrial ecosystems are the highest energy density systems on Earth. These systems are subsidized by oil, gas, coal, nuclear, hydropower, wind, geothermal, and other sources of energy besides direct solar energy. Once again, the density of energy within fossil fuels becomes clear, demonstrating the challenge of moving from a fossil-fuel-based economy to a biofuel-based economy.

Primary Production

Primary production is the amount of solar energy captured as chemical potential energy by photosynthesis. In concept, this is very simple. In practice, this is a very difficult parameter to measure. Gross primary production (GPP) is the total amount of carbon fixed per unit area per unit time (g/m^2-d, kg/ha-a, etc.). Net primary production (NPP) is the amount of organic carbon and other biomass that remains after the autotrophic organism (plant, algae) has met its metabolic needs (respiration, growth, reproduction). Net community production (NCP) is the amount of organic carbon and other biomass that remains from all autotrophs in a given area. This is typically the standing biomass in an area. Global NPP has been estimated at 1.1 E11 tonne (Mg) dry weight per year; terrestrial NPP produces about 6 E10 Mg dry weight per year (59 percent), while aquatic systems produce 5 E10 Mg dry weight per year (41 percent) (Odum, 1993; Pauly and Christensen, 1995).

Primary production is the single most important variable to know, understand, and design in an ecosystem. All energy within that ecosystem derived from primary production is termed autochthonous production, while energy introduced from outside the ecosystem boundary (from animals immigrating into the system, woody debris flow into the system, or other mass flux phenomena) is termed allochthonous. As demonstrated in Table 9-1, only a portion of the energy striking Earth's surface is converted to plant biomass. The amount of energy cascading through the trophic levels is dependent upon net primary production and conversion efficiencies of animals eating the plants. Typical trophic level transfer of energy (plant to herbivore, herbivore to carnivore, etc.) is 5 to 20 percent (Ricklefs, 2008). Organisms only digest a portion of the food they eat, passing the rest through as fecal materials to be consumed by detritivores. *Assimilation efficiency* is the amount of food assimilated relative to the amount ingested. Animal biomass is more energy dense and digestible than plant biomass; thus predator efficiency of conversion can range from 60 to 90 percent, while herbivore conversion rates range from as low as 15 percent for high-cellulosic material to as high as 80 percent for seeds. When designing ecosystems for trophic support of predators, the type of vegetation will directly dictate the density of prey, and thus must be designed accordingly.

Exploitation efficiency is the ingested food divided by prey production (kg biomass/ha-a). *Net heterotrophic production efficiency* is the ratio of the biomass accumulated through growth and reproduction over the assimilation biomass. *Gross heterotrophic production efficiency* is net production divided by ingestion. *Ecological efficiency* is consumer production divided by prey production (Ricklefs, 2005).

Ecological engineers design systems for a particular level of net community productivity based upon energy and mass flows through the system. As described in Chapter 4, the rate of primary productivity is biome-, ecoregion-, and place-specific. Understanding the ranges of productivity for a given area is critical for effective ecosystem services design. The most productive unsubsidized solar-powered ecosystems—tropical forests—produce about 1.75 kg/m^2-a biomass through net productivity, while subsidized solar-powered ecosystems

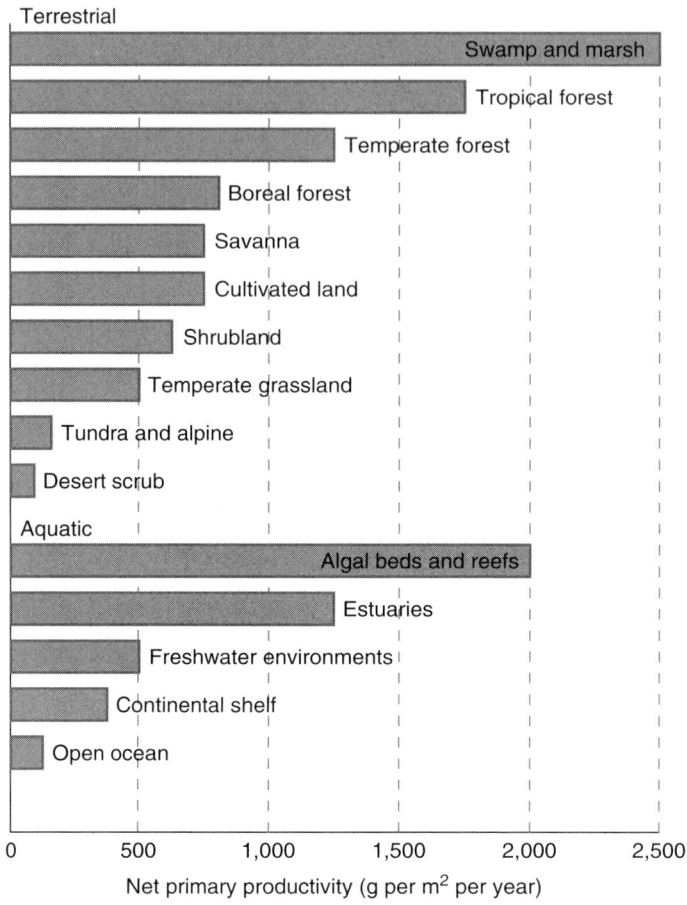

FIGURE 9-6. Net primary productivity and biomass accumulation ratios by biome. (*Source*: Modified from Ricklefs, 2009.)

such as wetlands (marshes and swamps) can produce as much as 2.5 kg/m²-a biomass (Figure 9-6). The *biomass accumulation ratio* is the stored biomass (kg/m²) divided by net productivity (kg/m²-a), expressed as years accumulated biomass. Thus, a forest has 20 years or more accumulated biomass, while cultivated land may only have 2 years accumulated biomass, and the open ocean less than 0.02 years (Figure 9-6).

The factors that affect primary production include light, temperature, water, nutrients, and space. Organisms in a place compete for these factors; the way they compete results in a set of life strategies for growth and reproduction that often becomes unique to the community within an ecoregion. Plants and algae adapt to light intensities, depending on the availability of light for the place in which they exist. Full sunlight can have energy as high as 500 W/m². Sun-tolerant plants reach their photosynthetic saturation point, or the point where increased sunlight no longer increases primary productivity, at as low as 50 W/m².

Many plants grow tall fast, in order to capture light, resulting in significant investment to support biomass; others have adapted to light limitation by increasing their photosynthetic efficiency, or the percent of sunlight converted to net primary productivity per year. The amount of sunlight necessary for a plant to meet its respiratory needs is its compensation point; in many plants this is between 1 and 2 W/m². The potential for designing high-energy systems using unsubsidized solar power is rarely sunlight limited, but more often water and nutrient limited. The energy density of a desert is very high, for example, yet the net primary productivity is very low.

Design parameters for primary production must include limiting factors as conditions for performance of the system. Water is perhaps the most important limiting factor for primary productivity globally (see Chapters 4 and 5). After water, nutrients are most critical in limiting primary production. Nutrient requirements for plants are generally grouped into macro- and micronutrients. Macronutrients (besides atmospheric carbon, hydrogen, and oxygen) are nitrogen (N), phosphorus (P), potassium (K), calcium (Ca), and magnesium (Mg). Micronutrients include sulfur (S), iron (Fe), manganese (Mn), boron (B), copper (Cu), chlorine (Cl), zinc (Zn), molybdenum (Mo), selenium (Se), and silicon (Si). In terrestrial plants, the availability of these critical nutrients in soils governs the primary producer community composition. The influence of primary production on community structure will be described in Chapter 10, "Designing Community Structure."

Designing Trophic Levels

The process of designing ecosystem services starts with understanding place, as indicated in Chapter 3. Design criteria for ecosystem services include redundancy and complexity for the purpose of resiliency in processes upon which those services depend. The goal for ecological engineers should be to design as much trophic complexity into a system as possible. The first estimator of that complexity is the number of trophic levels (n).

The variables required to estimate the number of trophic levels include net primary productivity (NPP), average ecological efficiency (EFF), and average energy flux of top predator populations. If plants are trophic level 1, the energy available to trophic level n is:

$$E(n) = NPP \ Eff^{n-1} \qquad (9.2)$$

where Eff is the geometric mean of ecological efficiencies across levels (Ricklefs, 1993). The number of trophic levels supported by a level of NPP can therefore be estimated as:

$$n = 1 + \frac{\log[E(n)] - \log(NPP)}{\log(EFF)} \qquad (9.3)$$

This relationship is an estimate at best, and thus should be used to assess threshold design criteria, rather than to support final design conditions. In general, the higher n within an ecosystem, the more stable the ecosystem will be. *Ecosystem stability* is a function of its ability to resist changes in response to external influences (*constancy*) and its ability to return to some predisturbed state after disturbance (*resilience*). Both of these ecosystem characteristics are critical for successful ecological engineering design. Biome-specific examples of these processes are presented in Table 9-3.

Even low-NPP systems can produce as many as seven trophic levels, with recycling of mass and energy throughout the system. The complexity of trophic design should not be underestimated. Marczak et al. (2007) demonstrated that many ecosystems should be considered open (subsidized) systems rather than unsubsidized, based on a review of 115 datasets from 32 studies of food web effects of resource subsidies. Gamfeldt et al. (2005) found that increased consumer richness resulted in increased consumer biomass and the most diverse prey assemblage, reinforcing the theory that biodiversity enhances trophic stability.

The primary productivity required (PPR) to support a trophic structure can be estimated using these relationships. Pauly and Christenson

TABLE 9-3 Approximate Values for Net Primary Productivity, Ecological Efficiency, and Trophic Level Numbers for Example Biomes

Biome	NPP (MJ/m² a)	Predator Ingestion (kJ/m² a)	Ecological Efficiency Eff (%)	Number of Trophic Levels (n)
Open ocean	2.1	0.42	25	7.1
Coastal marine	33.5	4.61	20	5.1
Temperate grassland	8.4	13.0	10	4.3
Tropical forest	33.5	17.2	5	3.2

Source: Modified from Ricklefs, 1993.

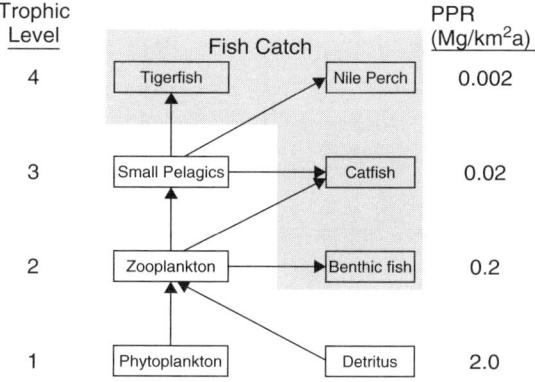

FIGURE 9-7. Primary productivity required (PPR) to support freshwater fisheries, using 10 percent trophic efficiency. (*Source*: Modified from Pauly and Christensen, 1996.)

(1995) evaluated PPR for fisheries based on almost 50 trophic models. They found that mean energy transfer efficiency across trophic levels was approximately 10 percent (Figure 9-7). The PPR to support global fisheries was estimated at 8 percent of global aquatic primary productivity. Fisheries PPR ranged from as low as 2 percent in open oceans to as high as 35 percent in freshwater systems (Pauly and Christensen, 1996).

Best estimates to date put human appropriation of NPP (HANPP) at 23.8 percent of potential NPP (Haberl et al., 2007). Human impact on ecosystems across trophic levels through land use change is evident throughout the biosphere. Coastal marine ecosystems, especially estuaries, have consistently been demonstrated to be the most complex, dynamic, and productive ecosystems in Earth's biosphere. Lotze et al. (2006) estimated that human impacts have depleted more than 90 percent of important coastal and estuary species, and destroyed more than 65 percent of seagrass and wetland habitat. The potential for recovery of services from these critical ecosystems should be a high priority for ecological design. Ecological engineers, restoration ecologists, and systems ecologists still have much work to do before ecosystem trophic levels can be designed explicitly. These general guidelines should serve as bookends to design criteria, with general ranges of inputs.

MASS FLOW THROUGH ECOSYSTEMS

Ecosystem boundaries can be conceptualized as the scale necessary for mass and energy flows and cycles to be accounted for. Because

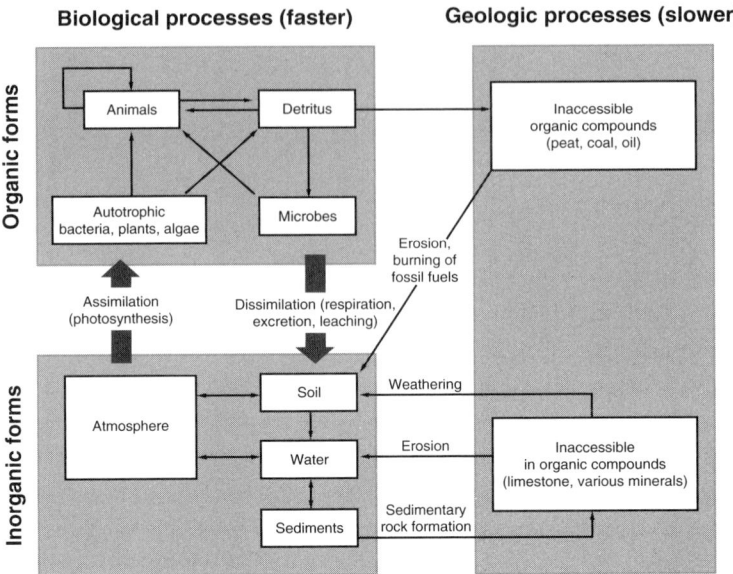

FIGURE 9-8. Compartmental model of mass cycling through ecosystems. (*Source*: Modified from Ricklefs, 2008.)

the biosphere and all its elements are open thermodynamic systems, energy moves across these boundaries continuously. However, for all intents and purposes, the biosphere is a closed system with relation to mass. Mass moves across ecosystem boundaries over time (flux) in organic forms and inorganic forms mediated by relatively rapid biological processes and by relatively slow geologic processes (Figure 9-8).

Cycling of mass through biotic and abiotic elements of the biosphere is termed *biogeochemical cycling*. The rate of mass flux is a critical component for assessing and designing ecosystem services. Many of those services explicitly regulate, facilitate, or retard mass flux. The predominant components of concern in mass flow through ecosystems are water, carbon, nitrogen, and phosphorus.

Hydrologic Cycle

Water cycles throughout the biosphere as gas, liquid, and solid. This cycle is global in scope but very local in impact. The timing of precipitation over land determines biome characteristics and habitability for human beings. Small changes in timing of precipitation mean the difference between flood and drought, famine and plenty. The hydrologic cycle is familiar, yet many do not appreciate the relatively small amount of liquid fresh water on Earth relative to the total volume, only 2.53 percent. Of that small amount, 1.56 percent is frozen in the Antarctic glaciers, and 0.17 percent is frozen in the Greenland glaciers. Groundwater constitutes 0.76 percent, while lakes, rivers, and wetlands only compose 0.008 percent (Gleick et al., 2009). These numbers are a bit deceptive, in that they do not account for the rate of cycling of water throughout a system; this hydrologic cycle is the most ecologically and economically important geochemical process on Earth.

Climate governs the processes of water evaporation, atmospheric transport, and precipitation. These processes are driven by thermal regimes, wind, and ocean currents. The global atmospheric water balance is (Oki et al., 1995):

$$\frac{\partial w}{\partial t} + \frac{\partial Wc}{\partial t} = [-\nabla_H \vec{Q}] + [-\nabla_H \vec{Q}_c] + (E - P) \qquad (9.4)$$

where

  W is precipitable water (vapor in the atmospheric column).
  Wc is liquid and solid water in the atmospheric column.
  ∇_H is horizontal divergence.
  $\vec{Q}$ is the vertically integrated water vapor flux.
  $\vec{Q}_c$ is the vertically integrated solid and liquid flux.
  E is evaporation.
  P is precipitation.

Ignoring solid and liquid phase water in the atmosphere, the relationship simplifies to:

$$\frac{\partial W}{\partial t} = -\nabla_H \vec{Q} + (E - P) \qquad (9.5)$$

Similarly, water balance for river systems can be represented globally as:

$$\frac{\partial s}{\partial t} = -[-\nabla_H \vec{R}_O + [-\nabla_H \vec{R}_U] - (E - P) \qquad (9.6)$$

where

  S represents storage in the river.
  $\vec{R}_O$ is overland flow (runoff).
  $\vec{R}_U$ is underground flow (interflow and groundwater).

Assuming that atmospheric and basin storage do not change on an annual basis (setting equations 9.5 and 9.6 equal to 0), then:

$$-\nabla_H \vec{Q} = (P - E) = -\nabla_H \vec{R}_O \qquad (9.7)$$

For a given basin, water vapor convergence $(P - E)$ is equal to the runoff from the basin over an annual basis, $-\nabla_H \vec{R}_O$ (km³/a). Thus, for a basin, evaporation can be estimated as:

$$E = P - \nabla_H \vec{R}_O \tag{9.8}$$

As a first approximator, this approach can be very useful in ecological engineering design. Most basins have average annual flow and precipitation data, but evaporation data are difficult to collect, and often not available. This approach can be used to allocate annual evaporation across a basin based upon land cover, slope, soil type, and other known variables.

Changes in climate will have dramatic impacts on how water moves throughout the biosphere. Most climate change models predict that increasing global energy will result in increased extremes of rainfall—more floods and more droughts. The global water budget has shifted as well, with global atmospheric water vapor content increasing (Bates et al., 2008). In general, Equation 9.8 seems to be accurate; where precipitation increases, evaporation also increases. This relationship is directly related to soil moisture, a determinant in NPP. White et al. (1999) estimated that a one-day increase in growing season in the northern forest would increase NEP by 1.6 percent, GPP by 0.5 percent, and evapotranspiration (ET) by 0.2 percent. There is mounting evidence that global soil moisture is increasing, representing a net increase in landscape storage (Bates et al., 2008). If these observations are indeed trends, the biosphere from the northern forest to the subtropical biomes could generate increased NPP in the coming century.

Ecological engineers must understand the implications of potential changes to climate, and thus weather patterns, for the ecosystem services they are designing. The Intergovernmental Panel on Climate Change (IPCC) analyzed global precipitation records over the period 1901–2005 and determined that there have been regional changes in precipitation over the past 100 years, with some areas seeing net increases in precipitation and others decreasing (Bates et al., 2008). In general, the high latitudes and large parts of the U.S. are experiencing an increase in annual runoff, while parts of southern Europe, southernmost South America, and West Africa are experiencing decreases (Bates et al., 2008).

Carbon Cycle

Carbon exists in the biosphere in four major compartments: atmosphere, hydrosphere (oceans), land, and living things (Figure 9-9). A

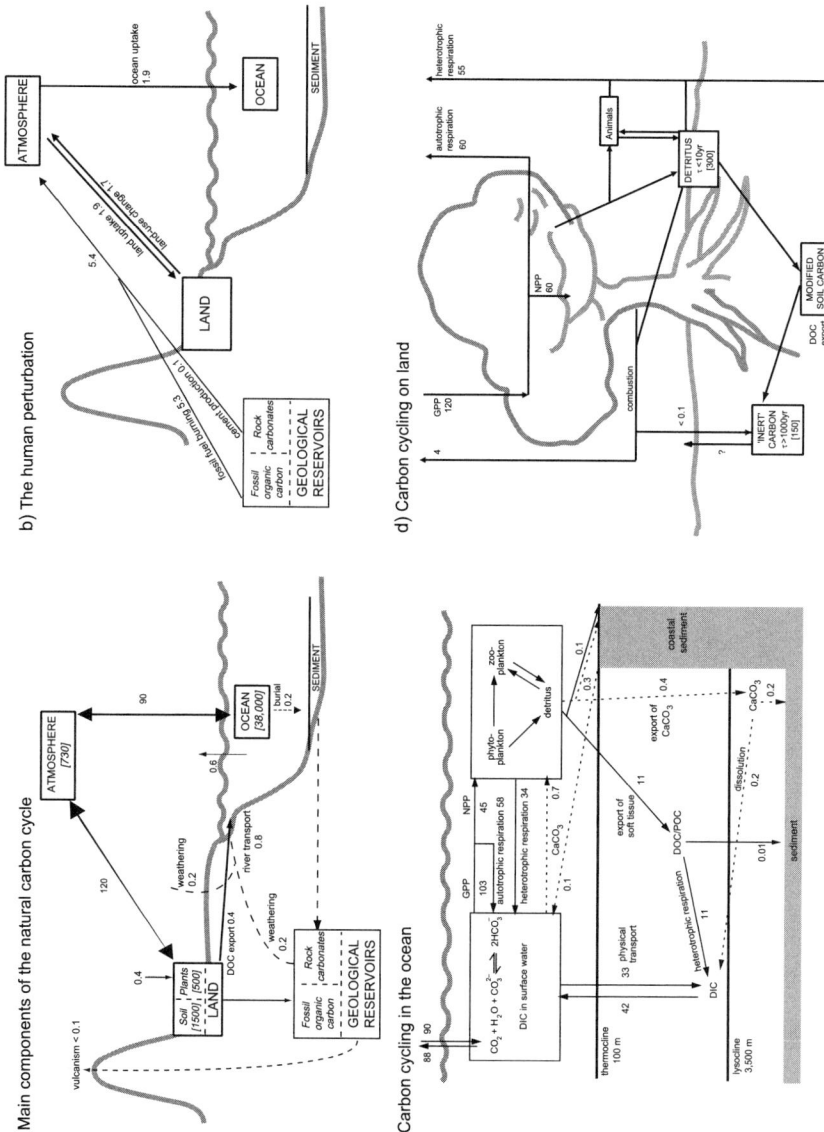

FIGURE 9-9. Global carbon cycle. (*Source*: Modified from IPCC TAR3.)

common misperception is that the atmosphere is the largest reservoir of carbon on Earth; however, the largest category of carbon reservoir is coal, oil, and natural gas (Table 9-4). Much of that carbon was sequestered into the land during past epochs. The IPCC estimates that carbon in the atmosphere increased from approximately 280 ppm during the pre-industrial era (circa 1770) to 379 ppm in 2005 (Solomon et al., 2007). This increase is responsible for an increase in radiative energy in the atmosphere of 1.7 W/m^2. Extraction and combustion of carbon-based fuels accounted for 60 percent of this increase, with the remainder from the effects of changes in land use. About 45 percent of the increased CO_2 has remained in the atmosphere, while 30 percent has been dissolved in the ocean, and the remainder has been assimilated by terrestrial autotrophs (Solomon et al., 2007). The rate of CO_2 pulse to the atmosphere from fossil fuel combustion has increased over the past 20 years by as much as 30 percent (Table 9-5).

The IPCC estimated that half the increased mass flux of CO_2 into the atmosphere can be assimilated within a couple of years, 30 percent

TABLE 9-4 Carbon Distribution in the Global Carbon Cycle

Global Reservoir	Estimated Mass (billion tonnes)
Atmosphere	578–766
Terrestrial plants	540–610
Soil organic matter	1500–1600
Ocean	38,000–40,000
Coal, oil, natural gas	10,000,000–25,000,000
Marine sediments and sedimentary rocks	66,000,000–100,000,000

Source: Modified from Pidwirny and Gulledge, 2009.

TABLE 9-5 Global Changes in Carbon Budgets (positive values are CO2 fluxes [Gt C/a] into the atmosphere, and negative values are sinks from the atmosphere; NA indicates data were not available)

Carbon Source/Sink	1980	1990	2000–2005
Atmospheric increase	3.3 ± 0.1	3.2 ± 0.1	4.1 ± 0.1
Fossil carbon dioxide emissions	5.4 ± 0.3	6.4 ± 0.4	7.2 ± 0.3
Net land-to-ocean flux	−1.8 ± 0.8	−2.2 ± 0.4	−2.2 ± 0.5
Net land-to-atmosphere flux *Partitioned as follows:*	−0.3 ± 0.9	−1.0 ± 0.6	−0.9 ± 0.6
Land use change flux	1.4 (0.4 to 2.3)	1.6 (0.5 to 2.7)	NA
Residual land sink	−1.7 (3.4 to 0.2)	−2.6 (−4.3 to −0.9)	NA

Source: From Solomon et al., 2007.

can be assimilated over 30 years, and the remainder will cycle in the atmosphere for thousands of years (Solomon et al., 2007). The rates of cycling are controlled by mass exchange rates between major components of the global carbon cycle (Figure 9-9a). Carbon cycles through the biosphere through three primary processes:

1. Assimilation and dissimilation through photosynthesis and the biosphere
2. Physical exchange with the abiotic components of the biosphere
3. Carbonate cycling through deposition and dissolution of carbonate compounds

Carbon cycling on land begins at NPP (Figure 9-9b and d). Earth's biosphere captures an estimated 1.1 E11 Mg dry weight per year, about 17 percent of the carbon in the atmosphere (6.4 E11 Mg). Once assimilated on land or in the ocean, carbon is ingested, sequestered or excreted, metabolized, or mineralized, and a portion is returned to the atmosphere as carbon dioxide (CO_2) (Figure 9-9c and d).

Ecological engineers design carbon sequestration ecosystem services provided by increasing NPP and soil sequestration. Carbon sequestration in soil as dissolved organic carbon (DOC) and particulate organic carbon (POC) can exceed the rates of detritus respiration. This process also increases other ecosystem services, including nutrient cycling, water storage, soil fertility, refugia, and many others. The interconnected nature of ecosystem services design becomes apparent when approached from the perspective of fundamental processes.

Nitrogen Cycle

Nitrogen (N) is predominantly found in the form of nitrogen gas (N_2), composing approximately 78 percent of Earth's atmosphere. Unlike carbon, N is absent from parent material, but enters soil through microbial fixation and, to a lesser extent, in precipitation via lightning conversion to nitrate (NO_3^-) (Figure 9-10). In unsubsidized ecosystems, N is in limited supply, and often prohibits growth of autotrophs, especially in aquatic systems. Human-subsidized ecosystems have high N inputs, often resulting in increased loads to waterbodies and eutrophication. The dead zones located at the deltas of many major rivers around the world, including the Gulf of Mexico, have been attributed to increased N loads and subsequent oxygen depletion in deeper waters.

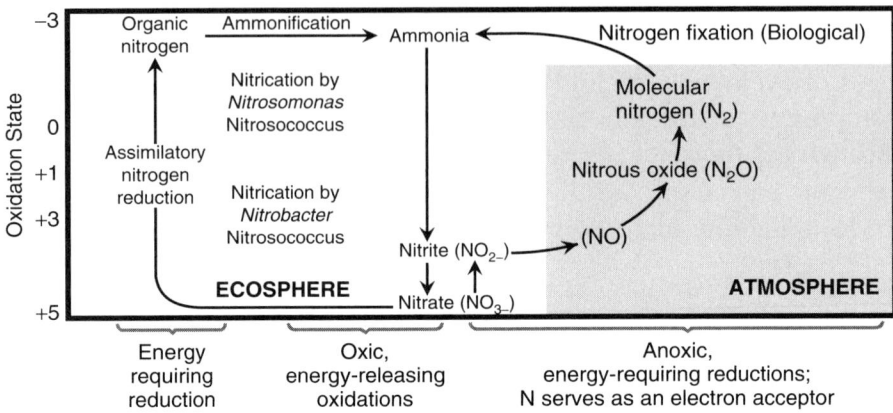

FIGURE 9-10. Global nitrogen cycle. (*Source*: Modified from Ricklefs, 2008.)

The process of N cycling is mostly biologically mediated. Nitrogen is fixed into ammonium (NH_4^+) by specialized bacteria using sugars and other organic compounds to provide the energy for conversion. This process is performed by free-living bacteria in soils, such as *Azotobacter*, and bacteria symbiotically associated with plants, such as *Rhizobium*. Plants can take up nitrate and ammonium, but assimilate N to the highest Gibbs free energy form, the reduced form (NH_4^+). This form is the basic component for the amine group on amino acids, building blocks for proteins. Proteins are metabolized and discharged into the environment (a process called ammonification). All organisms perform ammonification; this process keeps high-energy ammonium molecules cycling in an ecosystem. Wastewater treatment facilities that treat human municipal and industrial wastes have ammonium in their wastewater. In aquatic ecosystems ammonium can be toxic, and is thus regulated as a pollutant for wastewater dischargers.

Specialized microorganisms metabolize this high-energy molecule to its oxidized form (NO_3^-) in a process called nitrification. In soils and freshwater systems, *Nitrosomonas* performs the first oxidative step, converting ammonium to nitrite ($NO2^-$). In marine systems, this step is performed by *Nitrosococcus*. The final step is performed by *Nitrobacter*, resulting in nitrate molecules. In marine systems, this step is performed by *Nitrococcus* (Ricklefs, 2008). These are oxidative processes, and are commonly performed in wastewater treatment plants with very high energy inputs. Ecological engineers have been very successful at designing treatment wetlands as subsidized ecosystems that perform this process as well (Mitch and Gosselink, 2000).

In anoxic conditions, facultative anaerobic bacteria such as *Pseudomonas denitrificans* can reduce nitrates to nitric oxide (NO), and ultimately to nitrogen gas (N_2), which returns to the atmosphere. This process, called denitrification, is also exploited in wastewater treatment, where anaerobic conditions are maintained to remove N from aquatic waste streams. This process occurs in wetlands and other reduced environments. In anoxic systems, this process is incomplete, resulting in release of nitrous oxide (N_2O) into the atmosphere. This is particularly common in soils that are near field capacity or otherwise mildly oxygen depleted. Nitrous oxide is a very potent greenhouse gas, with the global warming potential (or carbon dioxide equivalent impact) of 310 (IPCC, 1997).

Nitrous oxide emissions into the atmosphere are increasing by as much as 0.3 percent per year, largely from agricultural application of nitrogen fertilizers (Mosier et al., 1998). Reducing nitrous oxide emissions has become a critical strategy in reducing the impact of greenhouse gas emissions. Ecological engineers will need to develop design strategies for agricultural production systems that increase nitrogen use efficiency while reducing emissions of nitrous oxide.

Phosphorus Cycle

Compared to carbon and nitrogen, phosphorus (P) has a relatively simple global cycle (Figure 9-11). Phosphorus is a critical element in all

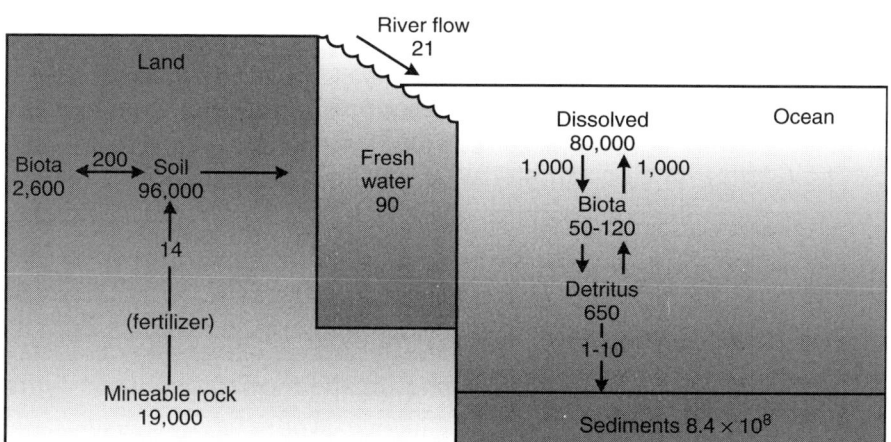

FIGURE 9-11. Global phosphorus cycle. (*Source*: Modified from Ricklefs and Miller, 2000.)

living things; it is assimilated by plants in its oxidative form as phosphate ions (PO_3^-). Phosphorus has no gas phase under normal redox conditions; it is only transported in the atmosphere as dust. Phosphorus occurs as inorganic salts with sodium, calcium, iron, and magnesium in soil and parent material. In highly weathered soils, P leaches out of the system, resulting in phosphorus-limited primary producer systems. Thus, plant growth on younger soils tends to be nitrogen limited, while plant growth on older soils tends to be phosphorus limited (Vitousek, 2004).

Like nitrogen, phosphorus is extracted as a fertilizer and applied to increase NPP in agricultural and other human-supplemented ecosystems. Phosphorus moves across the landscape otherwise as mineralized parent material, eroding from uplifted areas, flowing through rivers attached to sediment, and finally arriving at the ocean, where it is deposited as sediment, only to recycle over geologic time as parent material.

> Ecology tries to understand the interactions between living things and their environment. Every living thing represents an equation of give and take.
>
> —Aldo Leopold

References

Bates, B., Z. Kunzewicz, S. Wu, and J. Paluticof, eds., Climate Change and Water, Technical Paper of the Intergovernmental Panel on Climate Change, IPCC Secretariat, Geneva, Switzerland, 2008.

Brown, M.T., M. Wackernagel and C.A.S. Hall. Comparative estimates of sustainability: economic resource base, ecological footprint and emergy. in C.A.S. Hall (ed). *Quantifying Sustainable Development: The Future of Tropical Economies*. Academic Press. pp. 695–713. 2000.

Dale, V., and R. Haeuber, *Applying Ecological Principles to Land Management*, Springer Publishers, New York, NY, 2001.

Gamfeldt, L., H. Hillebrand, and P. Jonsson, Species richness changes across two trophic levels simultaneously affect prey and consumer biomass, *Ecology Letters*, 8: 696–703, 2005.

Gleick, P., H. Cooley, M. Cohen, M. Morikawa, J. Morrison, and M. Palaniappan, *The World's Water, 2008–2009*, Island Press, Washington, DC, 2009.

Haberl, Helmut, Karl-Heinz Erb, Fridolin Krausmann, Veronika Gaube, Alberte Bondeau, Christof Plutzar, Somone Gingrich, Wolfgang Lucht, and Marina Fischer-Kowalski, Quantifying and mapping the global human appropriation of net primary production in Earth's terrestrial ecosystem, *Proceedings of the National Academy of Sciences of the USA*, 104: 12942–12947, 2007.

Hegger, M.; Fuchs, M.; Stark, T.; Zeumer, M. *Energy Manual: Sustainable Architecture*. Basel: Birkhauser. 280 p. 2008.
Hof, J., and M. Bevers, *Spatial Optimization for Managed Ecosystems*, Columbia University Press, New York, NY, 1998.
Holyoak, M., M. Leibold, and R. Holt, *Metacommunities: Spatial Dynamics and Ecological Communities*, University of Chicago Press, Chicago, IL, 2005.
IPCC, Guidelines for National Greenhouse Gas Inventories, Intergovernmental Panel on Climate Change/Organization for Economic Cooperation and Development, Paris, France, 1997.
Lotze, H., H. Lenihan, B.Bourque, R.Bradbury, R.Cooke, M.Kay, S.Kidwell, M.Kirby, C.Peterson, and J.Jackson. Depletion, Degradation, and Recovery Potential of Estuaries and Coastal Seas. *Science* 312(5781): 1806. June 6, 2006.
Marczak, L, B.Ross, M.Thompson, and John S.Richardson, Meta-analysis: Trophic level, habitat, and productivity shape the food web effects of resource subsidies, *Ecology*, 88(1): 140–148, 2007.
Mitch, W., and J. Gosselink, *Wetlands*, 3rd ed., John Wiley & Sons, Hoboken, NJ, 2000.
Mitsch, W.J. and S.E. Jørgensen, *Ecological Engineering and Ecosystem Restoration*. John Wiley and Sons, Inc., New York. 2004.
Moriarty, P., and D. Honnery, What energy levels can the Earth sustain? *Energy Policy*, 37(7): 2469–2474, July 2009.
Mosier A., C.Kroeze, C.Nevison, O.Oenema, S.Seitzinger, and O.van Cleemput, Closing the global atmospheric N2O budget: Nitrous oxide emissions through the agricultural nitrogen cycle, *Nutrient Cycling in Agroecosystems*, 52: 225–248, 1998.
Naiman, R., and H. Décamps, *The Ecology and Management of Aquatic-Terrestrial Ecotones*, UNESCO, Paris, France, 1990.
Naiman, R., H. Décamps, and M.McLean, *Riparia*, Elsevier Press, Boston, MA, 2005.
Odum, E., *Ecology and Our Endangered Life Support System*, Sinauer Associates Press, Sunderland, MA, 1993.
Odum, H.T., *Environmental Accounting: EMERGY and Environmental Decision Making*, John Wiley & Sons, Hoboken, NJ, 1996.
Oki, T., K. Musiake, H. Matsuyama, and K. Masuda, Global atmospheric water balance and runoff from large river basins, *Hydrological Processes*, 9: 655–678, 1995.
Pasitschniak-Arts, M., and F. Messier, Effects of edges and habitats on small mammals in a prairie ecosystem, *Canadian Journal of Zoology*, 76(11): 2020–2025, 1998.
Pauly, D. and V. Christensen. Primary Production Required to Sustain Global Fisheries. *Nature* 374:255–257, 1995.
Pidwirny, Michael, and Jay Gulledge, Carbon cycle, in Cutler J. Cleveland, ed., Encyclopedia of Earth, Environmental Information Coalition, National

Council for Science and the Environment, Washington, DC, first published in the *Encyclopedia of Earth* December 10, 2006, last revised September 16, 2009, www.eoearth.org/article/Carbon_cycle (accessed January 29, 2010).

Ricklefs, R., *The Economy of Nature*, 6th ed., W.H. Freeman and Co., New York, NY, 2008.

Ricklefs, R., and Miller, G., *Ecology*, 4th ed., W.H. Freeman and Co., New York, NY, 2000.

Ricklefs, R., and D. Schluter, eds., *Species Diversity in Ecological Communities*, University of Chicago Press, Chicago, IL, 1993.

Solomon, S., D. Qin, M. Manning, Z. Chen, M. Marquis, K.B. Averyt, M. Tignor, and H.L. Miller, eds., Contribution of Working Group I to the Fourth Assessment Report of the Intergovernmental Panel on Climate Change, Cambridge University Press, Cambridge, UK, and New York, NY, 2007.

Verry, E., C. Dolloff, and M.Manning, Riparian ecotone: A functional definition and delineation for resource assessment, *Water, Air, & Soil Pollution*, 4(1): 67–94, 2004.

Vitousek, P., *Nutrient Cycling and Limitation*, Princeton University Press, Princeton, NJ, 2004.

White, M., S.Running, and P. Thornton, The impact of growing-season length variability on carbon assimilation and evapotranspiration over 88 years in the eastern US deciduous forest, *International Journal of Biometeorology*, 42(3): 139–145, 1999.

10
Designing Community Structure

> Is it not enough for you to feed on the good pasture? Must you also trample the rest of your pasture with your feet? Is it not enough for you to drink clear water? Must you also muddy the rest with your feet?
> —Ezekiel 34:18

INTRODUCTION

Ecosystem services are the product of the interactions of populations in a place, a process called community structure. The diversity of life in a site determines the scale and effectiveness of ecosystem functions over the range of expected conditions for an ecoregion. Ecological design uses the principles of restoration ecology to create community structures that provide ecosystem services. Restoration ecology considers how regional processes determine species composition, thresholds of environmental and habitat characteristics that favor species survival, and the manner in which biotic interactions shape community structure (Menninger and Palmer, 2006).

Hierarchical Processes

Ecosystem processes are affected at different scales by processes unique to the scale. Ricklefs and Schluter (1993) described ecological communities as a hierarchy of structures interconnected by processes unique to each level (Figure 10-1). On a global scale, the biosphere processes that govern community structure are biotic exchange and mass extinction. These processes drive species formation at the biome level. Species dispersal over ecological and geographic time defines ecoregional characteristics. All of these processes occur over very long periods, usually hundreds of generations. The more local landscape processes include habitat dynamics and selection. Metapopulations interact across landscapes and within ecoregions based on patch and corridor dynamics. Local populations interact based on community structure characteristics such as competition,

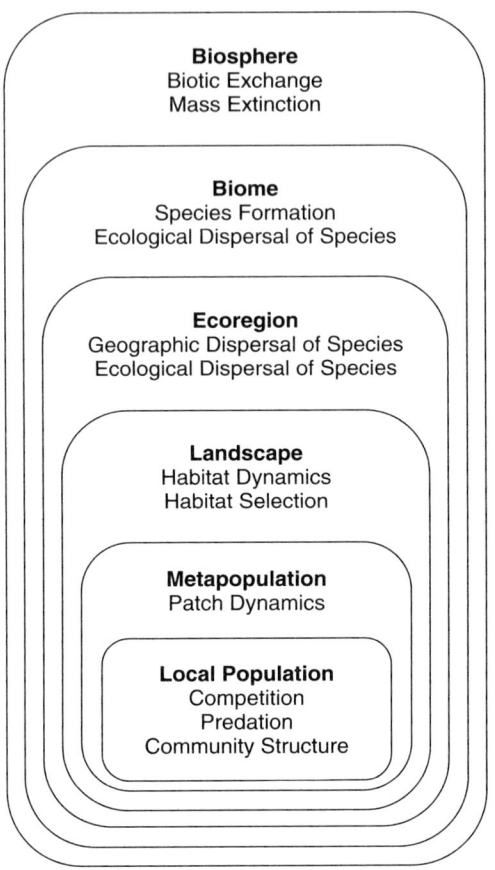

FIGURE 10-1. Hierarchical community structure processes. (*Source*: Modified from Ricklefs and Schluter, 1993.)

predation, mutualism, and other population dynamics. Ecological engineering design is predominantly a site or local activity within the context of landscape, and therefore metapopulation, processes.

Types of Restoration Design

Ecological engineering design projects can be classified from a community structure perspective as introduction, reintroduction, and augmentation designs (Falk et al., 2000, Chapter 3). *Introduction* designs are those whereby species or genotypes not currently present at the project site are introduced. Introduced species are collected off-site, and are generally from outside the ecoregion of the site. Some ecologists advocate for only using resident (indigenous) species in design;

however, some systems are too disturbed or altered for local populations. In other cases, local propagules may be too rare or threatened to be available for redistribution to project sites. Introduced (non-native) species should only be used in an ecological design when conditions at the site have been altered to the degree that indigenous species cannot adapt, or introduced species may serve as transitional organisms to provide food, shelter, or other habitat services for indigenous species while they are being established. Indigenous species have evolutionary histories and established community interactions that cannot be replicated easily (if at all) with introduced species. Invasive species (exotics with high reproductive rates, no known predators, and the capacity to expand geographically) should never be used in ecological engineering design. The risk of unintended release and ecological havoc is too great.

Reintroduction designs reestablish species or genotypes that are not currently present at a site but occurred there some time previously. Reintroduction designs are especially common in stream restoration projects where some activity downstream or upstream (reservoirs, unpassable grade changes) has extirpated a species of fish that was common to the system. These design projects are for sites that may be highly disturbed, but remain strongly connected to metapopulations from multiple trophic levels. Reintroductions may seem simple, but their success depends upon a source of adequately robust individuals from the target population to repopulate the project site. These translocated individuals must be adequately genetically similar to the extinct population to survive under the extremes of the site. In many cases, that means they must come from the vicinity (same watershed or ecoregion). Translocating species across biomes is rarely a good idea. There are many examples of reintroduction designs—wolves in Yellowstone National Park, buffalo on the southern Great Plains, brook trout and salmon in the rivers and streams of the Pacific northern coast. Many times, these projects are not predictive in their design, but rather empirical; the goal of ecological engineering is to be able to quantitatively simulate and predict outcomes from these projects prior to implementing them.

Augmentation designs are the most common in ecological design; the population of a species is restocked with individuals from the same species in another area. These types of projects require remnant resident populations with adequately intact habitat to support population growth, and local populations from which to translocate new individuals to the site. It is imperative that the ecotechnology team in any of these design strategies have adequate competency to understand the life strategies of the target species. Even simple aquatic organisms such as crayfish are incredibly territorial; introduction of new individuals without adequate preparation of habitat may result in feeding rather than breeding behavior.

BIOTIC INTERACTIONS

Community structure is the product of the interaction of organisms in a place over long periods of time. The interactions between individuals and populations can be characterized further as *assemblages* and *guilds* within an ecosystem. Assemblages are taxonomically related groups of populations within the same ecoregion or site. Guilds are groups of populations within an ecoregion or site that use resources in a similar manner. The number of species in a site, called species richness, and the relative abundance of individuals within each species determine the complexity of the community structure and function (Ricklefs and Miller, 2000).

Ecosystem community structures respond to resource limitations over time and space. Species composition changes along a continuum of changing landscapes within an ecoregion or even biome. This *continuum concept* of community structure describes the variation of ecosystem characteristics across longitudes and latitudes. At the ecoregion and larger ecosystem scales, characteristics do not often change abruptly (coastal edges being an obvious exception), but are gradients across which the relative abundance of key species changes. The changing conditions across gradients are generally shifts in scarce critical resource availability across trophic levels. The way organisms respond to these changes in resources can be characterized broadly as competing strategies for accessing scarce critical resources.

Community Interactions

Energy available for use by individuals at a site derives from the primary productivity of the site. Species richness and diversity at a site are also affected by the frequency and intensity of disturbance events. The productivity-stability hypothesis states that over a given area, species diversity is a product of the net primary productivity (NPP) and stability (infrequency of disturbance events) of a site (Ricklefs and Miller, 2000). What is not known is which is the dominant force in this process, NPP or disturbance. This "bottom-up" theory of trophic interactions may not always be the best predictor of ecosystem structure; in some cases the top predators may control species composition in a "top-down" process. Either way, ecosystem processes are consumer-resource interactions. According to Liebig's law of the minimum, each population in an ecosystem increases until the supply of some resource (the limiting resource) is no longer available in adequate supply to meet the demand of the consumer (Ricklefs and Miller, 2000). The way organisms interact across trophic levels to access resources can be categorized as competition, consumption (predation, parasitism, herbivory, and detritivory), and commensalism.

Organisms can compete for scarce resources within species (intraspecific) and between species (interspecific). Both types of competition are constants in individuals across all ecosystems. Gause's principle of competitive exclusion suggests that over time two species that occupy the same niche (use the same resources at the same time in the same way) will either drive one to extinction or develop separate life strategies and divergent niches (Odum, 1993). The two outcomes from long-term competition are competitive exclusion or coexistence. The ecotechnology team must have the competency to anticipate these potential competitive processes.

Competition

Competition occurs as interference, exploitation, and competitor-driven apparent competition. Many ecosystem restoration and design projects have been undone by introduced plants that exhibit unanticipated allelopathy, a plant's exclusion of others from its root zone by emission of phytotoxic chemicals. This process is very common in plants as diverse as black walnut (*Juglans nigra*) and sorghum (*Sorghum bicolor*). Exploitation competition occurs when one organism uses a resource to the exclusion of others. The most common types of exploitation competition are intraspecific and interspecific plant competition for nutrients and water. Apparent competition occurs when a consumer induces competitive pressure on a species that gives another species an advantage. An example of apparent competition is two plants with different grazing pressure; the plant that is preferred by the grazer can be more easily outcompeted for resources by the other plant (Ricklefs, 2008).

Nutrient use efficiency (NUE) for plants is the amount of biomass generated per unit of nutrient cycled (Ricklefs and Miller, 2000). Plants occupying the same location compete for soil nutrients (N and P predominantly) and when scarce, water. The availability of nutrients in soil (soil fertility) was described in Chapter 7. Soil fertility is among the easiest variables to control at a site. Ecological engineers can adjust the soil

N and P ratios to promote species that are desirable and to reduce the competitive advantage of less-desired and undesired species.

Consumption

Predator interactions with prey often become very specialized within a biome. Predator-prey interactions may dominate community structure. If a predator depletes its prey, the predator's numbers will also decline, resulting in a population increase in the next lower trophic level. The resulting coupled oscillation of populations may become the most dominant impact on the food chain. This process can cascade down the food web according to top-down predation theory (Falk et al., 2006). Herbivory is a type of predator-prey relationship, whereby the prey is a plant and the predator is an herbivore (Odum, 1993). Many plants produce chemicals to dissuade herbivory; these include alkaloids such as nicotine, caffeine, and morphine; terpenoids such as latex; and many other compounds that human beings have yet to exploit.

Parasitism is another form of consumption, whereby the parasite consumes part of the living prey (host) (Ricklefs, 2008). Parasites are commonly much smaller that the host; many are microorganisms that cause disease (pathogens). Parasites, like predators, can be generalist (exploit a large number of hosts) or specialists (exploit only a few hosts, which is called host specificity). Parasites reproduce much more rapidly than their host, and often co-opt the life cycle of the host for distribution. In general, the limiting effects of both parasites and predators are reduced, and the regulating effects enhanced, when both predator/parasite and prey/host have evolved over time in a geographically dispersed metacommunity. The predators/parasites that killed their prey/hosts often died with them, while prey/hosts that could reciprocally adapt with the predators/parasites persisted. The catastrophic predator/parasite impact is most commonly associated with recent interactions, where a species is introduced or translocated to another location. This is an ongoing concern with global climate change, as traditionally tropical diseases begin to move beyond historic boundaries. It is also a warning to ecological engineers about introducing new species into a site.

Commensalism

Commensalism is a beneficial interaction whereby one species benefits and the other is not affected. If two species benefit from interactions but are not dependent on each other for survival, the relationship is *cooperative*. If the relationship becomes co-dependent for survival of the participating species, it is called *mutualism*. These relationships can be trophic, defensive, and dispersive (Ricklefs and Miller, 2000). Trophic mutualism involves interactions between species to obtain energy and nutrients from the environment, usually through specialized exchanges. Defensive mutualism occurs where species exchange habitat for defense; it includes cleaning symbiosis, whereby one species consumes parasites from another. Dispersive mutualism is predominantly associated with animals aiding pollination and seed distribution of plants (Ricklefs and Miller, 2000).

METAPOPULATIONS

Metapopulations are populations from one area or patch linked with another area or patch through immigration and emigration of individuals. The population within a patch in the context of metapopulations is called a subpopulation. The concept of metapopulations is critical to understanding species dynamics in fragmented ecosystems. Ecological engineers design in this framework, and must understand how populations interact and function under these circumstances. The relationships between species and the areas or size of patches and population viability are the critical design variables.

Species-Area Relationship

Organisms live in a place; the characteristics of place are the most important predictors of the viability of a species in an area. The species-area relationship between the number of species and habitat area has been described as:

$$S = cA^z \quad (10.1)$$

where
  S is species richness.
  A is area of habitat within a patch.
  c and z are constants associated with each habitat.

A log transformation of S and A gives a linear relationship:

$$Log\ S = Log\ c + z \log A \quad (10.2)$$

where
  c is the log of the Y intercept.
  z is the slope.

The values for z fall between 0.20 and 0.35 most of the time; a conservative value for design is 0.30 (Ricklefs and Miller, 2000). High values of z reflect more isolated conditions for populations, thus the need for larger areas. Thus, area of habitat can be designed at a preliminary level based upon the number of species desired in a patch.

Minimum Viable Populations

The ability to design a viable habitat is dependent on the viability and stability of the community being designed. The stability of a community is not just a function of the area, but also of the connectedness and viability of species within an area. The minimum viable population (MVP) is the smallest population of the species that can sustain itself in the face of environmental variation (Ricklefs and Miller, 2000). The effective population size (N_e) is the size of a population (N) that theoretically has the same amount of genetic drift as the total population. N_e is usually less than N because some of the organisms in the population do not reproduce as effectively as others, and some immigrate out of the population prior to reproduction. In general, the effective population size is a function of the reproductive success of males and females, and is therefore a function of the ratio of sexually mature male and female populations that are reproducing (Ricklefs and Miller, 2000):

$$N_{males} = \frac{N_m K_m - 1}{(K_m + V_m/K_m) - 1} \quad (10.3)$$

$$N_{females} = \frac{N_f K_f - 1}{(K_f + V_f/K_f) - 1} \quad (10.4)$$

where
- N is the number of breeding males or females.
- K is the average number of offspring produced by each in their lifetimes.
- V is the variance in the number of offspring produced by each in their lifetimes.

The overall effective population size is:

$$N_e = 4 \left[\frac{1}{N_{males}} + \frac{1}{N_{females}} \right]^{-1} \quad (10.5)$$

When the effective population is less than 500, genetic diversity may be compromised. The MVP therefore is in the $N_e = 500$ range, with

N depending on the reproductive success of the species (Falk et al., 2006).

Minimum Viable Metapopulations

The minimum viable metapopulation (MVM) is the minimum number of local populations that interact as metapopulations necessary for long-term persistence of the species (Falk et al., 2006). The habitat patches and area required to support an MVM are defined as the minimum amount of suitable habitat (MASH). In general, this number is 15–20 patches for a given population. Fragmented habitat below this threshold will result in local population declines, and with no repopulation, extinction.

Source populations are those that produce surplus individuals that can populate less productive (sink) habitats. These propagules are critical for buffering population disruptions, especially in highly fragmented MASH systems with low patch size. Corridor connectivity is required between source habitats and sink habitats if populations are to be maintained. For example, for a population of birds the MVM might be 100 individuals, but if a few unrelated individuals are reintroduced every two years, this number could be as low as 75 because of increases in N_e.

Metapopulations are the unit of ecosystem population design. The MVM is the lower threshold for design success. For a given species at a site, the MASH is the minimum design area for persistence of a species. Ecological engineers rarely design for just one species, however. Indicator species should be selected that represent the larger community in their sensitivity to patch fragmentation.

REGIONAL PROCESSES

Regional processes govern local communities. These processes are a function of the structure of the local communities; they include the number of species within a watershed (local diversity, or alpha diversity) and the number of species within the larger ecoregion (regional diversity, or gamma diversity). The difference in species between habitats in an ecoregion is the beta diversity (Ricklefs, 2008). The Sørensen similarity index provides a comparison of the species in two communities:

$$Sorensen\ similarity = \frac{C}{(S_1 + S_2)/2} \quad (10.6)$$

where
> C is the number of species held in common.
> S_1 and S_2 are the average number of species within each community.

This index ranges from 0 (completely dissimilar) to 1 (completely the same) (Ricklefs, 2008). The diversity of species within a watershed or ecoregion is dependent on regional processes that impact habitat and MVM directly.

Species Pool

The *species pool* is the composite group of species living in an area; watershed species pools collectively make ecoregion species pools, though they may not contain the same species. The niche concept dictates that communities are composed only of species that can coexist in the same habitat. Some species in one watershed or site may not be able to survive in another watershed or site within the same ecoregion. The restrictions of distribution may be the result of niche competition at the other site due to invasive species, disturbance or loss of critical habitat or MASH between one site and another, or the presence of predation, grazing, or other consumer pressures at one site and not another. This process of differential distribution of species across different habitats is called species sorting. Often sites will have *core* species that are common and abundant, and *satellite* species that are less abundant. When a species is reduced in abundance or goes extinct at a site, other species can move into the evacuated niche, in a process called ecological release. This phenomenon can be observed in sites with significant disturbance events, where opportunistic organisms emerge early after the disturbance and establish high densities. Ecological engineers should focus designs on core species across trophic levels first, anticipating that satellite species will follow, presuming there are propagules and corridors.

Dispersal

Dispersal is the process of distribution of species across boundaries and barriers. Many organisms have complex dispersal strategies. Plants disperse seeds by wind, by insects, birds, and animals through attachment and ingestions, and by water. Animals disperse through emigration (movement out of a population) and immigration (movement into a population) (Ricklefs and Miller, 2000). Dispersal distance for animals is typically estimated as the standard deviation (s) of the normal distribution of the probability of the distance an organism will move from a release point. The dispersal distance s_t is the standard deviation estimated over the life span t of an organism. The population of individuals within a circle whose radius is $2s$ is the neighborhood size. For snails, s may be as low as 5 m, and neighborhood size greater than 7,500 snails. For birds, s may be as high as 15,00 m, and neighborhood size may range from 150 to greater than 7,500 birds (Ricklefs and Miller, 2000).

The dispersal across patches may have preferential densities for higher-quality habitats early in a dispersal cycle. However, over time, the higher-quality habitats will become resource depleted, and lower-quality habitats will become more populated as organisms immigrate to them. This *ideal free distribution* of organisms suggests that within a metapopulation patch gradients will eventually even out with regards to individual fitness (Ricklefs, 2008). This concept does not consider that organisms will fight for superior habitats, and dispersed organisms may not have even probabilities of dispersing across all habitats.

Colonization Sequence

The sequence with which organisms inhabit a location can affect the species composition of the location. This founder effect is pronounced in frequently disturbed sites, where pioneer species (often thought of as weeds) establish vegetative cover and organic root mass over bare soil, protecting soil from erosion and establishing the foundation for the succession of plants to recover similar populations as in predisturbance conditions. In other cases, founders may out-compete indigenous species after disturbance events, displacing and in some cases replacing the previous species.

Dispersion

The dispersion of individuals of a species is their spatial distribution within a patch or habitat. Some organisms are clumped, others are evenly spaced. Even spacing of individuals tends to derive from interactions between individuals competing for shared resources. Clumped or aggregated distributions often derive from protection behavior such as ungulate herding or schooling fish. Clumped distributions can also form as the result of clustered resources where all individuals share a common space, such as *Plecoptera* under a rock in a stream, or snails on a periphyton patch. Dispersal mechanisms can affect dispersion as well. Dispersal of seeds in clumpy patterns can result in clumpy plant distribution; even dispersal results in even plant distribution. Finally, individuals may disperse across a site in a random manner. Randomly distributed populations are often estimated with the Poisson distribution to predict the probability of x individuals in a space (Ricklefs and Miller, 2000):

$$P(\chi) = \frac{M^x e^{-M}}{x!} \tag{10.7}$$

where
   M is the mean number of individuals per space.
   e is the base of the natural logarithm.

ENVIRONMENTAL AND HABITAT IMPACTS

Ultimately, the ability of an organism to survive and persist in an area is dependent on the abiotic and biotic characteristics of the habitat. Ecological engineers design habitat as an integral element of designing ecosystem services. The elements of environment that are most critical are the extremes. Very few organisms are affected by averages. All organisms are affected by extremes. In aquatic environments, temperature, dissolved oxygen, pH, and velocity are among the most common extremes of concern. In terrestrial systems, extended rain-free periods, lowest daily temperatures, highest daily temperatures, soil pH, and many other threshold extremes determine whether a species can occupy a location. Design considerations for habitat must control extremes to below impact thresholds for core species.

Abiotic Filters

Abiotic characteristics of habitat can limit a species' ability to survive at a site. Environmental stress occurs when conditions tax an organism beyond its physiological tolerance, resulting in behavioral and physiological dysfunction (Falk et al., 2006). These characteristics may include light, wind, temperature, soil characteristics, water chemistry, flow, topography, rainfall, and others. Habitats often exhibit gradients from benign to harsh, with benign habitats dominated by predator-prey relationships, and harsh habitats dominated by competition for resources (Falk et al., 2006). Thus, the abiotic characteristics filter species composition at a site, influencing the species pool. Abiotic filters are a key design component. Predators are more susceptible to environmental stress than prey, suggesting that predators could serve as design sentinels for a site.

Disturbance Regimes

Disturbance in ecosystem terms applies to habitat characteristics and composition. Natural disturbance regimes are common across the landscape, and include fire, windfalls, landslides, and floods. *Succession* is the sequence of changes initiated by disturbance. Plant communities in a region often converge to a common species pool with similar distributions, referred to as the *climax community*. While the phenomenon is common, the term "climax" implies directional succession, and this is not an accurate characterization of the process. Climax communities are the product of converging responses to common abiotic filters acting on common species pools. Different species pools or disturbance regimes result in different climax species. This is simply the process of open systems with varied composition responding to environmental gradients in similar manners (Ricklefs, 2008).

Succession sequences post-disturbance for many ecosystems are predictable. The transition from a burned forest to a meadow to a forest, for example, or from an old field to a forest, involves many successional sequences, each called a *sere* (Ricklefs, 2008). Primary succession is the establishment of communities in previously unpopulated habitats that have been formed or created from disturbance or human activities. Secondary succession is more common, and is the displacement of species after a disturbance event, over time. Breaks in canopy can lead to rapid growth of understory trees (dogwoods, serviceberry, and redbuds, for example) that then enhance the habitat of other species through food and refugia support.

Disturbance events are affected by spatial, temporal, and magnitude factors. Localized disturbance events have less impact on ecosystems than broad events, often enhancing habitat diversity and therefore species diversity. Broad events can reduce habitat diversity, thus reducing and in some cases eradicating population numbers. Frequent disturbance events become defining characteristics, resulting in disturbance-dominated ecosystems, such that their primary species pool is dictated by the frequency and intensity of disturbance events. Floodplain forests are examples. Infrequent events often become cycle reset processes; examples include forest fires that convert closed-canopy forests to open meadows. Loss of disturbance through fire suppression, flood control, and other human activities alters community structure dramatically, sometimes interrupting reproductive or other life processes of species, leading to local extinction.

Intermediate levels of disturbance create the highest habitat diversity and therefore most rich communities. Disturbance-adapted communities are most appropriate in many cases for restoration design; the species tend to be generalists, have broad niche plasticity, and are less sensitive to abiotic filters. Disturbance can and should be used as a management tool for ecosystem design. Controlled burning of meadows and woodlands can eradicate invasive species, recycle nutrients, and enhance diversity. Pulse flooding of riparian wetlands can return sediment to the upper benches, reseed annual grasses, and repopulate benthic organisms. Pulse flooding in rivers can redistribute gravel bars, grade substrate, scour pools, and create critical habitat for fish and benthic organisms.

Habitat Heterogeneity

The diversity of habitat types in an area is its habitat heterogeneity. Human impacts on ecosystem services are largely associated with homogenization of habitat. Ecosystems at some scale exhibit high heterogeneity; even monoculture cropping systems may have high heterogeneity at the soil substrate level. Macro-habitat diversity is the dominant characteristic of most ecosystems. The loss of macro-level heterogeneity is very destructive to species diversity and composition. Designing ecosystem services is very much an act of designing habitat heterogeneity. In fact, for many ecosystems, just replicating the pattern of diversity observed in nonhuman-dominated elements of the ecoregion will restore ecosystem services to high function.

Topographic heterogeneity is the pattern of elevation over a specific area. Topographic heterogeneity in streams is a measure of riffle, pool, and run composition, for example. Many stream systems have been altered by channelization or hydrologic regime modification, resulting in loss of riffles and pools. Restoring these topographic characteristics is the first stage of stream restoration. Topographic heterogeneity affects abiotic patterns and ecosystem processes, distribution of organisms, organism habitat use, and reproduction success (Falk et al., 2006).

Macro-level topographic heterogeneity restoration and design activities include restoring river channels, floodplains, and meanders, and reconnecting tidal creeks to salt marshes. Meso-level examples include constructing pits and mounds in rangelands, deserts, and woodlands, and creating microcatchments in arid lands. Micro-level topographic heterogeneity restoration examples include creating small mounds in pastures, prairies, and forests (<20 cm), creating mound-pool topography in forest riparian wetlands (<0.6 m), creating microtopography in freshwater wetlands, and restoring large woody debris to streams (Falk et al., 2006).

> One is put in mind of Shakespeare's warning that "virtue, grown into a pleurisy, dies of its own too-much." Be that as it may, the forest landscape is deprived of a certain exuberance which arises from a rich variety of plants fighting with each other for a place in the sun.
> —Aldo Leopold

References

Falk, D., M. Palmer, and J. Zedler, eds., *Foundations of Restoration Ecology*, Island Press, Washington, DC, 2006.

Menninger, H.M., and M.A. Palmer. 2006. Restoring ecological communities: from theory to practice, pp. 88–112, in *Foundation of Restoration Ecology*. Island Press, Washington DC.

Odum, E., *Ecology and Our Endangered Life Support System*, Sinauer Associates Press, Sunderland, MA, 1993.

Ricklefs, R. *The Economy of Nature*, 6th ed., W.H. Freeman and Co., New York, NY, 2008.

Ricklefs, R., and Miller, G., *Ecology*, 4th ed., W.H. Freeman and Co., New York, NY, 2000.

Ricklefs, R., and D. Schluter, eds., *Species Diversity in Ecological Communities*, University of Chicago Press, Chicago, IL, 1993.

11

Ecosystem Control and Feedback Systems

> From time immemorial, man has desired to comprehend the complexity of nature in terms of as few elementary concepts as possible.
> —Abdus Salam

INTRODUCTION

Ecosystem control mechanisms can be characterized as within populations and between populations. Ecosystem services are the product of the interactions of populations in a community. The resiliency of an ecosystem is directly related to the density and redundancy of those interactions. High species diversity and habitat diversity help to enhance the complexity of trans-species interactions. These interactions result in very complex dynamics that may seem unpredictable, but in fact complex systems can be very predictable in their responses to disturbance.

Complex systems are characterized by nonlinear interactions between components, with both positive and negative feedback systems. The interaction of the elements of complex systems results in feedback systems that elicit both positive and negative responses. The cumulative interactions of the elements in complex systems result in chaotic behavior, where ability to predict a state change is dependent upon previous state changes. If the interactions and previous state conditions within an ecosystem are known, the probable outcome of disturbance is very predictable.

Predicting the potential impact of climate-induced changes on ecosystems illustrates this complexity. Scholze et al. (2006) used a complex dynamic global vegetation model to predict the risk of global ecosystem change in response to climate change. They evaluated three global climate change scenarios: <2°C, 2–3°C, and >3°C (Figure 11-1). They found high risks of forest loss in Eurasia, eastern China, Canada, Central America, and Amazonia (Table 11-1).

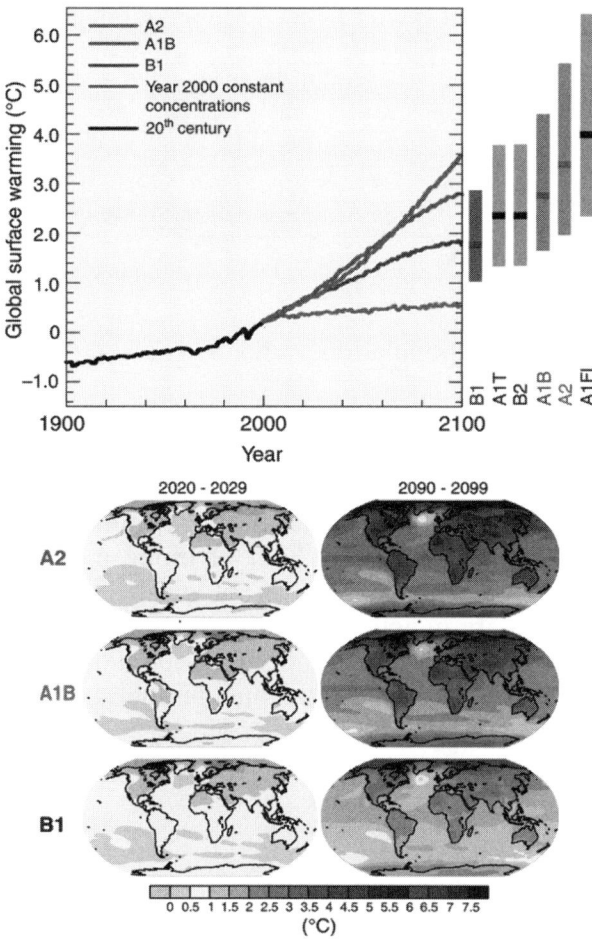

FIGURE 11-1. Median (solid line) and range of global land-atmosphere carbon flux for three levels of global warming. (*Source*: Modified from IPCC, 2007.)

A 3° increase in temperature would have a 56 percent probability of a 5 percent loss in tropical Latin American forests, and a 100 percent probability of a 5 percent loss in boreal northern latitude forests.

These predictions were based on simulations of very complex processes across the globe, and have very high uncertainty associated with them. However, our understanding of the mechanisms behind these phenomena allows for some level of confidence about these risk analyses. Ecosystems have higher-order emerging properties that can confound predictions (Sinclair and Byron, 2006). The nonlinear effects across trophic levels can produce counterintuitive responses. Ecosystem control and feedback processes will be explored within

TABLE 11-1 Probability of Percent Losses in Forest Vegetation under Three Global Climate Change Scenarios

Global Climate Temperature Increase Scenario	Tropical Latin America			Boreal Northern Latitudes		
	5%	10%	20%	5%	10%	20%
T >3°C	56	38	12	100	88	31
2°C<T <3°C	25	20	0	100	70	10
T<2°C	19	19	0	75	44	0

Source: From Scholze et al., 2006.

populations, between populations, and as tools for management of ecosystems.

POPULATION CONTROL PROCESSES

Ecosystem services are the product of the composition of the biota within a site. Populations are the unit of biotic management at site, watershed, and ecoregion scales. Ecological engineers designing ecosystem services are in actuality designing habitats to support stable populations. The dynamics of individual populations are affected by resource limitations and spatial and temporal competition from other species within communities. Reproductive strategies, population growth rates, and resource limitations are key variables in ecosystem control and feedback mechanisms.

Reproductive Strategies

Reproductive strategies within populations have a dominant effect on the way populations respond to environmental stress. In general, the number of individuals in a metapopulation (N) is dependent on the rate of population growth (r) and the carrying capacity of the patch (K):

$$\frac{dN}{dt} = rN\left(1 - \frac{N}{K}\right) \tag{11.1}$$

The reproductive strategies of species can be classified as reproductive rate–limited (r-selected) or carrying capacity–limited (K-selected). Species that are r-selected reproduce very rapidly, with many offspring, low survivability, and very little parental investment in raising the offspring. These organisms tend to be small, widely dispersed, with very fast generation times. Species that are K-selected reproduce more slowly, have fewer offspring, generally have some level of parental investment beyond reproduction, and live longer than r-selected organisms.

Of course, these generalizations do not hold up for many organisms. The white oak tree (*Quercus alba*) is long-lived, large, slow growing, and slow reproducing, yet produces copious amounts of seeds (acorns), which are dispersed broadly. For many animals, this categorization holds true at some level; rodents reproduce rapidly, are relatively small, and grow to reproductive age quickly. But compared to a cockroach, rodents are big, slow, and impotent. Discussions of reproductive strategy are context-specific, usually comparing species' trophic interactions. Populations that are K-selected tend to be relatively buffered from environmental stresses and are usually resource-limited. In contrast, populations that are r-selected often have adequate resources for growth, but are limited by predation or other consumption processes.

Survivorship

All organisms in populations die. The rates of death directly affect the probability that the population will survive. Survival over time is generally expressed as:

$$S(t) = e^{-kt} \tag{11.2}$$

where
    k is the rate of population decline.

Survivorship (l_x) is the product of the probabilities of survival of all the individuals in a population. Survivorship can be expressed as:

$$Log_e l_x = -\sum_{i=0}^{x-1} K_i \tag{11.3}$$

The expectation of further life is the weighted average of the survival periods between i and $i+1$ (Ricklefs and Miller, 2000):

$$e_x = \frac{1}{l_x} \sum_{i=x}^{\infty} (i - x)(l_i - l_{i+1}) \tag{11.4}$$

The expectation of further life is a key parameter in life tables for species (Ricklefs and Miller, 2000). Dynamic life tables can be constructed to follow a cohort group introduced to a habitat or born at the same time through their life spans; a static life table evaluates the survival and fecundity of a population at a given point in time

(Ricklefs, 2008). In many cases, ecological engineers need to determine the survivorship of a population at a point in time. Given the number of individuals of age x alive at time i, survivorship is:

$$l_x = \frac{n_x(i)}{n_o(i-x)} \tag{11.5}$$

Growth Rates

As Equation 11.1 shows, population growth rate is a product of the intrinsic growth rate (r), the population size, and the effects of resource limitations (crowding) as N approaches K. In order to describe the population at time t, this relationship can be integrated as the logistic growth equation:

$$N(t) = \frac{K}{1 + be^{-r_0(t-1)}} \tag{11.6}$$

where

e is the natural log root.

b is a constant related to the initial population at time zero:

$$b = \frac{K - N_0}{N_0} \tag{11.7}$$

The logistic growth curve is an S-shaped curve with the inflection point at $K/2$ (Figure 11-2). Population size and growth rate are density-dependent for many organisms (Ricklefs, 2008). Density-dependent

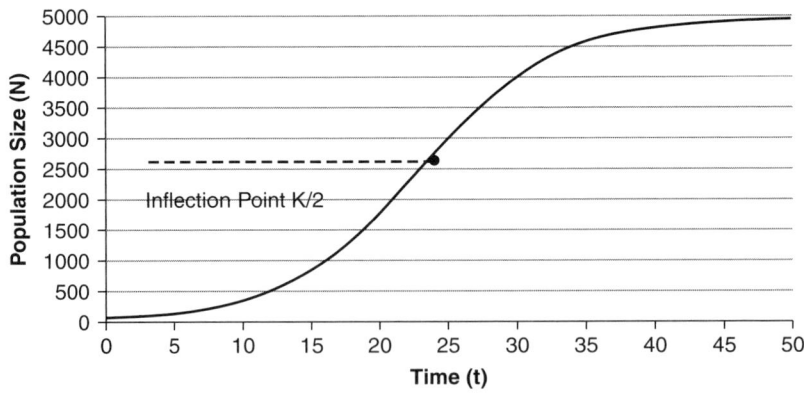

FIGURE 11-2. Logistic growth curve, with the inflection point at K/2.

factors include food supplies, refugia, predator pressure, parasites, and disease. Density-independent factors include temperature, precipitation (or lack thereof), and disturbance events.

For metapopulations occupying a habitat composed of a series of patches, the rate of extinction e is related to patch area (A) (Ricklefs and Miller, 2000):

$$e = e_0 e^{-bA} \qquad (11.8)$$

where

e_0 and b are species-specific parameters.
e^{-bA} is the natural log of $-bA$.

Migration rate between patches (m) is dependent on patch isolation distance (D) for most species. The proportion of patches that are occupied by a species in equilibrium can be estimated as:

$$\hat{p} = 1 - \left(\frac{e_o}{m_o}\right) e^{-bA+aD} \qquad (11.9)$$

where

b and a are species and patch density proportion, respectively.

The interplay between population density and resource limitations within a population creates internal feedback controls on population growth. When a population colonizes a patch, the population growth rate may exceed the resource base within the patch, resulting in overpopulation of the patch. This process can be described by the logistic model (Ricklefs and Miller, 2000):

$$N(t+1) = N(t) e^{r[1-\frac{N(t)}{K}]} \qquad (11.10)$$

where the population overshoots patch carrying capacity then adjusts, resulting in a series of dampened oscillations about K (Figure 11-3). Those feedback systems become even more complex when community-level processes are considered (Ricklefs and Miller, 2000).

COMMUNITY CONTROL PROCESSES

Community control processes are generally resource-consumer related, with population feedback systems between species. These processes are difficult to design explicitly because they are the product of the species that occupy a space. Understanding community control

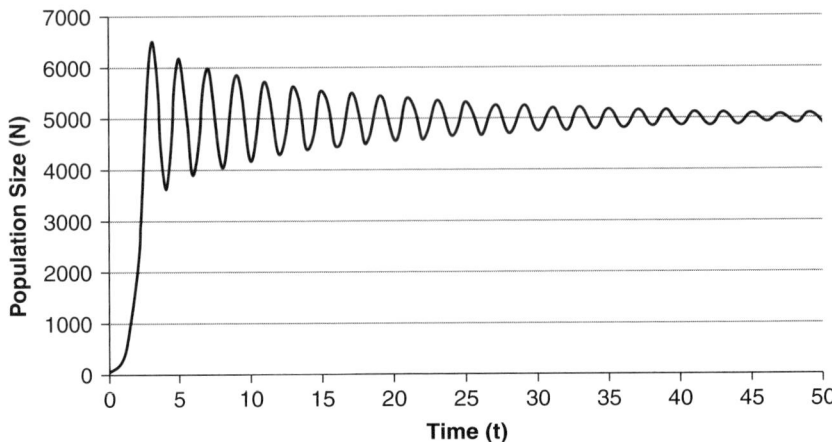

FIGURE 11-3. Discrete logistic population model, with $r = 1.95$, $K = 5000$, and $N_0 = 50$.

processes can enhance the ability of ecological engineers to construct robust, stable, and resilient systems that provide ecosystem services. The frontier of ecological engineering is in community control processes.

Plants and Nutrients

For plants, the limiting resources are generally water, nutrients, and light. When plants are fertilized with nitrogen through natural or human activities, the nutrients enhance the pre-fertilization cycling of the compound. Nitrogen and phosphorus respond differently in the cycling. Vitousek (2004) reported that nitrogen fertilization of forests in Hawaii did not increase decomposition or regeneration rate (Figure 11-4). Phosphorus additions did increase decomposition and regeneration rates (Figure 11-5). It is likely that the microbial community responsible for cycling processes was phosphorus-limited in the nitrogen fertilization treatment. Nutrient limitations occur when supply is less than demand, and can be induced in one population by competition through the nutrient cycle (Figures 11-4 and 11-5).

Plant uptake, microbial growth, and turnover drive nutrient demand, while decomposition and mineralization (and fertilization) drive supply. Microbial decomposers are both supplier and demander of nutrients (Vitousek, 2004). Increasing one nutrient in a system can induce limitations of other nutrients across the system. For example, if carbon is limiting decomposer community growth, and carbon is added to

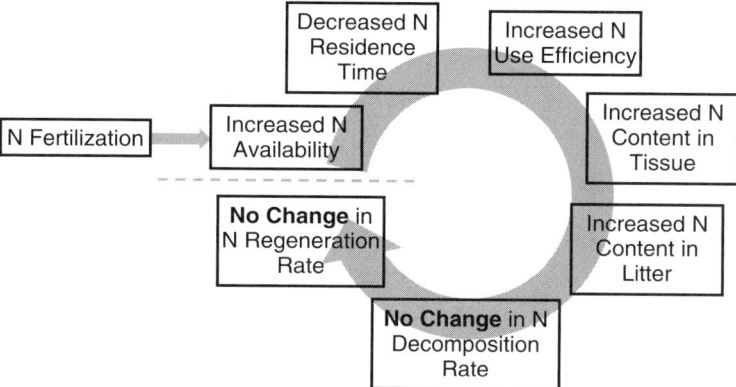

FIGURE 11-4. Nitrogen cycle response to fertilization. (*Source*: Modified from Vitousek, 2004.)

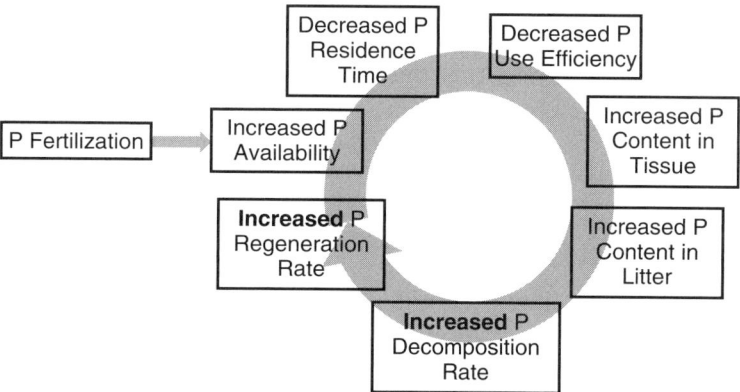

FIGURE 11-5. Phosphorus cycle response to fertilization. (*Source*: Modified from Vitousek, 2004.)

the soil through conservation tillage or other practices, plant available nitrogen and phosphorus can decline as a result of increased decomposer demand and competition for those nutrients.

Resource Competition

Resource-consumer competition occurs within and between species. The coefficient of competition for species i with respect to j (a_{ij}) is the effect of the growth rate of one on the other due to competition for resources (Ricklefs and Miller, 2000). The number of individuals of species j can be estimated with regards to the number of individuals in species i competing for a resource (Ricklefs and Miller, 2000):

$$\frac{dN_j}{dt} = r_j N_j \left(1 - \frac{N_j}{K_j} - \frac{a_{ij} N_j}{K_j}\right) \qquad (11.11)$$

where
> r is the intrinsic population growth rate.
> N_j is the number of organisms j.
> N_i is the number of organisms i.
> K_j is the carrying capacity for organism j.
> K_i is the carrying capacity for organism i.
> a_{ij} is the coefficient of competition for species i and j.

Over time, competing species evolve mechanisms to co-exist, or one goes extinct. At some threshold, the equilibrium population size of species i, the carrying capacity is influenced by species j (Ricklefs and Miller, 2000):

$$\hat{N}_i = K_i - a_{ij} N_j \qquad (11.12)$$

where
> $\hat{N}_i$ represents the equilibrium population of species i relative to species j.

The equilibrium population size of species i is in effect the carrying capacity of population i minus the population of species j times the competition coefficient. Thus, the term $a_{ij} N_j$ can be thought of as the equivalent j population of Ni, since it is a carrying capacity abstraction.

FEEDBACK PROCESSES

Feedback processes occur when some element in a circular phenomenon influences another in a positive or negative manner. A positive feedback system provides resources or accelerates processes that enhance another element of a system, while negative feedback systems restrict or reduce processes or rates. Geese graze on aquatic plants, for example. In the process, they digest the plant material and defecate nutrients into the water. These nutrients fertilize algae, but also plants that the geese are grazing on. The rate of nutrient cycling back to the plants is increased, so the process of grazing enhances the growth of the plants being grazed on (Figure 11-6).

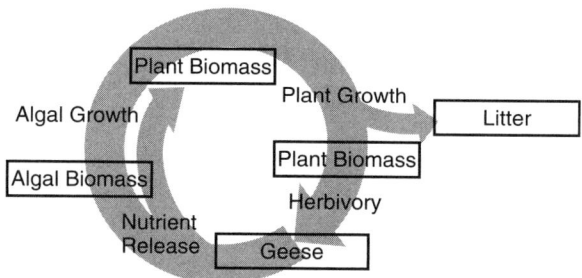

FIGURE 11-6. Positive feedback system. (*Source*: Modified from Ricklefs and Miller, 2000.)

Atmospheric Feedback Loops

Recall that ecosystem responses tend to be nonlinear; Heimann and Reichstein (2008) reported that the terrestrial biosphere interactions with greenhouse gasses are more complicated than previously considered. They postulated three feedback mechanisms:

1. *The permafrost thaw*: The potential thawing of permafrost could increase microbial metabolism, releasing large amounts of CO_2 and CH_4 (Figure 11-7a).
2. *The microbial priming effect*: Increased carbon and energy sources stimulate microbial decomposition of latent soil carbon in grassland soils (Figure 11-7b).
3. *The nitrogen cycle interaction*: Increased CO_2 concentrations increase photosynthesis and growth, which in turn reduce nitrogen availability to decomposers, increasing fungal use of lignin, increasing the rate of CO_2 released through decomposition. This positive feedback may be overcome by subsequent increases in N_2 fixation through increased carbon content in soils and associated photosynthesis and growth (Figure 11-7c).

Atmospheric chemistry may also affect the community composition of plant communities (Lerdau, 2007). Isoprene emissions from vascular plants, especially oak trees (*Quercus* sp.), increase their ability to tolerate oxidative stress from atmospheric ozone contamination. Only about one-third of plants emit isoprene; maples, birches, and hickory species do not. Isoprene in the atmosphere reacts with nitrogen oxide emissions from soils, increasing ozone levels. Ozone stresses the nonprotected species more than the oak trees, resulting in a positive feedback system that could select for oaks over other trees in

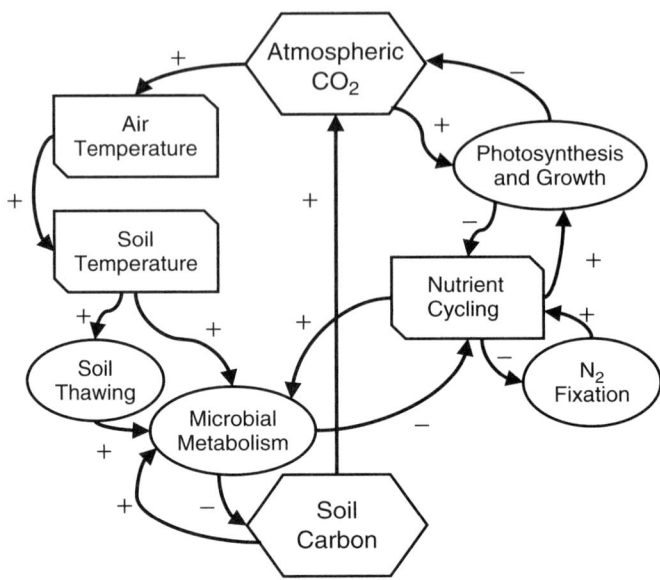

FIGURE 11-7. Potential feedback loops associated with climate change. (*Source*: Modified from Heimann and Reichstein, 2008.)

disturbed systems with elevated ozone concentrations from human-induced atmospheric pollution (Figure 11-8).

Soil Feedback Loops

Soil mineralization is critical for nutrient cycling and fertility (Figures 11-4 and 11-5). Nutrient cycling is mediated through microbial activities; Barot et al. (2006) showed that earthworms also accelerate mineralization and enhance plant growth. This process can result in increased primary productivity of the community, with the positive feedback of increasing soil organic matter and therefore decomposer (earthworm) biomass.

Daufresne, Hedin, and Tilman (2005) proposed that soil nutrient losses through dissolved organic pathways can alter the ratio of nutrients available in the soil matrix, influencing competition among plant species. They demonstrated that recycled nutrients have a time delay in availability that can lead to population oscillations (Figure 11-3). The role of the detrital nutrient pool is critical in plant systems, where losses occur through leaching and conversion to complex organic forms. The plant-soil-detritus system represents a feedback system

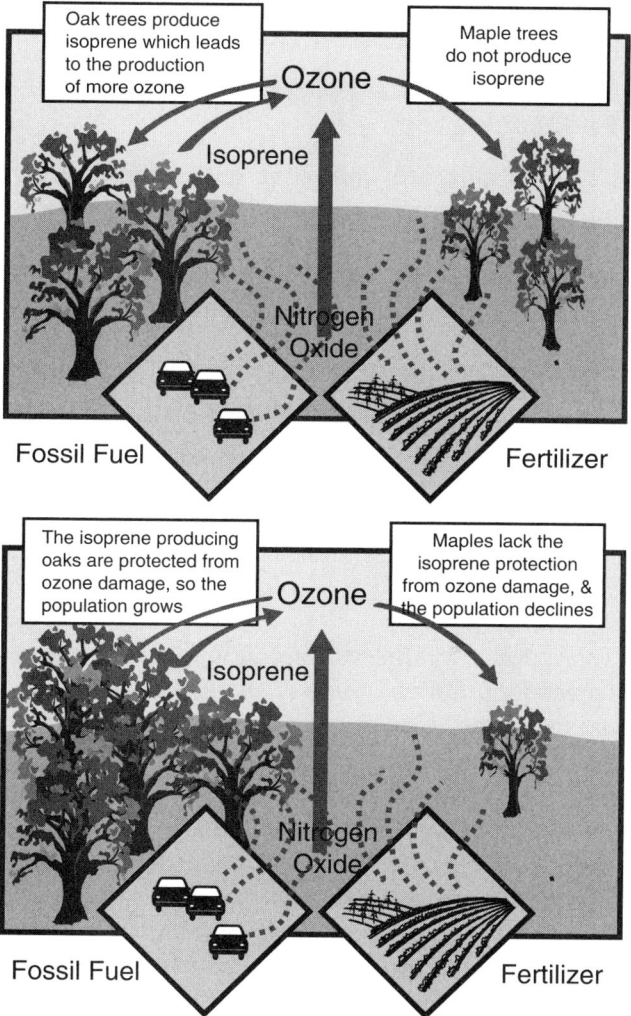

FIGURE 11-8. Potential feedback loops associated with isoprene and oak trees. (*Source*: Modified from Lerdau, 2007.)

for competition through exclusion of nutrients from one species by another.

Coughenour and Chen (1997) suggested that soil-plant-water-atmosphere feedback systems would respond to increased temperatures with lower NPP at current atmospheric CO_2 levels (350 umol/mol). The feedback process governing NPP under increased temperature is soil moisture and plant water status. Increased atmospheric CO_2 levels did increase NPP, although the relative increases in NPP found when atmospheric CO_2 levels were increased were largely due to increased water use efficiency.

Consumer Feedback Loops

Grazer-plant interactions are complex, with many subtle and some overt feedback loops. The grazing optimization hypothesis states that NPP increases with moderate grazing (Mazancourt et al., 1999). The proposed mechanisms for this improvement in community productivity are the combined benefits of the increased nutrient cycling from grazer defecation, and the plant community benefits of persistent but minor disturbance. In grasslands, herbivores can consume between 15 and 60 percent of annual aboveground NPP (Mazancourt et al., 1999). In the Serengeti savanna, that number can be as high as 94 percent, but averages 60 percent. In spite of this dramatic insult on plants, when compared with ungrazed areas (grazer exclusions), the grazed systems have higher NPPs. Most grasslands are nitrogen limited; increasing nitrogen cycling appears to be a significant positive feedback system for these plant communities.

Similarly, predator-prey interactions show positive indirect effects of predators on prey populations (Gude et al., 2006). In predator-ungulate-plant-soil systems, wolf (*Canis lupus*) predation on elk herds (*Cervus elaphus*) showed no impact on elk group size. Wolf predation appears to stimulate elk movement, reducing grazing pressure on grasslands where wolves hunt. This increased movement also reduces browsing on woody plant stands, which increases cover for stalking activities.

Feedback mechanisms in community trophic responses to stress take several forms. Hypoxia (low oxygen concentration) in bays and estuaries occurs when high-nutrient (predominantly nitrogen) water discharges into low-nutrient systems, resulting in increased microbial respiration. Oxygen cannot diffuse fast enough from the atmosphere to offset the increased consumption, and is thus depleted. Altieri and Witman (2006) developed a foundation species feedback model to study the impact of environmental stress (hypoxia) that integrates individual, population, community, and ecosystem levels. The model was based on observations at Narragansett Bay, Rhode Island, a hypoxic estuary. They showed that seasonal hypoxia reduced starfish predator pressures on mussels, but that the hypoxia stress was too great for the mussels to benefit from the predator refuge. A negative feedback system emerged whereby mussel stress and mortality decreased filtration feeding effectiveness by as much as 75 percent, resulting in increased algal density and longer hypoxic periods (Figure 11-9).

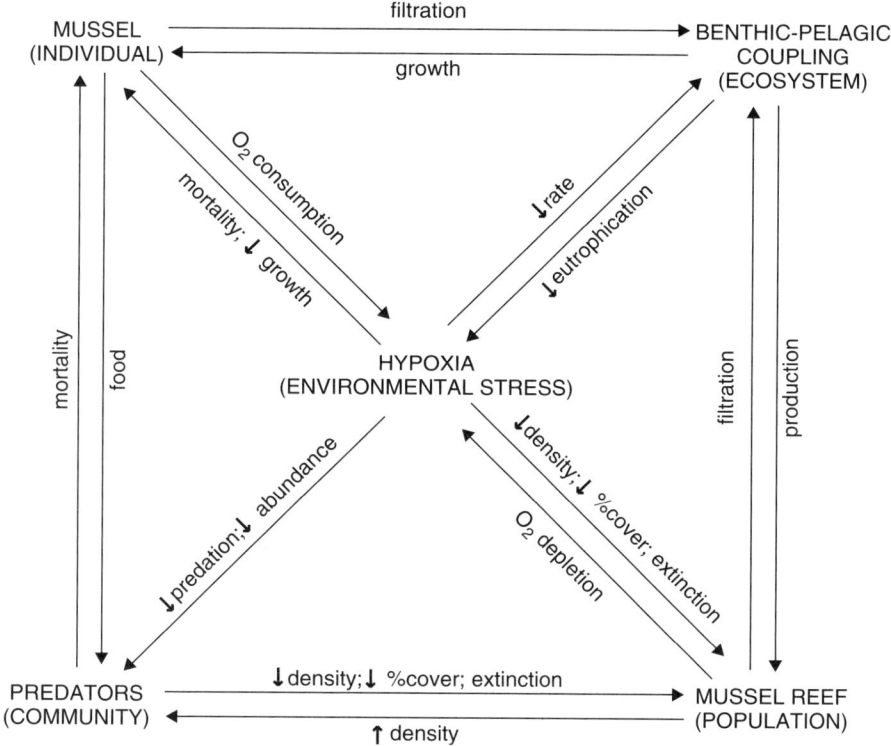

FIGURE 11-9. Estuarine trophodynamic feedback model. (*Source*: Modified from Altieri and Witman, 2006.)

DESIGNING ECOSYSTEM COMPLEXITY

Ecosystems are complex systems; the properties of complex systems must be integrated into any ecosystem design, if the effort is to be successful. Kay and Regier (2000) summarized the properties of the complexity characteristic of ecosystems:

1. *Nonlinear*: The system cannot be understood by its component parts, but rather as a whole, because the interactions of the components result in nonlinear and nonintuitive responses of the system, called *emergent properties*.
2. *Hierarchical*: The elements of ecosystems are nested within elements, creating a holarchically nested system. Holarchical organization is a subset of hierarchical systems, where each element (holon) is a complete system yet is connected to the whole and cannot be considered independent of it.

3. *Internal causality*: Each holon behaves with purpose, and each interconnection results in emergent behavior through positive and negative feedback, autocatalysis, and hysteresis. Autocatalysis is the self-initiating characteristic of a process, and hysteresis is the tendency of a process to respond on the basis of its history, or to have memory.
4. *Self-organization*: The interactions of holons in the nested system result in a form of self-organization that emerges where energy utilization is optimized within the range of possible behaviors. This range is limited by the complexity of holons, also characterized as biodiversity.
5. *Dynamically stable*: Ecosystems are not homeostatic; they are dynamically stable at a given equilibrium point. However, ecosystem processes can change rapidly and catastrophically (from a holonic perspective) in response to external forces.
6. *Multiple steady states*: Ecosystems do NOT have a preferred system state. Current system states are the product of previous conditions; changing current conditions will alter the successive states. Ecosystems cannot be restored to previous states, because the process of changing states alters them irreversibly.

These properties are as critical to ecosystem design as precipitation, slope, and vegetation. Ecosystems respond to change by exhibiting self-organizing, holarchic, and open (SOHO) attributes (Kay and Regier, 2000). A holarchic system is one where holons (individual components) interact reciprocally across scales. The elements of ecosystems (individuals, populations, species, communities, guilds, assemblages) interact with each other at multiple environmental scales (site, patch, watershed, ecoregion, biome, and biosphere).

The goal of the ecological engineer is to design ecological integrity from which ecosystem services emerge. Conceptual models of ecosystem elements provide a framework for identifying holons, as described previously. The context for the interactions between holons should be described, and, where possible, quantified. These relationships may include competition for resources, influences on survivability, and predator-prey interactions. The scales of interaction should be explicitly mapped in this process, so that emergent properties can be anticipated. The process of analysis and deliberation, described in Chapter 3, is one mechanism to ensure that multiple perspectives from the ecotechnology team as well as the affected community are considered.

Self-Organization

The processes of energy transformation between and within holons and complex multiscale interactions in ecosystems result in a predictable range of organization of individuals within a space. This self-organization tendency is the basis of many of the emergent properties associated with ecosystems (Odum, 1988). Ecosystem holons exhibit the propensity to remain in a limited state, referred to as an attractor (Kay and Regier, 2000). This attractor state is often confused with homeostasis, but the concept that ecosystems are static is dangerously flawed (everything is changing). Ecosystems move to equilibrium points based upon the conditions that constrain holonic behavior.

Ecosystems do exhibit resilience—that is, they can return to an attractor state after disturbance—but only if the disturbance does not alter a critical holonic interaction. Ecosystem design for restoration often means understanding the disturbed condition attractor state, and the thresholds of impacts necessary to shift the system to a preferred attractor state. Ecosystems can change attractor states (energy levels, structures, stability) very rapidly. From an ecosystem services standpoint, this is both a valuable and potentially catastrophic property. Ecosystem flips have been observed in many systems where altering a single holon results in a cascading impact on the trophic dynamics of the system.

The simplest form of self-organization emerges when organisms compete for scarce resources. Guttal and Jayaprakash (2007) observed that banded topographic vegetation patterns in arid ecosystems were the result of plants growing at the bottom of slopes, where water accumulated and infiltrated. This simple observation may be self-evident, but the consequences of that physical stratification become complex quite quickly. Plants grow where the water is. Rodents live where the plants are. Predators prey on rodents. This predator-prey-consumer-producer relationship is geographically constrained by water infiltrating into soil at the foot of small slopes in arid areas. The patch dynamics that emerge can be very complex. Changing the way water moves across this landscape could change these interactions in dramatic ways (everything is connected).

Boyer and López-Corona (2009) investigated the impact of frugivorous (fruit-eating) animals on seed dispersion and fruit tree success. They found that animals that fed on randomly located patches dispersed seeds and thus modified the distribution of patches on the landscape. The survival probability of a seed to fruit-bearing age increases with the distance to its parent patch and decreases with the size of the colonized patch. Animals that forage on fruits are deterministic, in that they have memory and thus forage from patches within minimal traveling distances. Foragers disperse seeds along a trajectory between forage sites, creating emerging patches.

The memory of the foragers, acting as holons with discrete operational motives, creates heterogeneous seed deposition (Boyer and López-Corona, 2009). The impact of this holonic interaction is a complex geospatial and temporal pattern where patches with a small initial size increase in biomass over time and reach a maximum NPP. This maximum NPP is an attractor state that corresponds to a self-organized critical state; however, in areas outside these corridors, fruit-bearing biomass sharply decreases, even though resources and space are available for colonization.

Self-organization is a tricky process to design, but can be observed and simulated. Salt marsh responses to sediment loads in coastal estuaries illustrate the difficulty. A salt marsh sediment budget (deposition minus erosion) can be represented as (van de Koppel et al., 2005):

$$\frac{\partial S_x}{\partial t} = l_{max}\left(1 - \frac{S_x}{K_x}\right) - e_{max}\frac{a}{a+p_x}\tau(x)S_x - d_s\frac{b}{b+P_x}\frac{\partial S}{\partial x}S_x \quad (11.13)$$

where
  l_{max} is the maximal sediment accumulation rate.
  S_x is the maximal sediment depth.
  K_s is the maximal sediment elevation (mean high water level).
  e_{max} is the maximum erosion rate of sediment by water.
    a is the level of P at which erosion is limited by 50 percent.
  P_x is the standing crop of plants at point x.
  $\tau(x)$ is the bottom shear stress (usually ranging from 0 at the landward edge to 1 at the seaward edge).
  d_s is a conversion coefficient.
  b is plant density.
  $\frac{\partial s}{\partial x}$ is the slope of the sediment.

Density-dependent vegetation dynamics were calculated as:

$$\frac{\partial P_x}{\partial t} = r\left(1 - \frac{P_x}{K_P}\right)\frac{S_x}{c+S_x}P_x - dP_x - d_p\frac{\partial S}{\partial x}P_x \quad (11.14)$$

where

r is the intrinsic growth rate of the plant.
K_p is the maximal plant standing crop.
c is a half-saturation constant.
d is plant senescence (mortality).

Thus, $d_P \frac{\partial s}{\partial x} P_x$ is the plant-specific mortality due to wave damage.

The results from van de Koppel et al. (2005) illustrated that as a result of self-organizing properties with sediment, the tension between sediment capture and plant erosion often resulted in vegetation collapse (Figure 11-10). Salt marshes with high sediment input approach a critical state where sediment depth in the tidal flats creates steep approaches to the seaward edge. This results in a potential flip where a disturbance event induces a cascade of vegetation collapse and severe erosion of the cliff edge, which can result in destruction of the salt marsh.

The notion that self-organization creates intelligent outcomes is therefore clearly not accurate. Rather, ecosystems exhibit SOHO properties, and must be designed with consideration of the implications of those properties. Ecosystem elements (holons) are not machines, nor are they parts of machines. Each component of the ecosystem, down to the organism, is an independent agent with discrete motives and life strategies. These organisms act and react, controlling aspects of their

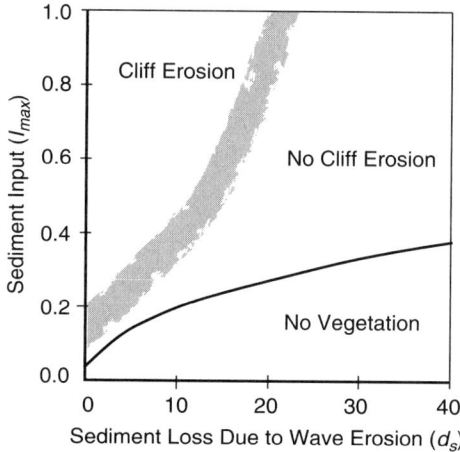

FIGURE 11-10. Salt marsh erosion as a function of sediment input. (*Source*: Modified from van de Koppel et al., 2005.)

environment that they can control. In order to design the processes that yield ecosystem services, ecological engineers have to understand the behavior of each holon in the context of the ecosystem.

> The engineer has respect for mechanical wisdom because he created it. He has disrespect for ecological wisdom, not because he is contemptuous of it, but because he is unaware of it.
>
> —Aldo Leopold

References

Altieri, H. and J. Witman. Local Extinction of a Foundation Species in a Hypoxic Estuary: Integration of Individuals to Ecosystem. *Ecology* 87: 3, 717–730, 2006.

Barot, S., A. Ugolini, and F. Brikci, Nutrient cycling efficiency explains the long-term effect of ecosystem engineers on primary production, *Functional Ecology* 21: 1–10, 2006.

Boyer, D., and O. López-Corona, Self-organization, scaling and collapse in a coupled automaton model of foragers and vegetation resources with seed dispersal, *Journal of Physics A: Mathematical and Theoretical*, 42: 434014, 2009.

Coughenour, M., and D. Chen, Assessment of grassland and ecosystem responses to atmospheric change using linked plant-soil process models, *Ecological Applications*, 7(3): 802–827, 1997.

Daufresne, T., L. Hedin, and D. Tilman, Plant coexistence depends on ecosystem nutrient cycles, *Proceedings of the National Academy of Sciences*, 102(26): 9212–9217, 2005.

Falk, D., M. Palmer, and J. Zedler, eds., *Foundations of Restoration Ecology*, Island Press, Washington, DC, 2006.

Gude, J., R. Garrott, J. Borawski, and F. King, Prey risk allocation in a grazing ecosystem, *Ecological Applications*, 13(1): 285–298, 2006.

Guttal, V., and C. Jayaprakash, Self-organization and productivity in semi-arid ecosystems: Implications of seasonality in rainfall, *Journal of Theoretical Biology*, 248: 490–500, 2007.

Heimann, M., and M. Reichstein, Terrestrial ecosystems carbon dynamics and climate feedbacks, *Nature*, 451(17): 289–292, 2008.

IPCC, 2007: *Climate Change 2007: Synthesis Report. Contribution of Working Groups I, II and III to the Fourth Assessment Report of the Intergovernmental Panel on Climate Change* [Core Writing Team, Pachauri, R.K and Reisinger, A. (eds.)]. IPCC, Geneva, Switzerland, 104 pp.

Kay, J., and Regier, H., Uncertainty, complexity, and ecological integrity: insights from an ecosystem approach, in P. Crabbe, A. Holland, L. Ryszkowski, and L. Westra, eds., *Implementing Ecological Integrity: Restoring Regional and Global Environmental and Human Health*, NATO

Science Series, Environmental Security, Kluwer Academic Publishers, Dordrecht, The Netherlands, pp. 121–156, 2000.

Lerdau, M., A positive feedback with negative consequences, *Science*, 316: 212–213, 2007.

Mazancourt, C., M. Loreau, and L. Abbadie, Grazing optimization and nutrient cycling, *Ecological Applications*, 9(3): 784–797, 1999.

Odum, E., *Ecology and Our Endangered Life Support System*, Sinauer Associates Press, Sunderland, MA, 1993.

Odum, H., Self-organization, transformity, and information, *Science*, 25(242): 1132–1139, 1988.

Ricklefs, R., *The Economy of Nature*, 6th ed., W.H. Freeman and Co., New York, NY, 2008.

Ricklefs, R., and Miller, G., *Ecology*, 4th ed., W.H. Freeman and Co., New York, NY, 2000.

Ricklefs, R., and D. Schluter, eds., *Species Diversity in Ecological Communities*, University of Chicago Press, Chicago, IL, 1993.

Scholze, M., W. Knorr, N. Arnell, and I. Prentice, A climate change risk analysis for world ecosystems, *Proceedings of the National Academy of Sciences*, 103(35): 13116–13120, 2006.

Sinclair, A., and A. Byron, Understanding ecosystem dynamics for conservation of biota, *Journal of Animal Ecology*, 75: 64–79, 2006.

van de Koppel, J., D. van der Wal, J. Bakker, and P. Herman, Self-Organization and vegetation collapse in salt marsh ecosystems, *The American Naturalist*, 165(1):E1–E12, 2005.

Vitousek, P., *Nutrient Cycling and Limitation*, Princeton University Press, Princeton, NJ, 2004.

12

Stream Restoration Design

> No man steps in the same river twice, for it is not the same river and he is not the same man.
>
> —Heraclitus

INTRODUCTION

Restoration is returning something to a former position or condition. A functioning stream is one where the physical, biological, and chemical characteristics support the expected ecosystem services. As Heraclitus noted, rivers change constantly, even without human intervention (everything is changing). Degraded streams cannot really be restored. Once a stream system is changed, it never goes back to its exact condition prior to the change. However, ecological function may be reestablished. The objective of stream restoration is restoration of ecosystem function and services.

Streams provide a number of ecosystem services. Provisioning services include water supply for consumption uses (domestic, agricultural, and industrial) and nonconsumptive uses (power generation, transport), and provision of aquatic organisms for food and medicines. Regulatory services include maintenance of water quality and buffering of flood flows. Cultural services include recreation, tourism, and spiritual values. Streams also provide supporting services, including nutrient cycling, sediment transport, primary production, predator/prey relationships, and ecosystem resilience. When any of these services are degraded or lost, "restoration" of those services may be desired.

There are several reasons to consider stream restoration, including: naturalization of channelized stream reaches, restoring natural flow after dam removal, repair of rapidly eroding stream banks, and improving habitat and refugia. Streams are sometimes enhanced to improve fisheries or some other specific function. Enhancements may be made for a single desired species such as brook trout or smallmouth bass, or the enhancement may take an ecosystem approach. The ecosystem approach assumes that the desired species will come if the propagules are available and the stream is healthy.

Stream restoration can be extraordinarily expensive. Full channel restoration cost on the order of $200 (2010) per foot is normal. Examples of failed restoration attempts are plentiful. Careful design can reduce the risk of failure but cannot totally eliminate risk. A range of options for restoration should be considered. Included in the range should be the option of doing nothing. Streams are evolving systems; if the underlying cause of disturbance is removed, they may heal themselves in time. Self-healing may or may not be possible within the constraints applied to the stream by society.

The art of stream restoration requires careful attention to assessment, alternative development and selection, construction oversight, monitoring, and maintenance. Because of the complexity of stream systems, and the interaction of physical, chemical, and biological elements, stream restoration is not the purview of any single profession. Engineers, hydrologists, biologists, restoration ecologists, geologists, geomorphologists, and horticulturalists may all be needed for successful design. The occurrence of floods, droughts, fires, and the like is unpredictable—except that they will inevitably occur sometime. Provision should be made to adaptively manage the project during and after such events to make repairs and adjustments as necessary.

ASSESSMENT

The first consideration before beginning stream restoration is the end goal of what a restored stream should be. A functioning stream clearly needs to handle a wide range of flow. Of special importance is the bankfull flow, generally considered to be the channel-forming flow, but the stream must also be resilient to both floods and droughts. The stream must be competent to transport sediment load both from the watershed and the channel. A functioning stream is stable, existing in a state of dynamic equilibrium where erosion balances deposition. A functioning stream should simulate natural stream geometry, including plan, profile, and cross section. Finally, the stream should provide a variety of habitats to support the full range of aquatic and riparian biodiversity. Assessing the stream with respect to the fully functioning condition is the first step in restoration.

All streams exist within the context of their watershed (everything is connected). The watershed provides both water and sediment to the stream. The amount and form of precipitation in the watershed, along with the watershed's infiltration and runoff properties, drive stream hydrology. Geology, topography, soils, land cover, and land use determine the morphology of the stream. Changes of land use

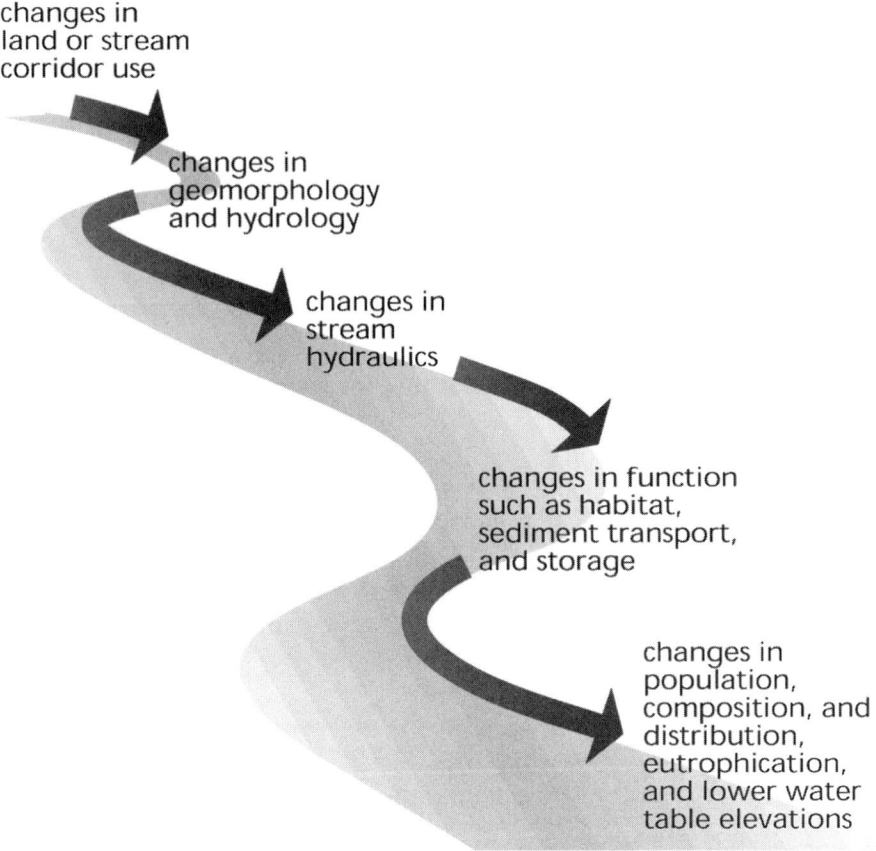

FIGURE 12-1. Connectivity of stream corridors. Disturbance to the stream corridor system causes alterations to stream corridor structure and functions. (*Source*: From *Stream Corridor Restoration: Principles, Processes, and Practices*, by the Federal Interagency Stream Restoration Working Group [FISRWG], October, 1998.)

in a watershed cause changes in watershed hydrology (Figure 12-1). Streams adjust to the new flow regime. The context of the watershed and its past and future trend must be considered in a stream restoration design.

In a fully functioning stream, the natural processes of erosion and deposition are roughly balanced. Erosion on the outside bend of a bank is matched by deposition on the inside bend. The stream only slowly moves across its floodplain. Disturbance of the dynamic equilibrium will propagate throughout the stream system. Because of the connectivity of stream systems, disturbances may be local, or they may be systemic. The restoration of a single stream bank risks failure if the problem is systemic in nature. The effect of a restoration project on upstream and downstream reaches must also be considered.

Watershed characterization was covered extensively in Chapter 5, "Defining Place: The Watershed." Significant questions to ask are:

- What are the watershed area, topography, relief, aspect, soils, and geology?
- What are the extents of different land uses and land covers?
- Are there significant trends in land use and land cover?
- What is the history of the stream?
- Are previous alterations working their way through the system?
- What conditions exist that may constrain restoration activities?

These questions may be answered through examination of topographic maps and aerial photographs.

In-stream processes are related to the position along the stream corridor from headwaters to mouth (Figure 12-2). A typical stream is erosional in the headwaters, transports sediment through mid-orders, and becomes depositional in higher orders. Is the restoration project in the erosional, transportation, or depositional zone of the valley? Stream ordination (see Chapter 5) places the proposed project in context with respect to stream processes. A long profile of the stream channel reveals the stream's general slope and local irregularities to the normal concave upward profile. Stream ordination and longitudinal profiles can be compiled from topographic maps, or from geographic information system (GIS) datasets.

A review of the history of a watershed and its stream channel may reveal past changes that are currently impacting the stream's dynamic equilibrium. Lane (1955) determined that the product of stream discharge and slope is directly proportional to the product of sediment load and sediment size. This relationship can be used to qualitatively evaluate a stream's response to a modification. For example, when a stream is channelized, the slope increases because length is decreased and drop is constant. When the slope increases, either the product of sediment load and sediment size must increase or discharge must decrease.

A stream in dynamic equilibrium generally frequently spills onto its floodplain and spreads across the valley floor adjacent to the channel. When a stream floods, the average depth of the channel is reduced and, as a result, the stream's power also is reduced. When a channel is disturbed, the stream undergoes a fairly predictable series of adjustments. Simon's (1989) stream evolution model describes this adjustment. The stream erodes into the channel bed and becomes incised. At some point, stream power becomes greater than the banks can withstand, resulting in bank erosion and widening of the channel. Eventually, eroded material accumulates and aggrades the channel. The channel will then form a new stable channel in the aggraded material. The former floodplain now exists as a terrace above the active floodplain in the evolved channel. Identification of the terraces and the stage of stream evolution increases the designer's understanding of what the

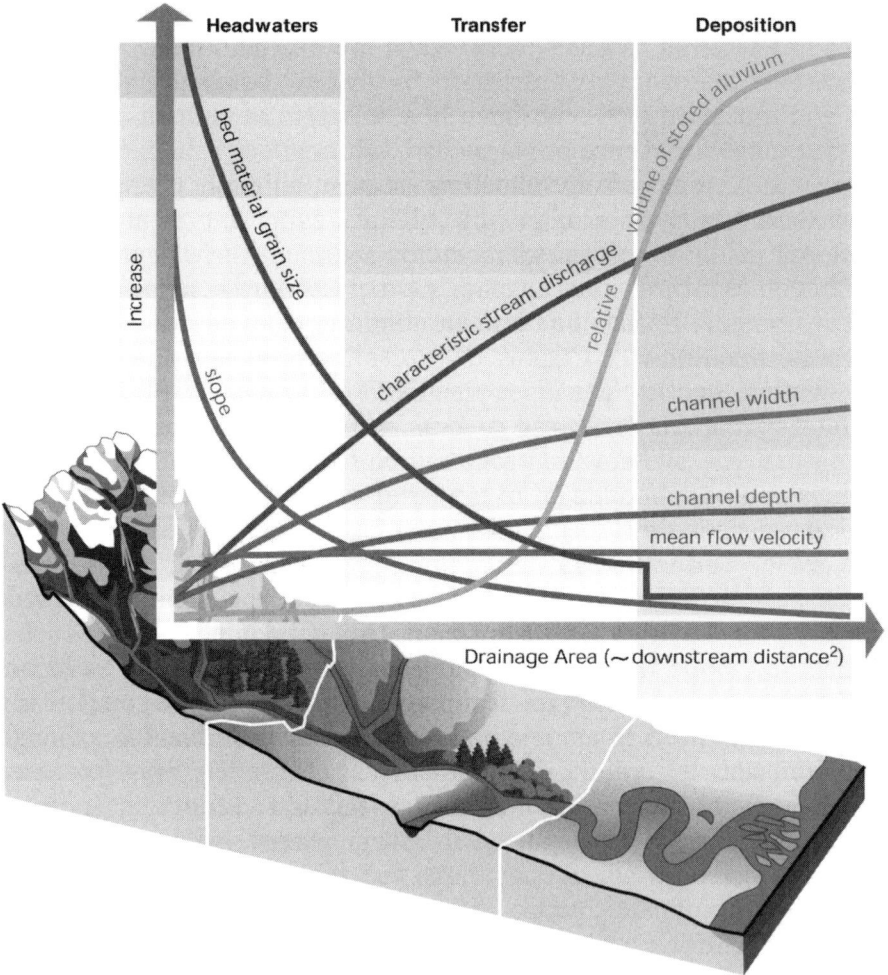

FIGURE 12-2. Longitudinal stream profile. From the headwaters to the mouth, the stream traverses three zones: an initial headwaters zone of erosion, a transport zone, and a depositional zone. (*Source*: EPA.)

stream is trying to do. The designer can then develop projects to bring the stream back into dynamic equilibrium.

There are several different types of streams. Streams may be straight or meandering, single or multiple channel, pool and drop or pool and riffle, and so forth. Similar streams exhibit similar characteristics. Streams classified by type can be evaluated with respect to reference streams of that same type. Dave Rosgen's (1996) stream classification system is a unified process for describing a stream in terms of valley and channel morphology. Streams are classified into seven types, A through G, based on the number of stream threads (single or multiple channels), entrenchment ratio (the degree of incision into the valley floor), width to depth ratio at bankfull flow, and sinuosity (channel length divided by valley length). Streams are then further classified by slope and channel material (clay, sand, gravel, cobble, boulders, bedrock). Channel classification has the obvious benefit of providing common nomenclature of channel types and characteristics. More importantly, the classification system provides a detailed protocol for assessing stream conditions with respect to reference predisturbance conditions, and provides the basis for effective restoration design. Readers should refer to Rosgen's book *Applied River Morphology* for a complete description of the classification system and its application to restoration.

Rarely does the stream restorer have the luxury of working in a totally natural system with freedom to do as he/she pleases. Restoration takes place within the constraints of society. Some constraints that impact restoration are the extent of existing drainage easements, existing structures, and Federal Emergency Management Agency (FEMA) flood insurance regulations. Most often, post-project flood elevations may not exceed pre-project elevations. Cities, states, and federal governments may all have permitting requirements. These constraints and permits need to be outlined at the start of the project.

The tools of stream restoration are hydrology, sedimentology, geomorphology, and habitat manipulation. The following sections provide a summary of those tools. Each discipline is extensive. Further study of the literature in each respective field is recommended.

HYDROLOGY

The obvious ecological service of streams is the transport of water from the watershed back to the ocean. The quantity of flow or discharge in a stream may vary several orders of magnitude between low flow and extreme floods. Discharge is frequently characterized as either storm flow or baseflow. Storm flow enters a stream as runoff during storm events. Baseflow is the discharge in the stream between storms.

Flood events are characterized by a flood frequency analysis. Flood frequency analysis gives the probability that a flood of a specific magnitude will be exceeded during any single year. If discharge data are available, a flood frequency analysis proceeds as follows:

- Identify the peak instantaneous discharge for each year (either calendar of water year).
- Rank the discharges from highest to lowest.
- Calculate the cumulative frequency or probability (P) for each event, using Weibull's approximation:

$$P = (rank \times 100\%)/(n+1)$$

where
  n is the number of records.

- Plot the peak discharge vs. probability on log-normal graph paper.

Probability of exceedance can be converted to flood frequency (T) by the equation:

$$T = 1/P$$

where
  T is the recurrence frequency in years.

For instance, a flood with a 20 percent probability of exceedance has a return frequency of $1/0.20 = 5$ years.

A flow duration curve depicts the percentage of time that a given flow is equaled or exceeded over a given period. To develop a daily flow duration curve:

- Compile all daily flows for the period of interest.
- Sort the data in order of decreasing flow.
- Assign a ranking number, m, using 1 for the highest flow and n for the lowest.
- Compute the probability of exceedance by the formula: $P = 100m/(n+1)$.
- Plot the exceedance probability vs. discharge on log-normal paper.

The stream restoration project will rarely be conveniently located at a stream gage. If the watershed is homogeneous, and gages exist within the watershed regression, equations can be developed to relate watershed area to flow. The flood frequency and flow duration curves (Figure 12-3) can then be developed from the regressed flows. If there are no gages in the watershed of interest, then regression against similar watersheds may be possible. In both cases, the regression equation should be validated by comparing to measured flows at the site of interest.

The channel-forming discharge is the discharge that is responsible for the dimensions of the channel. High-frequency floods are responsible for the dimensions of the channel. Low flows do not have the power to move sediment and hence cannot form the channel. Infrequent floods of high magnitude do move more sediment per hour than smaller floods, but they do not happen frequently enough to be the major player in channel formation. Most changes to the stream channel are caused by floods that occur roughly once per year. Bankfull discharge (Figure 12-4) is the flow at incipient flooding (Rosgen, (1996); Leopold, 1994); this is the discharge that restored channels are designed to carry.

There are several ways to determine bankfull discharge. Bankfull elevation can be observed directly in the field. Indicators of bankfull include:

- The top of the first depositional feature or gravel bar
- Change in vegetation
- Scour line
- Change in bank material

These features are not always readily apparent. Terraces can easily be misidentified as floodplains, and the like. Bankfull stage should be verified by multiple indicators. Other indicators of bankfull are:

- The flood frequency: Bankfull most often relates to the 1.4 to 1.6 year return storm, although some researchers report a much wider range of return.
- A break in the stage-discharge curve: At bankfull, the channel widens so there is a greater increase in flow for a given increment of depth.
- The flow at which the most sediment is transported over a long period of time: This point is the maxima in a plot of the product of flow duration and sediment discharge (see section on sedimentology).

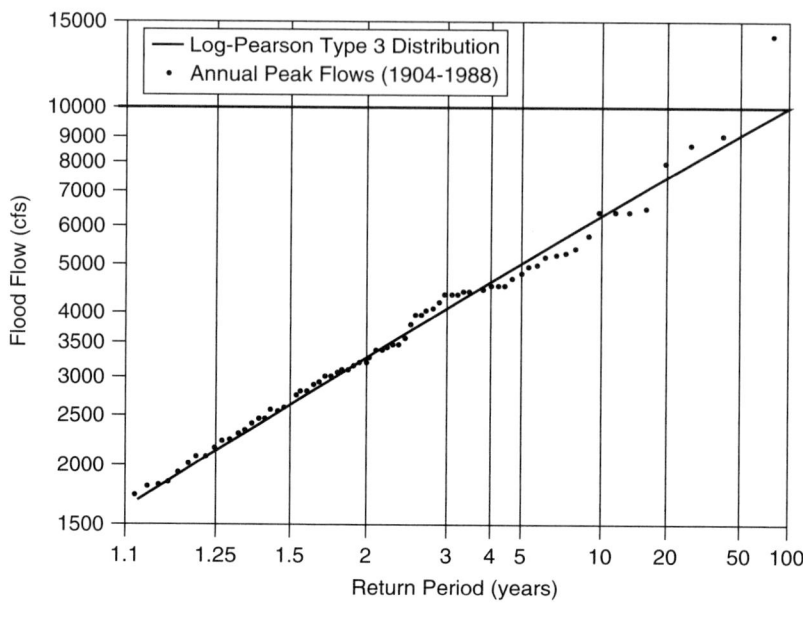
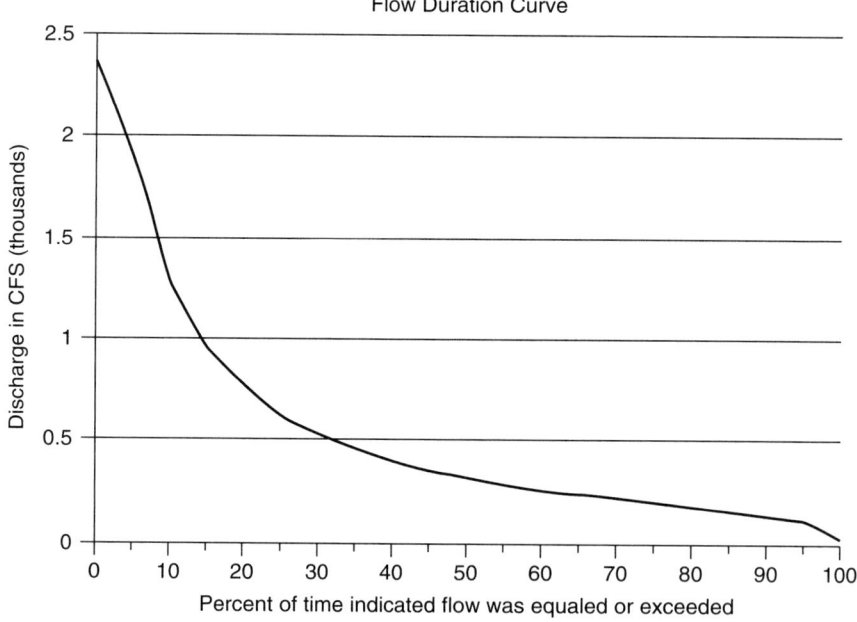

FIGURE 12-3. Typical flood frequency and flow duration curves.

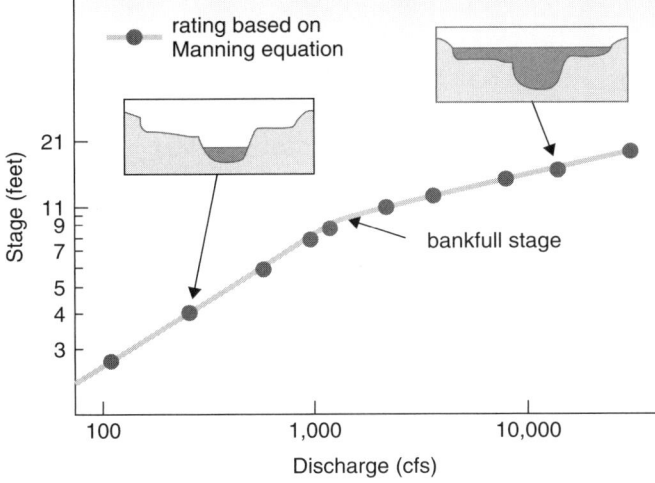

FIGURE 12-4. Determination of bankfull discharge. The discharge that corresponds to the elevation of the first depositional surface is the bankfull discharge. (*Source*: From *Stream Corridor Restoration: Principles, Processes, and Practices*, by the Federal Interagency Stream Restoration Working Group [FISRWG], October, 1998.)

The velocity of flow in open channels is determined by Manning's equation:

$$V = \frac{R^{2/3} S^{1/2}}{n} \qquad (12.1)$$

where
- V is the average velocity (meter per second, or m/s).
- R is the hydraulic radius (m), or the wetted perimeter divided by the area.
- S is the water surface slope (m/m)
- n is the coefficient of roughness.

For wide, shallow channels, the wetted perimeter can be estimated by average depth of flow. The area of the channel is found from field surveys. S is found by surveying the water surface slope over a distance equal to at least 20 bankfull widths. The coefficient of roughness, n, can be found several ways:

With discharge and velocity known at several stages, n can be back-calculated and a curve relating n to stage developed. If water depth (d) is greater than 3 times the median bed material size ($d50$), n can be estimated as:

$$n = 0.04 \, d_{50}^{1/6}, \text{ where } d \text{ is in meters.} \qquad (12.2)$$

If water depth is less than 3 times the median bed material size, then its velocity can be measured in order to back-calculate n (Newbury and Gaboury, 1993). When n is shifting from less than 3 to more than 3 times the median bed material size, refer to photographic guides or descriptive tables (see Ward and Trimble, 1995).

Manning's n is also affected by channel irregularity, variations in channel cross section, obstructions, vegetation, and meandering. A modified n corrected for variations in channel cross section can be determined by:

$$n = (n_0 + n_1 + n_2 + n_3 + n_4) \quad (12.3)$$

where
- n_0 is from channel material.
- n_1 is for channel irregularity.
- n_2 is for variation in cross section.
- n_3 is for obstructions.
- n_4 is for vegetation.
- m is an adjustment for meandering.

Ward and Trimble (1995) provides a table for the various ns.

Discharge is velocity times cross-sectional area, or:

$$Q = AV \quad (12.4)$$

Bankfull discharge can also be found using Manning's equation as follows:

- Find the cross-sectional area of the stream at bankfull depth.
- Use Manning to compute velocity.
- Multiply velocity by cross-sectional area to find bankfull discharge.

Restored channels are designed to handle bankfull discharge within their banks. Larger storms must be accounted for, as well as baseflow conditions. For extreme storm events, regulations will not normally allow any increase in flood stage. Typically, 5-, 10-, 25-, 50-, and 100-year floods are analyzed. Larger floods may have to be analyzed in sensitive locations. A very effective tool for analyzing water surface elevation in flood events is the River Analysis System (RAS) of the U.S. Army Corps of Engineer's Hydrologic Engineering Center (HEC) in Davis, California. A HEC-RAS model of the restoration reach is frequently required by local authorities before a permit for channel modification is approved. HEC-RAS performs one-dimensional hydraulic calculations for a full network of natural and constructed channels (Brunner, 2008). The model uses the energy, momentum, and continuity equations to solve for water surface elevations. Required data are cross sections at pertinent points in the stream, geometry of obstructions such as bridges and culverts, Manning's n, and discharge.

The process for hydrologic design for channel restoration is:

- Develop flood frequency and flow duration curves.
- Determine bankfull discharge.
- Design a channel competent to pass the bankfull discharge.
- Verify that flood stage is not increased for 5-, 10-, 25-, 50-, and 100-year floods.
- Accommodate baseflow conditions.

Streams transport sediment as well as water. The sediment load of a stream and the stream's competence to carry that load are the subject of the next section.

SEDIMENTOLOGY

Stream sediment transport is normally divided into suspended load and bedload. Suspended load is mostly washload, defined as the fine particle sizes that originate in the watershed and are not normally found in the bed materials of the stream. Suspended load is predominantly clays, but may include particle sizes up to fine sand. Bedload is the larger material that moves along the stream bottom by sliding, rolling, or saltating along the substrate surface (Waters, 1995). The suspended load of a river is generally controlled not by the river's transport capacity, but by the rate of supply from the basin. The source of bedload is dominantly a function of the stream's transport capacity (Knighton, 1998). Biologists are mostly concerned with suspended sediment concentration and the degree of sedimentation. However, sediment transport and channel stability must be considered first.

Tractive force is a general measure of average shear stress along a channel bed. Tractive force is a function of the unit weight of water, average depth, and channel slope:

$$\tau = \omega ds, \qquad (12.5)$$

where:
    τ is the tractive force (kg/m^2).
    ω is the unit weight of water (kg/m^3).
    d is the average depth (m).
    s is the slope of the water surface.

Since water weighs 1000 kg/m3, tractive force can be expressed as: $\tau = 1000ds$.

For non-cohesive bed materials greater than 1 cm in diameter, tractive force (kg/m2) equals the diameter (cm) of bed particles just starting to move, or the incipient diameter (Newbury, 1993). If the tractive force is 18 kg/m2, then particles on the streambed smaller than 18 cm diameter will be transported. This relationship may be used to evaluate the competence of a stream to move sediment. It may also be useful for sizing of stone for in-stream improvements.

The stream channel must be competent to transport its sediment supply through the restoration reach. Too little tractive force will cause deposition of material in the channel, while too much tractive force will result in erosion. Typically, the allowable tractive force is selected to transport the median size particle (d_{50}) expected in the restored stream reach.

Streambed particles include sand, gravels, cobbles, boulders, logs, and other debris that offer resistance to flow. A cumulative frequency curve of bed paving material is useful in determining the stability of a streambed. A simple way to sample bed paving material is to walk randomly through the reach, stopping every few steps to feel the bottom of the stream and measure the mean diameter of materials projecting into the flow (Newbury and Gaboury, 1993). Particles are measured in their longest, shortest, and intermediate diameters and averaged to get a representative diameter. Be sure the sample is representative of the reach—including riffles, pools, glides, runs, etc.—by estimating the percentage of each component and collecting a corresponding percentage of particles in each. The walk should include the entire bankfull width, both above and below the water surface. Rosgen (1996) recommends a slightly more rigorous approach of establishing cross sections at representative sections of the reach and collecting a specified number of samples equally spaced across each cross section. Either way, the diameter data are then plotted against frequency of occurrence in percent. The d_{50} particle in this plot is the one that is larger than 50 percent of the particles sampled, d_{84} is larger than 84 percent of particles, and so forth.

Tractive force is good for preliminary sizing of channels. Analysis of sediment transport will verify the competence of the channel. The U.S. Army Corps of Engineers' HEC-RAS program now contains a sediment transport component. This component simulates one-dimensional sediment transport resulting from scour and deposition over moderate time periods. The Bureau of Reclamation (U.S. Bureau of Reclamation, Technical Service Center, Sediment and River Hydraulics Group,

Denver, Colorado) currently supports the Sedimentation and River Hydraulics – One Dimension (SRH-1D) model that simulates changes to rivers and canals caused by sediment transport. A sediment rating curve at the influent boundary will be needed for either model to effectively simulate the system.

The best way to find sediment load is to measure the concentration at different flow levels. Suspended sediment can be measured fairly easily from water samples. Bedload must be measured with special bedload samplers. Because sampling is difficult and can be dangerous in floods, little data are available for high-flow conditions. There are several equations available for computation of bedload, including those by Einstein, Yalin, Yang, duBoys, and Schoklitch (Haan, 1995). These equations have specific assumptions that should be evaluated before deciding which to use.

The tools for managing bedload in streams are bankfull depth and channel slope. If the channel is degrading, then reducing bankfull depth will reduce stream power, and fewer particles will be transported. Reducing slope by lengthening the channel will also reduce stream power. Stream restoration requires replicating the geometry of natural systems in plan, cross section, and profile so that the stream is competent to carry both its hydraulic and sediment loads. The field of geomorphology provides the necessary knowledge to replicate natural conditions.

GEOMORPHOLOGY

Researchers have developed several empirical relationships among stream characteristics. Typically, these relationships take the form of power functions such as $X = aYb$, where X is a dependent variable, Y is the independent variable, and a and b are constants. Geometric relationships can help develop natural channel geometry. However hydrologic and sediment transport needs must be met. Empirical relationships are only a starting point in design.

Within a river system, width, depth, and velocity are frequently related to mean annual discharge and bankfull discharge. The functions take the form:

$$W_{bf} = eQ_{bf}^{f}; \quad D_{bf} = gQ_{bf}^{h}; \quad V_{bf} = iQ_{bf}^{j} \qquad (12.6)$$

where W, D, and V are width, depth, and velocity, respectively. The coefficients e, g, and i and the exponents f, h, and j are derived through regression. Data necessary to develop the regression are bankfull width and depth at several locations on the stream, water velocity,

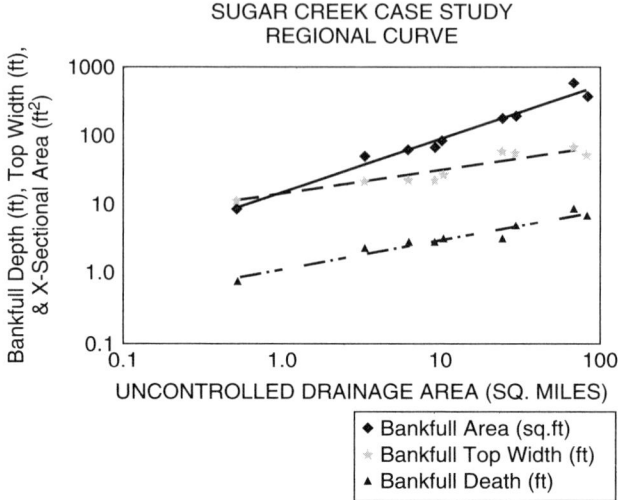

FIGURE 12-5. The relationship between bankfull channel characteristics and drainage area can be plotted regionally and used as a preliminary estimate of design cross section. (*Source*: Natural Resources Conservation Service, National Water Management Center.)

and stream discharge. Manning's equation is used to develop velocity and discharge. The power function can be developed with simple spreadsheet programs. This relationship may not hold between river systems nor between distinctly different geologic formations within a given river system. The relationship is tighter when streams of the same classification are compared.

The ratio of bankfull channel width, depth, and cross-sectional area to drainage area (Figure 12-5) is another useful relationship. These ratios are most useful if they are developed regionally.

Natural channels in erodible materials have a fairly predictable distribution pattern of pools, riffles, and sinusoidal meanders. This pattern is related to bankfull depth. A full meander wavelength includes two riffles and two pools (Figure 12-6). Meanders repeat themselves over a length equal to 7 to 15 bankfull widths. These ratios should be verified in project watersheds. Pools form at the meander bends, and riffles at the inflection point between bends. Sinuosity is the ratio of stream length measured along the thalweg, or flow way, divided by the valley length. More sinuous streams use more distance to cover a given length along the valley, and hence have less slope. Sinuosity modification is one tool for modifying a stream's erosive power.

Floodplains are the depositional area adjacent to stream channels. During storm events, when depth exceeds bankfull, the flow spreads

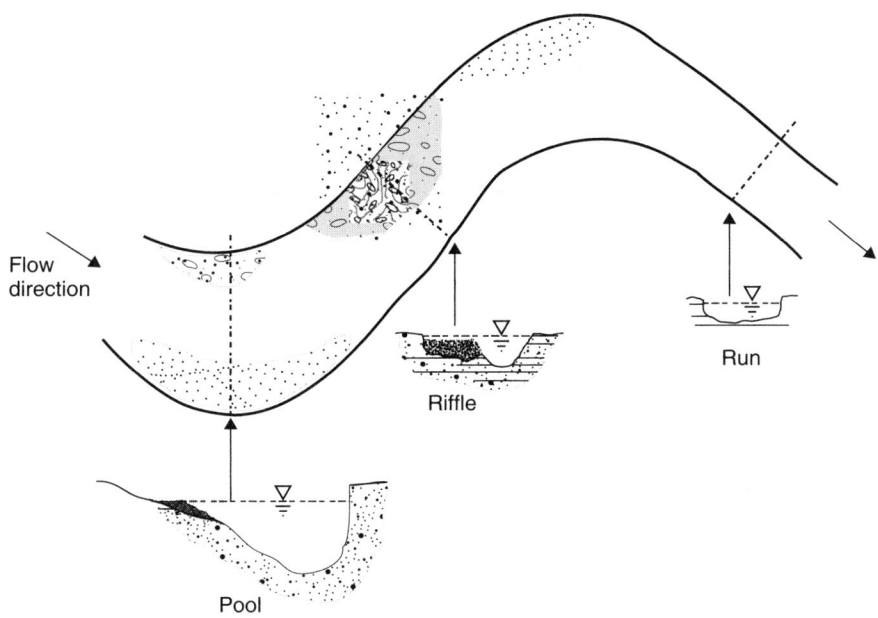

FIGURE 12-6. Typical stream meander pattern. (*Source*: EPA.)

over the floodplain. When the flow spreads, average depth decreases, and so does stream power. The ability of a stream to spread onto its floodplain provides relief from erosive conditions in the channel. Since the flow spreads over more area, velocity also subsides. With reduced velocity, suspended sediment settles, reducing the stream's pollutant load. The floodplain also provides refugia against high flows for many species. Maintaining or restoring the floodplain should be a priority of any channel restoration project. When a project is restricted to a given corridor, it may be advantageous to provide a two-stage channel. An inner stage is designed to carry bankfull discharge and sediment, and an outer channel is provided to simulate the floodplain. The width of the floodplain bench should be 3 to 5 times the bankfull width.

Localized streambank failure can be a source of excessive sediment in streams and also a problem for riparian or streamside landowners. Banks may erode as moving water removes soil particles, or by collapse. Collapse occurs when the strength of the bank is too low to resist gravitational forces (FISRWG, 1998). Frequently, collapse occurs as floodwaters recede. During the flood, hydrostatic pressure in the saturated soil of the bank is resisted by water in the channel. As the waters recede, the resistance to pressure in the bank is reduced, and collapse occurs. Rosgen (1996) provides an effective analysis of bank failure hazard.

Several tools exist to counter bank erosion. If the problem is erosion by moving water, then the flow of the stream can be moved away from the bank with wing deflectors or vegetative revetments. If the bank is failing due to collapse, then bank height can be reduced, bank angle or slope reduced, and toe protection provided. Adding root mass by vegetating the bank will also provide reinforcement and increase the soil's resistance to collapse. The Federal Interagency Stream Restoration Working Group (FISRWG, 1998) gives details on bank protection practices.

HABITAT

Habitat is a combination of the physical and biological characteristics of an area or areas that are essential for meeting the food and other metabolic needs and the shelter, breeding, and overwintering requirements of a particular species (FSSSWG, 2008). The scale of habitat for different species may vary from the interstitial space between gravels in a single riffle to entire river systems, and even oceans in the case of anadromous fish such as king salmon. Because aquatic communities are complex interconnected systems, it is seldom satisfactory to develop habitat for a particular species. Instead, an ecosystem approach considering the needs of all species is needed.

Different aquatic organisms need differing water flow conditions. Organisms may need shallow, fast water such as riffles or cascades, or they may need deeper, slower water such as a pool. Organisms may also select different habitat during different stages of their life cycle or during different times of the diurnal cycle. Natural streams consist of a variety of riffles, pools, glides, and runs that meet the flowage needs of the community. Glides are the shallow, slow parts of a stream leading up to a riffle. Runs are the relatively deep and fast parts of the stream coming out of a riffle. A channel design that incorporates channel complexity and simulates the natural stream conditions has many of the characteristics needed for habitat.

Fish and other stream biota also have microhabitat needs. Does the substrate provide spawning beds for fish and refugia for benthic organisms? Is there adequate overhead cover to provide refugia for larger species? Is there enough variability in the current to provide desired conditions for all species? One way to evaluate the variability of current in a stream reach is to plot on a sketch of the reach the different values of Froude's and Reynold's numbers for the different components of flow.

A given discharge in an open channel will be subcritical, slow, and deep, or supercritical, fast, and shallow, depending on channel slope and cross section. The state of flow, subcritical or supercritical, is characterized by Froude's number:

$$Fr = \frac{v}{(gd)^{1/2}}, \qquad (12.7)$$

where
  Fr is Froude's number.
  v is velocity in (m/s).
  g is acceleration f gravity (m/s/s).
  d is depth of flow (m).

When Froude's number is less than 1, flow is subcritical, and when greater than 1, flow is supercritical. Stream flow makes a smooth transition from subcritical to supercritical, but the transition from supercritical to subcritical is abrupt and turbulent. During low flow conditions, flow in pools is subcritical. At riffles, the flow transitions from subcritical to supercritical, and then the flow transitions back to subcritical below the riffle. It is the turbulence in the transition that entrains oxygen in the water. The pool and riffle pattern provides multiple habitats suitable to different species.

Reynold's number is a dimensionless number that describes the scale of turbulence of flow relative to a characteristic body length of depth. Reynold's number is the ratio of velocity of flow times the body length or depth to the kinematic viscosity of the fluid (Newbury, 1993). Reynold's number can be used to study the forces that act on fish or other organisms in a current.

$$\mathrm{Re} = \frac{VD}{v}, \qquad (12.8)$$

where
  V is the velocity (m/s).
  D is the depth (m).
  v is kinematic viscosity (m²/s).

Froude's and Reynold's numbers give insight into variability of flow in a stream. Froude's number indicates subcritical or supercritical flow. Reynold's number is a measure of turbulence. Variety in the numbers and the amount of change at transition points indicate variability of stream flow.

CONNECTIVITY

As with the hydrology and geomorphology of a watershed, everything is connected in the biota of a stream. In most river systems, there is movement of organisms both upstream and downstream. Benthic invertebrates are observed to continually drift downstream, but the community persists upstream. Some fish and amphibians are known to move up- and downstream with regularity, depending on seasonal needs or spawning cycles. Disturbance to the system can occasionally cause local extinctions of some species. When local extinctions occur, replacements are recruited from other stream reaches. Structures such as dams, culverts, and weirs may block passage of aquatic organisms and fragment habitat within the system.

Organisms use various strategies to move upstream. Some fish are strong swimmers and can move against swift currents. They will use eddies behind instream obstructions for resting and easing their way upstream, much as a kayaker does. Other organisms are not such strong swimmers and may need to crawl or swim along in back eddies along the shoreline, or move through the relatively slow current along the shore. Some species of mussels move upstream in their juvenile stage by attaching to the gills of fish and hitching a ride.

Culverts at road crossings frequently create fish blockage because they are smooth and frequently have shallow depth during baseflow. During higher discharge, water velocity may be too great for fish to move all the way through the culvert without exhausting themselves. Scour at the exit of the culvert may also cause a vertical grade adjustment that cannot be passed by organisms.

For fish, culverts can be designed for passage if the design provides water depth and velocity at low flow that will allow fish movement. The swimming speed of fish is classified as sustained (can be maintained indefinitely), prolonged (can be maintained for 200 minutes), or burst (can be maintained for 165 seconds). In some cases, instream obstructions are designed into the floor of the culvert to allow resting during the upstream traverse. Tables of the sustained speed and burst speed of some fish species are available (Newbury et al., 1993). For existing culverts, passage is frequently enhanced by partially backflooding the culvert with a grade control structure on the downstream side. The grade control structure must be designed for fish passage as well.

Designing a culvert for passage of the entire stream community is difficult. Even for fish, the prolonged and burst speeds of most species are not well documented. In addition, the culvert does not provide slow shallow zones along the edges to allow for movement of slower species and those that are not strong swimmers. Open-bottom culverts provide some opportunity for benthic organisms to move upstream, if they have sufficient zones of slow shallow water. Even with open-bottom culverts, the designer must be careful to ensure that the flow through the culvert is competent to move the sediment supplied by the upstream reach, or deposition will be a significant issue.

The U.S. Forest Service uses a stream simulation approach to road crossings (FSSSWG, 2008). In this approach, crossings are constructed to simulate the conditions of the stream in an upstream reference reach. Crossing structures may be larger with the simulation approach, but maintenance is greatly reduced because the channel through the crossing is self-regulating. The problem of fragmentation could be reduced if more agencies would adopt the stream simulation approach to crossings.

RIPARIAN CORRIDOR

Ecosystem functions of the riparian corridor include habitat, barrier, conduit, filter, source, and sink functions. A stream in dynamic equilibrium provides many more ecosystem services than transport of water and sediment. The riparian corridor is the vegetated zone along streams that forms the transition from aquatic to terrestrial habitat. Ecological functions provided by riparian corridors include (FISRWG, 1998):

- Habitat: Spatial structure that allows species to live, reproduce, feed, and move
- Barrier: The stoppage of materials, energy, and organisms
- Conduit: Transport of materials, energy, and organisms
- Filter: Selective penetration of materials, energy, and organisms
- Source: Production of materials, energy, and organisms
- Sink: Storage of materials, energy, and organisms

Healthy riparian corridors are known to reduce up to 50 percent of nutrients and pesticides, 60 percent of certain pathogens, and 75 percent of sediment from runoff into streams (NRCS, n.d.). Riparian corridors may be grass or wooded, but should match the native vegetation.

The required width of a riparian corridor is dependent on the ecological service desired. A generally accepted minimum width is 18 meters (Barbour et al., 1999). However, water quality protection may be obtained with corridors of 10 meters or less. If the corridor is expected to provide a conduit for movement of interior forest species, then the corridor should be wide enough to create interior forest conditions. In cases where it is desirable to provide passage for upland species as well as riparian species, the riparian corridor needs to include the floodplain and a portion of the upland fringe.

Considerations for riparian corridors other than width are their connectivity and the impact of edge habitat. Connectivity implies near continuous buffer. Gaps reduce the barrier effect and may interrupt movement along the corridor. Edges may be either abrupt of gradual. Gradual edges occur more often in natural settings and provide more diversity and encourage movement between the upland and the riparian area (FISRWG, 1998). Frequently, a three-zone buffer is recommended. Zone 1 is unmanaged native vegetation adjacent to the stream. Zone 2 is managed native vegetation, and zone 3 is a transition between native species and cropland or pasture. Riparian vegetation is an important part of the stream corridor. A stream cannot be truly restored unless the corridor mimics native conditions.

CONSTRUCTION

Stream restoration is a very specialized field, and adjustments to plans are to be expected. Do not expect general contractors to have the expertise to complete a stream restoration project without direct supervision by the project manager.

Even with good design and proper implementation, no one can predict nature. Stream restoration projects by their nature are evolving systems (everything is changing). The project needs a source of funding for follow-up inspection and maintenance for several years.

SUMMARY

Streams cannot be restored, but they can be returned to a functioning condition that provides desired ecological services. Assessment of stream condition should precede stream restoration efforts. Stream degradation can be local, or it can be systemic. If the damage is systemic, stream restoration should be done only in light of the systemic problem.

Bankfull or channel-forming flow occurs at the point of incipient flooding. The channel must be competent to transport bankfull flow and also transport the load of sediment provided by the watershed and upstream reaches. Hydrologic geometry relationships from reference reaches in the same river system or similar river systems may be used to make preliminary estimates of channel and cross-sectional geometry. The estimates must be analyzed with physically based models or methods.

Natural channel and cross-sectional geometry provide the foundation for aquatic habitat. Microhabitat conditions can be established by paying attention to substrate makeup, channel turbidity, and overhead cover. Maintaining connectivity throughout the stream network is essential for fully maintaining aquatic diversity in the system. The riparian corridor is integral to a healthy stream. Criteria for riparian zone width and vegetation depend on the ecological services expected. A multidisciplinary team is needed to assess all criteria in stream restoration projects.

> Some engineers are beginning to have a feeling in their bones that the meanderings of a creek not only improve the landscape but are a necessary part of the hydrologic functioning.
>
> —Aldo Leopold

Further Readings

Haynes, H.B. Noel, *The Ecology of Running Waters*, University of Toronto Press, Toronto, Canada, 1970.

Leopold, Luna B, *A View of the River*, Harvard University Press, Cambridge, MA, 1994.

References

Barbour, Michael T., Jeroen Gerritsen, Blaine D. Snyder, James B. Stribling, *Rapid Bioassessment Protocols for Use in Streams and Wadeable Rivers: Periphyton, Benthic Macroinvertebrates, and Fish*, 2nd ed., EPA 841-B-99-002, U.S. Environmental Protection Agency, Office of Water, Washington, DC, 1999.

Brunner, Gary W, *HEC-RAS River Analysis System Hydraulic Reference Manual*, U.S. Army Corps of Engineers Hydrologic Engineering Center, Davis, CA, 2008.

Bureau of Reclamation, *Erosion and Sediment Control Manual*, U.S. Department of the Interior, Bureau of Reclamation, Technical Service Center, Sedimentation and River Hydraulics Group, Denver, Co, 2006.

FISRWG, *Stream Corridor Restoration: Principles, Processes, and Practices*, Federal Interagency Stream Restoration Working Group, GPO Item no. 0120-A, SuDocs no. A 57.6/2:EN 3/PT.653, October, 1998.

FSSSWG, *Stream Simulation: An Ecological Approach to Providing Passage for Aquatic Organisms at Road-Stream Crossings*, United States Department of Agriculture, National Forest Service, National Technology and Development Program, Forest Service Stream-Simulation Working Group, San Dimas, CA, 2008.

Haan, C.T., B.J. Barfield, and J.C. Hayes, *Design Hydrology and Sedimentology for Small Catchments*, Academic Press, San Diego, CA, 1994.

Knighton, David, *Fluvial Forms & Processes*. Oxford University Press Inc, New York, 1998.

Lane, Emory W., The importance of fluvial morphology in hydraulic engineering, *Proceedings, American Society of Civil Engineers*, 81, Paper 745, 1955.

Leopold, Luna B., *A View of the River*, Harvard University Press, Cambridge, MA, 1994.

Newbury, Robert W., and Marc N. Gaboury, *Stream Analysis and Fish Habitat Design: A Field Manual*, Newbury Hydraulics Ltd. and The Manitoba Habitat Heritage Corporation, Gibson, BC, Canada, 1993.

NRCS, n.d., Buffer Strips: Common Sense Conservation, www.nrcs.usda.gov/FEATURE/buffers/ (accessed December 6, 2009).

Rosgen, Dave, *Applied River Morphology*, Wildland Hydrology, Lakewood, CO, 1996.

Simon, Andrew, A model of channel response in disturbed channels, *Earth Surface Processes and Landforms*, 14(1), 1989.

Ward, Andrew D., and Stanley Wayne Trimble, *Environmental Hydrology*, CRC Press, Boca Raton, FL, 1995.

Waters, Thomas F., *Sediment in Streams: Sources, Biological Effects and Control*, The American Fisheries Society, Monograph 7, Bethesda, MD, 1995.

13

Designing Ecosystem Services by Landform

> To waste, to destroy our natural resources, to skin and exhaust the land instead of using it so as to increase its usefulness, will result in undermining in the days of our children the very prosperity which we ought by right to hand down to them amplified and developed.
>
> —Theodore Roosevelt

INTRODUCTION

Each landform in the biosphere is host to a suite of ecosystem services characteristic of the processes unique to that landform. Landforms are arbitrary characterizations of the landscape, associated with the primary vegetative community or physical characteristics of a site. Landforms include agricultural lands, forests, grasslands, wetlands, coastal marshes and estuaries, and urban areas. These are not biomes, but are discrete parcels of land cover that can be designed at scales from very small (hectare) to very large (watershed). Landforms are interlaced on a landscape across scales. A given ecoregion might have all of these landforms distributed in different ways, with patches of meadows and woodlands dispersed across a predominantly urban landscape, for example. This system might include a stream with streamside wetlands. The complexity of the landscape is where the power of integrated ecological engineering design resides; there are almost always opportunities within a landscape to enhance ecological services by small incremental changes in landforms. The process of designing those changes is as complex as the landscape, and as resilient.

Ecosystem Services Design Process

Designing ecosystem services is the process of designing the landforms that encompass those services. This process requires a logical progression of design decisions based on competing demands at a site. The ecotechnology design team, working through the Ecosystem Services Advisory Committee (ESAC), should prioritize ecosystem services at a site. These ecosystem services can usually be provided within multiple landforms, so the next step in this progression is to prioritize landforms. The critical species for each ecosystem service and landform should then be defined, and the criteria for minimum viable population (MVP) calculated. These criteria include patch size, metapopulation connectivity, resource criteria such as net primary production (NPP) and net community production (NCP), and food web connections.

Optimizing the design of ecosystem services is more complex and subjective, because there are usually many ways to achieve design goals within a site. The strategy for species restoration (introduction, reintroduction, and augmentation) should be identified for each critical species. The species pool for each critical species should be characterized based on the restoration goals. With these criteria, the design process begins. Critical habitat should be mapped for each species. Minimum areas for patch sizes for each landform should be calculated. Maximum distances between those patches that support metapopulation interactions for each species should be identified.

Through this sequence of design activities, the Ecotechnology Design Team can inventory the landform options, overlay those options with current landforms, and develop a set of design strategies for a site. These design strategies should include the current ecosystem services sustainability index (ESSI), potential ESSI, and design ESSI for the site and at least one spatial level above the site (see Chapter 3). If the project site is a parcel within a watershed, ESSI should be characterized for both the parcel and watershed. If the project site is a watershed-scale project, these parameters should be calculated for the watershed and ecoregion. This process of prioritizing ecosystem services is often confused with the process of maximization.

Ecological services can be categorized by their utility as well as by the functions outlined in Chapter 2. These uses can be helpful in prioritizing ecosystem services design. *Direct uses* services are those that we consume or convert directly, and include food, fuel, feed, fiber, and water. *Indirect uses* include those services we benefit from indirectly, such as biodiversity, nutrient cycling, water treatment, and disinfection. *Option uses* include future discoveries such as pharmaceuticals and crop genetics. *Bequest uses* include intergenerational and sustainable development. *Existential uses* include aesthetic, cultural, and spiritual uses. Within each landform, each category can be inventoried and prioritized. Remember that everything is connected; the distinctions between landforms illustrate their similarities more than their differences.

AGRICULTURAL LANDS

Agricultural lands are the largest biome on Earth (see Chapter 1). This land use category is quite complex. Agricultural practices range from very low intensity range management to very high intensity polyculture. The range of energy subsidy is large, the geographic range is broad, and thus the net area productivity range is also large. Developing a list of ecosystem services for agricultural lands requires that the ecotechnology team understand the production systems they are analyzing, and the processes that are dominant in producing food, feed, fiber, and fuel.

Ecosystem stresses from agricultural production include water use (surface and groundwater, as well as rainfall), water pollution (sediment, nutrients, pesticides, and bacteria), and land conversion. These topics are addressed more explicitly in Chapter 16, in the context of sustainability metrics. The expanding biofuels economy is exacerbating many agricultural production stresses, because it represents a significant new production pressure from the landscape. Ecosystem services from agricultural landscapes represent a tension between increases in one category and decreases in another, with the land surface as the production limit. This trade-off has been referred to as "ecosystem dis-services" (Zhang et al., 2007), and represents a real cost of human activities on the landscape.

The categories of ecosystem services on the agricultural landscape include supporting, provisioning, regulating, and cultural (Figure 13-1). Provisioning services are most apparent, and include food (human consumption), feed (animal consumption), fiber (cotton, wool, hemp, rayon, etc.), and fuel. Supporting services include soil structure and fertility, nutrient cycling, water provision, and genetic biodiversity. The regulating services include soil retention, pollination, waste processing, pest control, refugia, water purification, and atmospheric regulation (greenhouse gases). Additional nonmarket services include water supply as an ancillary benefit from irrigation water development, soil conservation, climate change mitigation, wildlife habitat (refugia, feed, and water), and aesthetic landscapes. The ecosystem disservices include pest damage associated with large areas of land under monoculture, competition for water use, competition for pollinators, habitat loss, nutrient and pesticide pollution, soil erosion, and habitat fragmentation (Zhang et al., 2007).

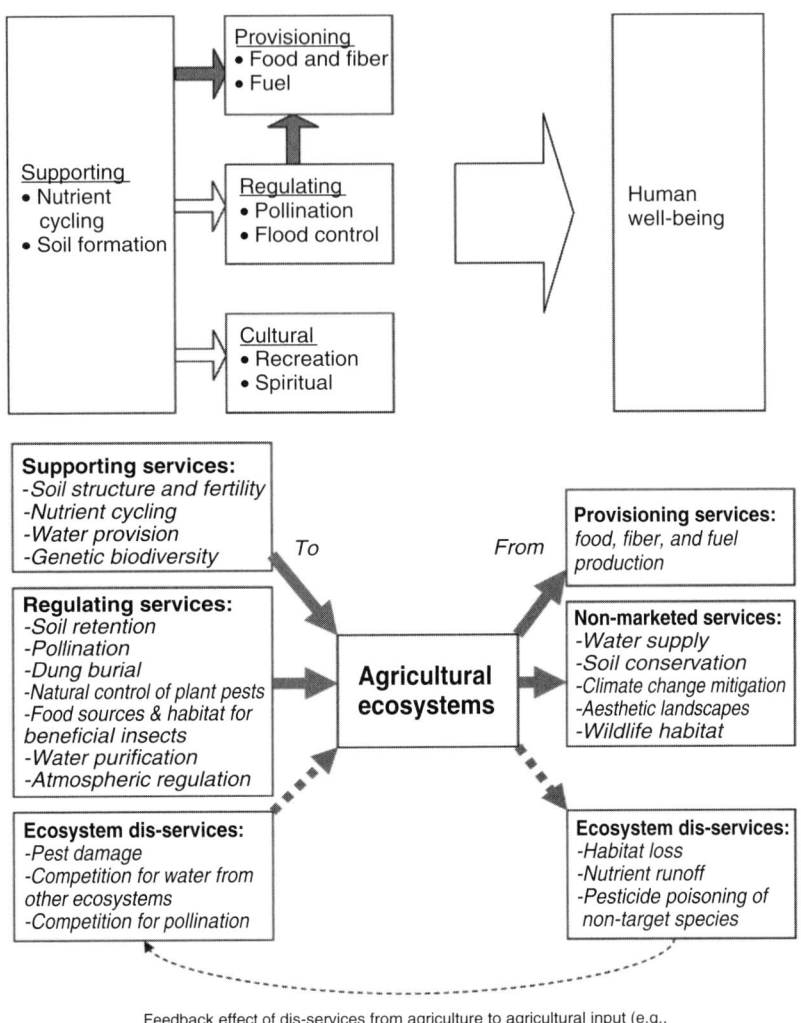

FIGURE 13-1. Ecosystem services from agricultural lands. (*Source*: From Zhang et al., 2007, with modification.)

Ecosystem services on agricultural lands are scalar functions. Processes and ecosystem guilds that drive ecosystem services differ at site (field or farm), watershed (landscape), ecoregion, and biome scales (Table 13-1). The density of services (process per square meter per unit time) is highly variable. Local measurements for each type of production system are critical for inventorying and designing those services. Additional consideration within a site should be given to the

TABLE 13-1 Ecosystem Services on Agricultural Lands across Multiple Scales

Ecosystem Service	Geographic Scale			
	Site	Watershed	Ecoregion	Biome
Soil Fertility and Formation	Microbes Invertebrates Cover crops Legumes	Hydrology Weathering Vegetative cover	Geology Topography	Climate
Pollination	Invertebrates Bees Moths Others Bats	Refugia for invertebrates Food supply Predators	Geology Topography	Climate
Pest Control	Predators Parasitoids Wasps Spiders Birds Bats	Refugia for predators Food supply	Geology Topography	Climate
Pest Damage	Insects Snails Birds Mammals	Invasive species	Invasive species Topography	Climate
Provisioning Food Feed Fiber Fuel	Land cover Soil fertility Water availability Pest pressure	Cumulative activities within the watershed	Invasive species Topography	Climate
Hydrologic flow	Vegetative cover Slope Soil type	Hydrology Weathering Vegetative cover	Topography	Climate
Water Treatment	Vegetative cover Soil infiltration Vegetation on edge of field Slope Soil type	Land use distribution across the watershed	Geology Topography	Climate
Climate Regulation	Vegetative cover	Vegetative influence on microclimate	Topography	Climate

Source: Modified from Zhang et al., 2007

differences in services provided by different intensities of agricultural production. Range management for grazers has a specific impact that differs from that of viticulture or orchard production, or row crop production. Agricultural ecosystem services derive directly from the soil; understanding soil processes is critical for assessing those services (see Chapter 7). Soil organic matter provides as much as 50 percent of plant nitrogen demand, and the fuel for microbial and invertebrate activities in the soil (Swinton et al., 2007).

The structure, composition, and function of agricultural landscapes provide context for indicators of ecosystem services (Dale and Polasky, 2007). Vegetative cover and distribution provide community structure for metapopulations. Regional topography and hydrology define corridors and connectivity. The challenge for the ecotechnology team is to determine those indicators that characterize the complexity of the system but have adequate simplicity for design and communication purposes (Figure 13-2).

Dale and Polasky (2007) recommended that ecological indicator criteria for measuring ecosystem services on agricultural landscapes must be easily measured. Remotely sensed indicators are most desirable from this perspective. Examples include vegetative cover, connectivity, and land use change. They recommended that indicators must be sensitive to changes in the system. Agricultural systems are characteristically in disturbed states. The soil is tilled, crops are cultivated for pest control, and crops are rotated within and between seasons. Ecosystem services indicators must be sensitive enough to changes in agricultural ecosystems to detect shifts in community structure and function, yet robust enough to filter the effects of continuous

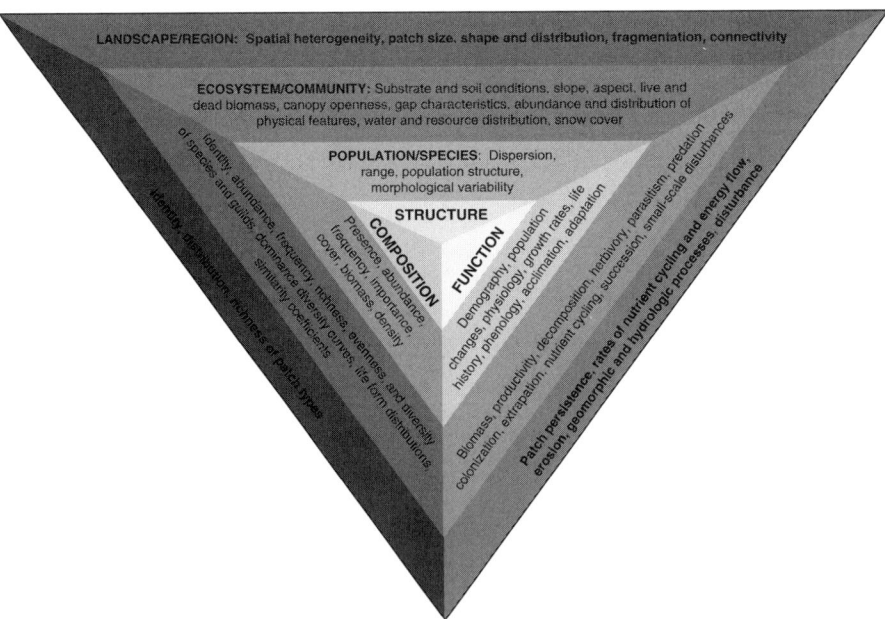

FIGURE 13-2. Ecological system indicators as a function of structure, composition, and function. (*Source*: Modified from Dale and Polasky, 2007.)

disturbance. This is one reason that measuring and assessing ecosystem services status in agricultural landscapes is so difficult.

Dale and Polasky (2007) recommended that indicators must respond to changes in the system in a predictable manner. Ecosystems are complex, nonlinear systems. However, they often behave in linear progressions over a range of perturbations. An indicator must measure those processes that show transition, rather than state change. State change indicators are valuable for assessment, but not very useful for predicting status, because by the time they have changed it is too late to manage the system for the prechange condition. The indicators must predict changes that are associated with management practices. The indicators must signify an impending change in key characteristics, as described previously, but with clear process connections to the management practices that can be controlled. The conceptual model approach described in Chapter 3 is critical for this assessment.

Finally, Dale and Polasky (2007) recommended that indicators must have known variability within the site and watershed. An indicator must be benchmarked to the range of possible conditions, and thus that range must be understood. Soil organic matter content variability in a watershed and across an ecoregion can inform the ecotechnology team about the status of a site relative to the landscape. The range of conditions across an ecoregion can be used to frame management objectives for a site, as previously described. The indicators can be used to assess current conditions and define future objectives in a design strategy.

FORESTS

While the agricultural biomes (crop and grasslands) are the largest on Earth, almost 80 percent of Earth's biomass is in forests. Nearly 50 percent of that biomass is in rainforests, which only cover 13 percent of Earth's ice-free land surface (Kindermann et al., 2008). Three-quarters of the forests on Earth are located in the tropical and boreal biomes. Tropical rainforests are the widest range forest type and are extremely diverse in tree species. They are defined as closed-canopy evergreen broadleaf forests that usually require a minimum constant temperature of 25 Celsius and 1.5 meters of annual rainfall. A majority of the rainforests are located in South America, Africa, and Asia (Millennium Ecosystem Assessment, 2005).

Perhaps more than half of all the world's plant and animal species live in tropical rainforests (Wilson, 1999). Tropical moist deciduous forests develop in areas having a dry season a third of the year, and range from closed-canopy forests to open savanna forests, depending on the length of dry season, natural fire frequency, and anthropogenic factors. Human activities in densely canopied rainforests result in open-canopy areas, which allows for more growth on the ground level and thus increased productivity on the edges of the forests, or the forest-meadow ecotone (Millennium Ecosystem Assessment [MEA], 2005). Tropical dry forests are present in regions with long dry seasons and contribute to a small percentage of tropical forest coverage.

Temperate and boreal forests are found around the subtropic region to the arid steppes and the sub-Arctic region. These forests encompass almost 2 trillion hectares of land and are located in 55 different countries (Millennium Ecosystem Assessment, 2005). Temperate forests can grow in temperatures varying from −30 to 30 Celsius with an annual precipitation of 1.5 meters distributed evenly throughout the year. Their canopies are moderately dense and allow sunlight to penetrate, resulting in diverse species populations on the floor. The ability for many species of plants to grow implies that the soil is fertile. Boreal forests represent the largest terrestrial biome after agriculture and are found in the broad belt of Eurasia and North America. Temperatures are very low, and precipitation is mainly in the form of snow, ranging from 40–100 cm annually. Because of these conditions, the soil is thin, lacking in nutrients, and generally acidic.

Between 2000 and 2005, approximately 7.25 million hectares of forest were destroyed each year worldwide, of which more than 1.5 million hectares were in Brazil (Figure 13-3). As a result of reforestation and developing forest protection policies, deforestation rates in Europe, North America, and China have slowed in the past 10 years. Deforestation results from agricultural expansion, infrastructure expansion, and wood extraction. Agricultural expansion involves the usage of forest area for permanent crops or pasture, and is the primary driver of deforestation, contributing to 96 percent worldwide (Geist and Lambin, 2002). Infrastructure expansion includes road development and expanding urban areas, and is the dominant driver in South America and in parts of the United States. Wood extraction pertains to industrial logging, and is the major issue in Asian countries, while the use of timber for fuel dominates in Africa.

Tropical rainforest standing biomass has been estimated at as high as 290 Mt/ha (Fearnside, 2008). This is more than 15 times the biomass extracted from the most productive maize fields on Earth. The diversity of ecosystem services provided by forested landscapes is predictably very high, as is the density of those services (Table 13-2). The specific

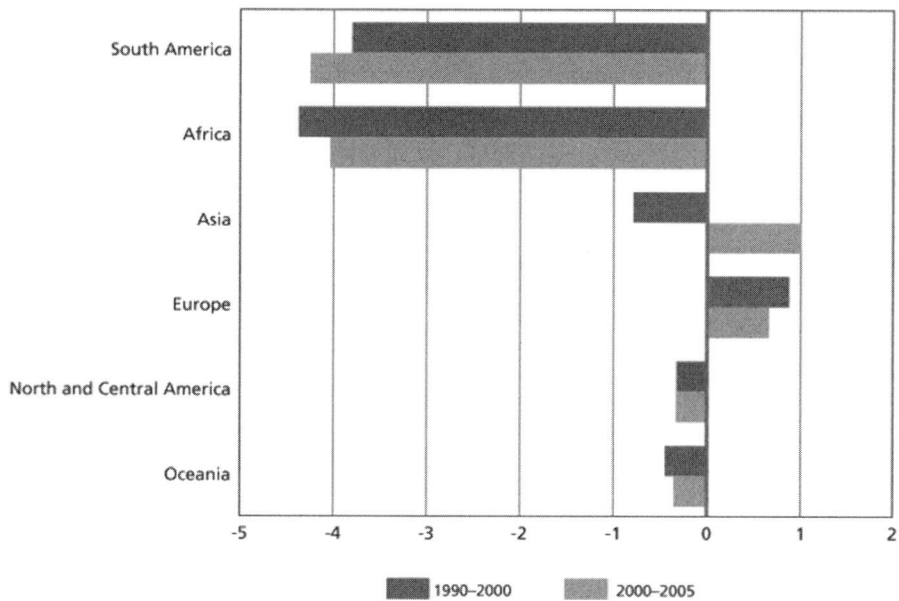

FIGURE 13-3. Net change in forest area by region (million hectares per year). (*Source*: Gleason et al., 2008.)

ecosystem services provided by a given site are a function of the characteristics of that forested ecosystem. The ecotechnology team will likely have to estimate the density and range of ecosystem services from existing literature. Care should be taken to use only data from within similar ecoregions in extrapolating forest ecosystem services data; estimates of even the most basic characteristics, such as standing biomass, can vary by a factor of three in a given forested ecosystem (Fairnside, 2008). Forests are relatively easy to design, and relatively popular to implement as landforms. The use of forested patches in urban areas can dramatically alter ecosystem services within urban landscapes, as described in this chapter.

GRASSLANDS

Grasslands (pasture and rangeland) and croplands constitute the largest biome on the planet (Foley et al., 2005). Rangelands are characterized by self-propagating plant communities, predominantly grasses, grass-like forbs, shrubs, and dispersed trees. Rangeland ecosystems are only partially subsidized ecosystems, with little direct management of plant communities other than occasional fire management

TABLE 13-2 Definition of Ecosystem Services Provided by Forests Based on Category

Category	Type	Definition
Provisioning Services	Biodiversity	Provide structural diversity, compositional diversity, and functional diversity; this is essential for sustaining ecosystem functioning.
	Wood and Non-wood products	Provide natural products such as paper, woodfuel, and industrial roundwood that humans rely on to survive
	Food	Provide a natural habitat, shelter, and food for animals. These animals are also hunted and consumed by people
Regulating Services	Climate Regulation	Trees are the largest source of carbon sequestration on the planet, which reduces global warming and provides shade which can reduce energy costs on a local scale.
	Soil and water protection	A tree's intricate root system provides protection of the soil, which stabilizes natural landscapes, thus reducing or preventing floods and landslides.
	Disease Regulation	Forests reduce the occurrence of standing water, which is a breeding area for mosquitoes, and therefore can reduce the prevalence of malaria.
	Pollination	Bees from nearby forests pollinate crops.
	Pest Regulation	Predators from nearby forests (e.g., bats, toads, and snakes) consume crop pests.
	Natural Hazard Regulation	Mangrove forests protect coastlines from storm surges. Also biological decomposition processes reduce potential fuel for wildfires.
Supporting Services	Nutrient Cycling	Transfers nitrogen, phosphorus, and carbon from plants to the soil.
	Water Cycling	Transfers water from the atmosphere through plants to the soil and vice versa.
Cultural Services	Recreational Services	Forests provide places to hunt, fish, hike, camp, and bird watch.
	Eco-tourism	Forests provide capital income through recreational services by bringing in tourists from other areas.
	Ethical Values	Old rainforests and the Redwood Forest provide spiritual, aesthetic, and other values people attach to them.

Source: From the MEA, 2005.

(Mazco and Hidinger, 2008). Pastures are more intensely subsidized systems, with inputs commonly consisting of fertilization, mowing, liming, and seeding. In the U.S., over 60 percent of the more than 310 million hectares of rangeland is on private lands (Mazco and Hidinger, 2008).

Like forests, grasslands are complex, diverse, and dynamic. Rangelands span many ecoregions, including grasslands, savannas, deserts, shrublands, alpine meadows, wetlands, and tundra (Mazco

TABLE 13-3 Definition of Ecosystem Services Provided by Grasslands Based on Category

Category	Type	Definition
Provisioning Services	Biodiversity	Provide structural diversity, compositional diversity, and functional diversity; this is essential for sustaining ecosystem functioning.
	Forage Production	Provide grasses and forbs for processing into hay for feeding livestock.
	Food	Provide a natural habitat, shelter, and food for beef and lamb for human consumption.
Regulating Services	Climate Regulation	Grasses are a source of carbon sequestration, which reduces global warming and provides organic matter to organisms in the soil. Grasslands increase soil organic matter content.
	Soil Protection	The root systems of grassland species provide erosion protection of the soil and increase water infiltration, reducing or preventing floods.
	Water Protection	Grasslands reduce overland sediment transport, and thus sediment, nutrient, and bacteria loads to rivers and streams.
	Pollination	Bees rely on meadow wildflowers and grasslands for food and shelter in order to pollinate crops.
	Pest Regulation	Birds from grasslands consume crop pests.
	Nutrient Cycling	Grasslands transfer nitrogen, phosphorus from plants to the soil and from soil to the plants.
Supporting Services	Water Cycling	Grasslands transfer water from the soil through plants to the atmosphere, reduce rainfall runoff, and increase infiltration into groundwater.
	Soil Formation	Grasslands mineralize top horizons of geologic material, creating soil.
Cultural Services	Recreational Services	Grasslands provide places to hunt, fish, hike, camp, and bird watch.
	Eco-tourism	Grasslands provide capital income through recreational services by bringing in tourists from other areas.
	Ethical Values	Grasslands provide spiritual, aesthetic, and other values people attach to them.

Source: From the Sustainable Rangeland Roundtable (SRR), 2008.

and Hidinger, 2008). The array of ecosystem services provided by grasslands is equally diverse. These services include provisioning, regulating, supporting, and cultural services (Table 13-3). Because of the geographic diversity of grasslands, as with forest and croplands, it is imperative that the ecotechnology team develop a broad suite of indicators of ecosystem services.

Designing grassland ecosystems requires definition of the indicators of successful design. Indicators for ecosystem services should comply with the criteria described previously by Dale and Polasky (2007). These indicators are both the assessment and design criteria for the ecotechnology team, and include:

1. NPP, a direct measure of provisioning services
2. Soil organic matter, a measure of carbon sequestration, nutrient cycling, and soil fertility
3. Area and extent of soil erosion, a measure of soil stabilization and water treatment
4. Water-holding capacity of the soil, a measure of infiltration and water-cycling services
5. Percent of water bodies within a watershed with impaired or degraded water quality, a measure of water treatment, or pollution, from the site
6. Fragmentation of rangeland or pasture communities and the larger landscape, a measure of refugia
7. Population status and geographic range of dependent species, a measure of biodiversity
8. Value of forage harvested from rangeland and pastures by producers and livestock, a measure of provisioning services
9. Number of domestic livestock on the grasslands, a measure of provisioning services
10. Soil compaction, a measure of soil tilth and water-cycling capacity.

These indicators represent the type of metrics that should be developed working with the Ecosystem Services Advisory Committee (see Chapter 3).

WETLANDS

Wetland ecosystems are the most productive systems on an area basis, providing a broad spectrum of ecosystem services at high spatial and temporal density (Table 13-4). Wetlands exist in every biome on Earth, covering about 9 percent of the globe (Zedler and Kercher, 2005). The definition of wetlands is difficult, since they encompass any area of land submerged under water for some period of time. Wetlands include areas referred to as swamps, bogs, marshes, mires, fens, billabongs, and numerous other parochial terms. Wetlands have

TABLE 13-4 Ecosystem Services Provided by Wetlands

Category	Type	Definition
Provisioning Services	Biodiversity	Wetlands are the most dense biodiversity landform after rainforests; they provide structural diversity, compositional diversity, and functional diversity; this is essential for sustaining ecosystem functioning.
	Fresh Water	Storage and retention of water for domestic, industrial, and agricultural use.
	Biochemical	Source of pharmaceutical and genetic material.
	Fiber and Fuel	Production of peat, wood (construction), fodder, fuelwood.
	Food	Provide a natural habitat, shelter, and feed for freshwater and marine animals used by humans for food, as well as wild game, fruits, nuts, and grains.
Regulating Services	Climate Regulation	Wetlands are a source and sink for greenhouse gases. They influence microclimates through heat sinks and sources associated with surface water and vegetation.
	Erosion Regulation	The root systems of wetland species provide erosion protection of the soil and increase water infiltration, reducing or preventing floods.
	Water Regulation	Wetlands are sinks for overland sediment transport, and thus dramatically reduce sediment, nutrient, and bacteria loads to estuaries, lakes, rivers, and streams.
	Natural Hazard Protection	Wetlands reduce peak flows from runoff events, reducing flooding and the transport energy of rivers and streams.
	Water Purification	Wetlands reduce sediment loads, treat and cycle carbon, nutrients, and bacteria, and sequester heavy metals.
	Pollination	Pollinating insects rely on wetland wildflowers and water for food and shelter in order to pollinate crops.
	Pest Regulation	Birds from wetlands consume crop pests.
	Nutrient Cycling	Grasslands transfer nitrogen and phosphorus from plants to the soil and from soil to the plants.
Supporting Services	Water Cycling	Wetlands are landscape sinks, and thus increase infiltration into groundwater.
	Nutrient Cycling	Wetlands cycle C, N and P through aerobic and anaerobic processes in the water column, soil, and biota.
	Soil Formation	Wetlands capture sediment, accumulate organic matter and nutrients, and create soil.
Cultural Services	Eco-tourism	Wetlands provide recreational services by bringing in tourists for bird watching, hunting, and fishing.
	Ethical Values	Wetlands provide spiritual, aesthetic, and other values people attach to them.
	Educational Services	Wetlands provide opportunities for people to observe the complex interactions of organisms across trophic levels, to learn about ecosystem functions, and human impacts on the land.

Source: MEA 2005, Lehner and Doll 2004.

features that are common across biomes (Mitch and Gosselink, 2000); these features include:

1. The presence of water (surface or root zone)
2. Unique soil conditions that differ from uplands (hydric soils)
3. Vegetation adapted to wet conditions (hydrophytes)

The difficulty in defining wetlands largely is due to the ubiquitous nature of water on the landscape. When is a field a wetland, or just a field after a rain, with water standing on the surface? These are not simple questions, and are addressed in great detail in *Wetlands* by Mitch and Gosselink (2000). They define a marsh as a frequently or continually inundated wetland characterized by emergent herbaceous vegetation adapted to saturated soil condition. By contrast, a bog is a peat-accumulating wetland that has no significant inflows or outflows and supports acidophilic mosses, particularly *Sphagnum*. A muskeg is a large area of peat bogs. Finally, a swamp is a wetland dominated by trees or shrubs. The most common freshwater wetland structures include marshes, swamps, riparian systems, tidal freshwater marshes, and peatlands. The most common marine and coastal systems include tidal saltwater marshes and mangrove swamps.

The U.S. Army Corps of Engineers regulates wetland removal and mitigation under federal statute, and so has developed a legal definition for wetland as well (Mitch and Gosselink, 2000). The regulatory definitions tend to change in response to legal challenges, so ecological engineers are encouraged to work closely with their local U.S. Army Corps of Engineers office or similar regulatory body when they suspect a site has wetlands concerns.

Ecological engineering in many respects evolved through the processes of designing wetlands for ecosystem services, especially wastewater treatment for biochemical oxygen demand reduction, disinfection, and nutrient removal. The process of designing wetlands for ecosystem services is as mature a discipline as designing the previous landforms, but has a more intensive hydrologic element. The ecotechnology team engaged in wetland design needs to ensure that the ecological engineer has experience in the process of wetland design for the ecosystem services of concern. Mitch and Jorgensen (2004), supplemented with Mitch and Gosselink (2000), provides a very thorough treatment of this design process.

Wetland ecosystems embody the ecological engineering axiom that everything changes. Wetlands are defined by hydrologic conditions. Alteration of hydrologic regimes (increased or decreased flows, altered timing of flows) impacts the integrity of the wetland system immediately (Zedler and Kercher, 2005). Since wetlands are landscape sinks with continuous nutrient loads and constantly disturbed regimes, they are prone to community composition shifts, especially from invasive species. Thus, wetland design must include aggressive metapopulation design to ensure that adequate propagules are transported to each node in the wetland design, in the event that critical species are weakened or eradicated from a site.

Wetland design should be incorporated into every ecosystem services design, when possible. Some have estimated that as much as half the wetlands on Earth have been destroyed or seriously altered in the past 1,000 years (Zedler and Kercher, 2005). Freshwater wetlands, especially emergent and riparian, are disappearing most rapidly, probably as a result of the arable soil associated with these sites. Global losses of wetland ecosystem services have been remarkable; in Asia, more than 500,000 hectares of wetlands are lost annually, usually through agricultural land conversion (Zedler and Kercher, 2005).

Much of the work of ecological engineers in the past decade has been associated with wetland restoration and mitigation. This process of restoring wetlands means that the ecotechnology team must be able to both index ecosystem services and diagnose impacts of site and watershed activities on wetland processes. Because wetlands are landscape sinks, they are also accumulators of toxics. Wetlands are prone to bioconcentrate metals and nonionic organic compounds (especially legacy pollutants such as polycyclic biphenols and dioxins). Wetlands across most biomes are degraded from a number of impacts (Zedler and Kercher, 2005); these can be categorized as:

1. Geomorphic and hydrologic regime alteration
2. Nutrient and sediment contamination
3. Plant community alteration
4. Climate change impacts

The cause and degree of impact determine the approach that will most probably be successful in restoring ecosystem services. No ecosystem can be restored, as explained in Chapter 8; however, functions can be recovered (Gleason et al., 2008). As wetland alteration increases, from unaltered to highly altered disturbance conditions, ecosystem services decrease (Figure 13-4). Water quality and biodiversity decline early in the disturbance gradient, followed by soil erosion reduction, carbon sequestration, and floodwater storage.

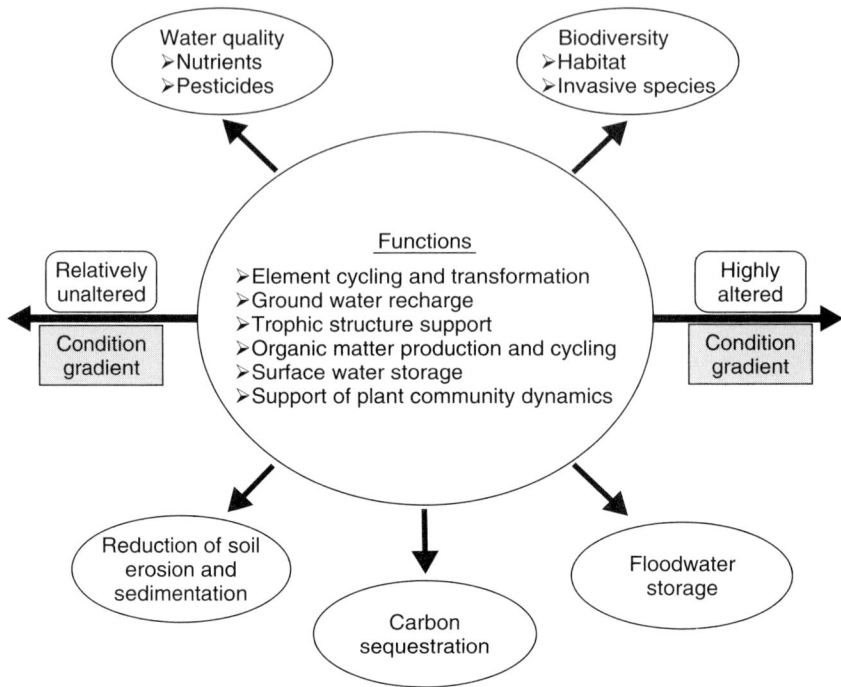

FIGURE 13-4. Wetland functions and ecosystem services impacts across a disturbance gradient, from unaltered to highly altered.

Geomorphic alteration has resulted in a global loss of about 35 percent of all mangrove swamps in the past 20 years (Millennium Ecosystem Assessment, 2005). This and other factors contributed to a loss of about 20 percent of coral reefs and degradation of another 20 percent globally (Millennium Ecosystem Assessment, 2005). Habitat change and nutrient loads are the largest impacts across inland tropical and temperate, as well as coastal, wetland systems. Finally, all coastal wetlands are subject to impact from changes in sea level as a consequence of global climate change.

URBAN AREAS

Urban areas are human-dominated built environments for the purpose of habitation for people. These are in contrast to rural areas, which may be human-dominated landscapes but not built environments (forests, agricultural areas, grasslands). Urban areas are net deficit areas for ecosystem services; they rely on outlying or peri-urban areas for supplemental resources, energy, water, and materials. Urban ecosystem

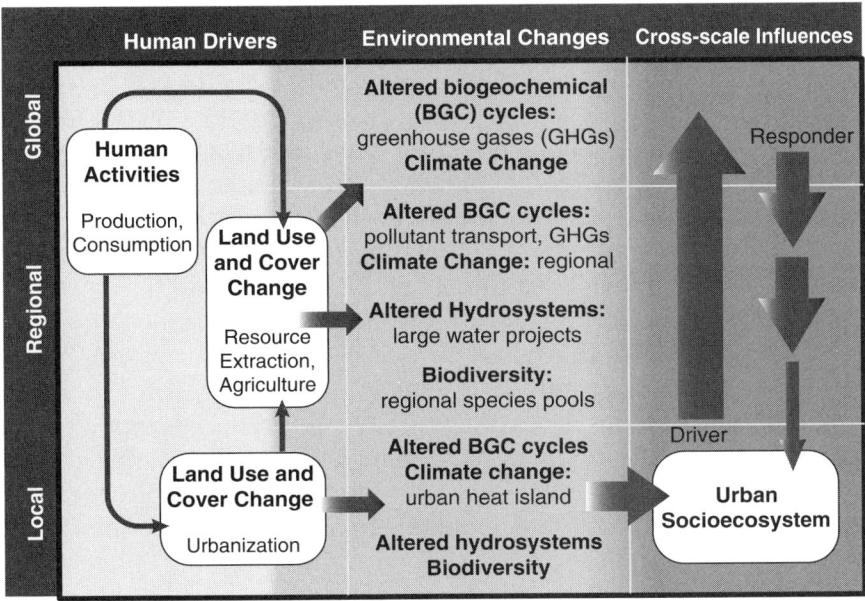

FIGURE 13-5. Urban ecosystems as drivers and responders to environmental change. (*Source*: Modified from Grimm et al., 2008.)

services footprints have been reported as being as high as 1,000 times the area of the city itself (Bolund and Hunhammar, 1999).

The deficit in services represents one of the greatest opportunities for innovation in ecological engineering. Urban ecosystems impact ecosystem services across multiple scales (Figure 13-5). The regional demand for goods and services to support urban areas has an ecoregion-level, and in some cases biome-level, impact. The predominant human drivers for these impacts are land use change and land cover change. These changes alter biogeochemical cycles, result in greenhouse gas emissions, particulate matter (PM) air pollution, heat island effects, stormwater runoff pollution, habitat fragmentation, and general disruption of ecosystem processes at all scales (Grimm et al., 2008).

For more than 10 centuries, urban dwellers have relied on rural communities for goods and services to support life, while rural communities received processed and manufactured goods, as well as markets for their products, from the urban communities. This system has become less efficient as humanity continues to urbanize, and fewer people produce the provisioning goods and services we need. Gutman (2007) suggests that urban communities need a new compact with rural communities. This new compact would have rural communities providing provisioning services, and urban communities providing higher incomes to rural communities by paying landholders for ecosystem services provided by the landscape. At a global scale, this model is being tested in Amazonia, where nongovernmental organizations (NGOs) provide funding to partner organizations or landholders for ecosystem services. In rural U.S. communities, landowners can participate in conservation reserve programs funded by the Natural Resources Conservation Service (NRCS), a direct form of compensation for indirect ecosystem services.

Urban landscapes can also provide ecosystem services. Urban ecosystem services require stakeholder adoption, since these are human-dominated ecosystems. The ESAC in urban systems must be representative of the community within which site design decisions have impacts. The ecotechnology team in urban design projects often is tasked with expanding understanding and appreciation of ecosystem services amongst stakeholders (Figure 13-6). Moving stakeholders from merely informed to empowered, and decision making from assessment to management, is a process that takes time and trust. The analysis and deliberation framework presented in Chapter 3, with collaborative learning as the method of engagement, has proven to be effective in this process. The first tier of analysis in urban ecosystem services design is the social assessment, whereby the people who benefit from and own or control the ecosystem functions from which ecosystem services will be derived are identified. This assessment also informs representation in the ADF (Cowling et al., 2008). The biophysical assessment for designing ecosystem services can proceed after this tier is complete.

The most challenging element of designing ecosystem services is the frustration associated with conventional urban design processes. Urban systems are the legacy of 10,000 years of incremental design experience. In most cities, urban design is not the dominant driver in urban expansion; instead, decisions are ad hoc and largely motivated by efficiency in infrastructure delivery (power, water, sewers, roads). In essence, we design our cities not for people, but rather for utilities. This is evidenced by the way cities are carved up by interstates, creating pedestrian barriers and fragmentation of human habitats. Similarly, streams and rivers are converted to concrete conveyances, fenced to prevent human access, resulting in additional fragmentation of human habitat. While human populations are very mobile, and metapopulation can move beyond these barriers, the costs can be significant, both economically and socially.

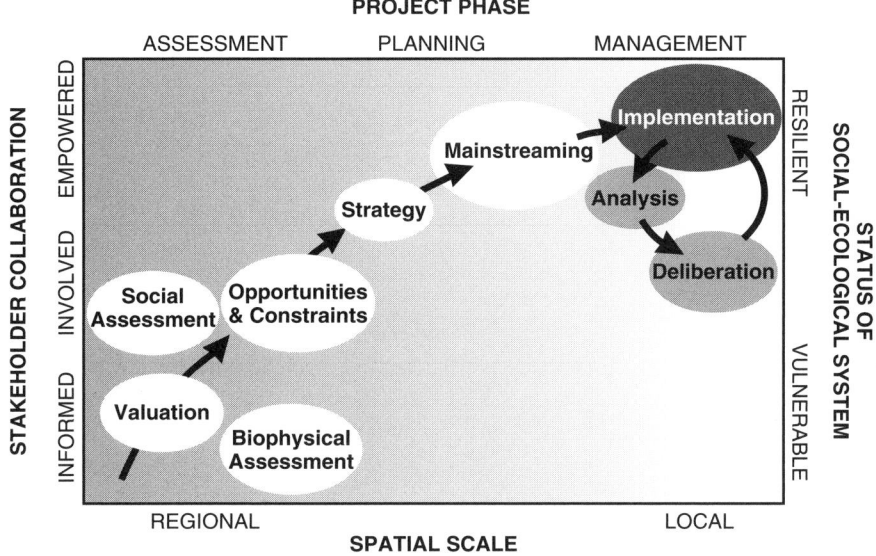

FIGURE 13-6. A process strategy for moving ecosystem services management from concept to implementation in a social framework. (*Source*: Modified from Cowling et al., 2008.)

Urban ecosystem services are landform dependent, and are often very fragmented. Designing effective ecosystem services in urban areas requires detailed attention to patch sizes and connectivity. Birds and rodents are the most common wildlife in urban settings, precisely because they can adapt to these limitations. Urban forests represent the most robust landform for enhancing urban ecosystem services, with urban meadows (often parks) and riparian areas (floodways) providing patches and corridors (Table 13-5).

In many cities in arid or semi-arid ecoregions, ecosystem services on the landscape may have actually increased as land use changed from agricultural rangelands to urban forests (Kreuter et al., 2001). These increases in ecosystem services are often not directly designed, and can require extraordinary energy subsidies to move water to otherwise arid locations. Cities such as San Antonio, Texas, Tempe-Phoenix, Arizona, Las Vegas, Nevada, and Los Angeles, California, are examples of landscapes with more ecosystem services per hectare than pre-urbanization. Kreuter et al. (2001) calculated that in San Antonio, Texas, from 1976 to 1991, the net value of ecosystem services declined only 4 percent, while rangeland decreased more than 65 percent, and urban land use increased almost 30 percent. The reason was that urban forests (woodlands) increased more than 400 percent during that period, offsetting

TABLE 13-5 Ecosystem Services Provided by Urban Ecosystems

Category	Type	Definition
Provisioning Services	Biodiversity	Provide structural diversity, compositional diversity, and functional diversity; this is essential for sustaining ecosystem functioning.
	Refugia	Provide residence, social spaces, and workplaces for people.
Regulating Services	Climate Regulation	Urban areas are a source of greenhouse gas emissions, which increase global warming; urban forests provide sequestration of carbon and reduction of ozone in traffic corridors.
	Water Protection	Urban systems can provide runoff treatment, infiltration, and storage if properly designed using low impact development.
Supporting Services	Water Cycling	Green infrastructure can enhance transfer of water from the soil through plants to the atmosphere, reduce rainfall runoff, and increase infiltration into groundwater.
Cultural Services	Recreational Services	Urban systems provide places for people to play. Parks, trails, greenways, and common lawns are some of the many urban spaces people use for recreation.
	Ethical Values	Urban places often provide spiritual, aesthetic, and other values people attach to structures and places.

the loss in ecosystem services provided by the rangelands. This illustrates the power of using ecosystem services as a design framework; it frees the ecological engineer from simply restoring systems to some approximation of what was, and allows the ecotechnology team to envision what can be.

Emerging urban design approaches provide novel connections for ecological engineers. Green infrastructure programs such as Chicago's green roof initiative provide financial incentives for innovation in design. The adoption by USEPA of low impact development (LID) technologies gives the ecotechnology team a pallet of design tools from which to draw, and reduces the barriers for implementation often found in urban ordinances. Estimates of building life spans suggest that 80 percent of the buildings that will exist in 2050 have not yet been built. Ecological engineers can design the urban future. The most compelling reason to do so is because we live there, our children will live there, and their children as well.

> I plead, in short, for positive and substantial public encouragement, economic and moral, for the landowner who conserves the public values—economic or aesthetic—of which he is the custodian.
>
> —Aldo Leopold

References

Bolund, P., and S. Hunhammar, Ecosystem services in urban areas, *Ecological Economics*, 29: 293–301, 1999.

Cowling, R.M., B. Egoh, A.T. Knight, P.J. O'Farrell, B. Reyers, M. Rouget, D.J. Roux, A. Welz, and A. Wilhelm-Rechman, An operational model for mainstreaming ecosystem services for implementation, *Proceedings of the National Academy of Sciences*, 105: 9483–9488, 2008.

Dale, V., and S. Polasky, Measures of the effects of agricultural practices on ecosystem services, *Ecological Economics*, 64(2): 286–296, 2007.

Falk, D., M. Palmer, and J. Zedler, eds., *Foundations of Restoration Ecology*, Island Press, Washington, DC, 2006.

Fearnside, P., Deforestation in Brazilian Amazonia and global warming, *Annals of Arid Zone*, 47(3– 4): 1–20, 2008.

Foley, J.A., R. DeFries, G.P. Asner, C. Barford, G. Bonan, S.R. Carpenter, F.S. Chapin, M.T. Coe, G.C. Daily, H.K. Gibbs, J.H. Helkowski, T. Holloway, E.A. Howard, C.J. Kucharik, C. Monfreda, J.A. Patz, I.C. Prentice, N. Ramankutty, and P.K. Snyder, Global consequences of land use, *Science*, 309: 570–574, 2005.

Geist, H.J., and E.F. Lambin, Proximate causes and underlying driving forces of deforestation, *BioScience*, 52(2): 143–150, 2002.

Gleason, R., M. Laubhan, B. Tangen, and K. Kermes, Ecosystem services derived from wetland conservation practices in the US prairie pothole region with an emphasis on the U.S. Department of Agriculture Conservation Reserve and Wetlands Reserve Programs, U.S. Geological Survey Professional Paper 1745, USGS, Washington, DC, 2008.

Grimm, N., S. Faeth, N. Golubiewski, C. Redman, J. Wu, X. Bai, and J. Briggs, Global change and the ecology of cities, *Science*, 319: 756–770, 2008.

Gutman, P., Ecosystem services: foundations for a new rural-urban compact, *Ecological Economics*, 62: 383–387, 2007.

Kindermann, G.E., I. McCallum, S. Fritz, and M. Obersteiner, A global forest growing stock, biomass and carbon map based on FAO statistics, *Silva Fennica* 42(3): 387–396, 2008.

Kreuter, U., H. Harris, M. Matlock, and R. Lacey, Change in ecosystem services valuation in San Antonio area, Texas, *Ecological Economics*, 39: 333–346, 2001.

Lehner, B., and P. Doll, Development and validation of a global database of lakes, reservoirs and wetlands, *Journal of Hydrology*, 296: 1–22, 2004.

Mazco, K., and L. Hidinger, eds., Sustainable Rangeland Ecosystem Services, SRR Monograph no. 3, Sustainable Rangeland Roundtable, Fort Collins, CO, 2008.

Millennium Ecosystem Assessment, *Ecosystems and Human Well-Being: Synthesis*, Island Press, Washington, DC, 2005, www.millenniumassessment.org/.

Mitch, W., and J. Gosselink, *Wetlands*, John Wiley & Sons, Hoboken, NJ, 2000.

Mitsch, W.J. and S.E. Jørgensen, *Ecological Engineering and Ecosystem Restoration*. John Wiley and Sons, Inc., New York. 2004.

Nelson, E., G. Mendoza, J. Regetz, S. Polasky, H. Tallis, D. Cameron, K. Chan, G. Daily, J. Goldstein, P. Kareiva, E. Lonsdorf, R. Naidoo, T. Ricketts, and M. Shaw, Modeling multiple ecosystem services, biodiversity conservation, commodity production, and tradeoffs at landscape scales, Frontiers in *Ecology and the Environment*, 7(1): 4–11, 2009.

Odum, E., *Ecology and Our Endangered Life Support System*, Sinauer Associates Press, Sunderland, MA, 1993.

Ricklefs, R., *The Economy of Nature*, 6th ed., W.H. Freeman and Co., New York, NY, 2008.

Ricklefs, R., and G. Miller, *Ecology*, 4th ed., W.H. Freeman and Co., New York, NY, 2000.

Ricklefs, R., and D. Schluter, eds., *Species Diversity in Ecological Communities*, University of Chicago Press, Chicago, IL, 1993.

Swinton, S.M., F. Lupi, G.P. Robertson, and S.K. Hamilton. Ecosystem services and agriculture: cultivating agriculture ecosystems for diverse benefits. *Ecological Economics*, 64(2): 245–252, 2007.

Wilson, E.O., *The Diversity of Life*, W.W. Norton and Company, New York, NY, 1999.

Zedler, J., and S. Kercher, Wetland resources: Status, trends, ecosystem services, and restorability, *Annual Review of Environmental Resources*, 30: 39–74, 2005.

Zhang, W., T. Rickets, C. Kremen, K. Carney, and S. Swinton, Ecosystem services and dis-services to agriculture, *Ecological Economics*, 64(2): 253–260, 2007.

14

Green Infrastructure Design

> A connected system of parks and parkways is manifestly far more complete and useful than a series of isolated parks.
> —John Olmsted and Frederick Law Olmsted Jr., 1903

INTRODUCTION

There exists a network of roads, utilities, water supply, and sewage disposal that makes normal human society possible. This network is the gray, or built, infrastructure of the community or nation. The built infrastructure exists at a huge cost to society, both for the installation of the infrastructure and for ongoing operation and maintenance. There also exists another network on which our society depends. That network consists of streams, lakes, wetlands, meadows, grasslands, and woodlands. This network also provides essential services to society, including water and air supply and purification, climate regulation, food and fiber, recreation, and spiritual rejuvenation (Figure 14-1). This network is called the green infrastructure. In contrast to the built infrastructure, green infrastructure exists as a part of the ecosystem and, if allowed, it will operate and maintain itself at no cost to society.

Resource conservationists now realize that preserving isolated natural landscapes will not meet their goals. To maintain biodiversity and ecological services, the green infrastructure will have to be maintained. They also realize that not all land can be saved, nor can all action be regulated. Thus, the field of green infrastructure planning has been born. Green infrastructure planning is more a tool for policy development than for engineering design. But engineering for ecosystem services is required, to fully implement a green infrastructure plan. The Conservation Fund (www.conservationfund.org/green_infrastructure) has advanced the concept of green infrastructure as a conservation-planning tool. Mark Benedict and Ed McMahon's *Green Infrastructure: Linking Landscapes and Communities* provides a detailed discussion of the green infrastructure planning process.

FIGURE 14-1. Gray infrastructure and green infrastructure networks both provide essential services to society. Gray infrastructure exists at high cost for installation, operation, and maintenance. Green infrastructure provides essential services at no cost to society. (*Source*: Image from Fayetteville Natural Heritage Association.)

The United States Environmental Protection Agency (USEPA) uses the term "green infrastructure" in a slightly different context. In the USEPA's literature, green infrastructure refers to using natural systems to mitigate the impacts of urban development. Special emphasis is given to mitigation of runoff during storm events through practices that mimic the natural hydrograph. Examples of green infrastructure practices include green roofs, green streets, rain gardens, rainwater harvesting, bioswales, and infiltration basins. A series of "Municipal Handbooks" (http://cfpub.epa.gov/npdes/greeninfrastructure/technology.cfm) developed by the USEPA provides information on use of green infrastructure techniques to manage wet weather flows in urban areas.

In this chapter, the principles and tools of green infrastructure planning are briefly described. Engineering tools to implement elements of the green infrastructure plan will be presented in subsequent chapters. For detailed information on implementing green infrastructure techniques, the reader is referred to the publications of the Conservation Fund and the USEPA.

THE GREEN INFRASTRUCTURE NETWORK

The green infrastructure is the framework upon which future conservation and development projects can build. The elements of green

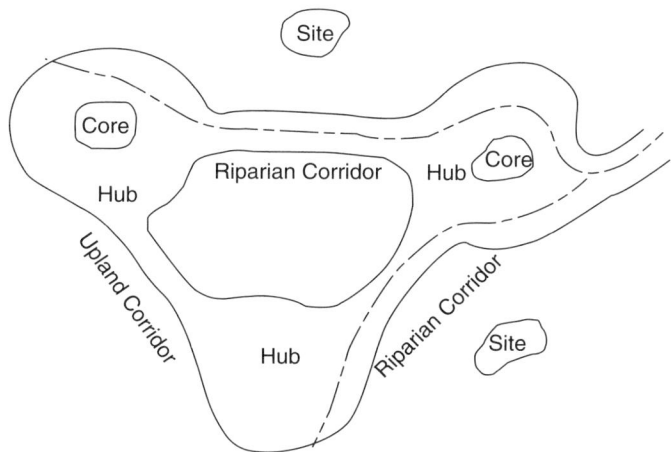

FIGURE 14-2. The green infrastructure network consists of intact cores, hubs, links, and sites. Taken together, the network meets the ecosystem service needs of the local community.

infrastructure are local woodlands, meadows, prairies, wetlands, lakes, streams, rivers, and other features that provide ecological services. Working lands—including farms and managed forests—also must be considered, as they provide services to the community as well. The green infrastructure network is a system of interconnected cores, hubs, links, and sites (Figure 14-2) that, taken together, preserve ecological services of the green infrastructure.

There are no hard and fast rules about the makeup of green infrastructure networks other than that they meet the goals of the community. An individual working in isolation cannot design the network. Design requires the coordinated work of multiple disciplines, including conservation ecology, biology, landscape architecture, hydrology, limnology, engineering, and so forth. One of the major benefits of developing a green infrastructure plan may be the interaction between disciplines. Developing a green infrastructure plan is also a messy, drawn-out process. It may take several years to generate consensus on the ideal network.

Cores. Cores of the green infrastructure network are intact natural areas. These natural areas are fully functioning ecosystems with habitat suitable for native species. The network will build on these cores to minimize fragmentation and maintain connectivity of the ecosystem. Cores may be woodlands, meadows or prairies, wetlands, lakes, or streams. Cores are frequently previously protected land such as interior areas of state or national forests, parks, or conservation areas.

The size of a core area is dependent on the goals of the green infrastructure plan. Some guidelines are: Bigger is better than smaller, rounder is better than linear, connected is better than segmented, closely segmented cores are better than more widely separated cores, and segmented cores that form a cluster are better than segmented cores that are linear. If the core is to provide habitat for interior dwelling species, then the core needs to be large enough to create interior habitat conditions. Local experts should be consulted regarding minimum sizing and conditions needed for suitable core areas.

Hubs. Hubs are larger areas that surround one or more cores. Hubs should be relatively undeveloped areas that provide buffer between the core areas and the surrounding areas. However, hubs may contain recreational lands and working lands, if those uses are consistent with the function of the core.

Cores and hubs are the bank of genetic material for the green infrastructure planning area. From these areas, individuals from different species can disperse into the surrounding areas. Genetic diversity is also maintained if individuals can move between cores with little danger.

Hubs need to be large enough to support populations of native species. As with cores, bigger, rounder, and connected hubs are better than smaller, linear, and fragmented. The buffer width provided by a hub around a core is quite variable. If the cores are selected to protect interior species, then the buffer needs to be wide enough to create interior forest conditions at the edge of the core.

Corridors. Corridors connect the hubs. The function of the corridor is to facilitate the movement of individuals and materials between the hubs of the network. The corridors need to provide for movement of aquatic, riparian, and upland species. A combination of riparian corridors and upland areas will be needed.

The width of a corridor is important, but as with hubs and corridors, there is no easy formula. Generally, wider corridors are better than narrower ones. Corridors should provide for movement of interior species. The minimum width should create interior conditions within the corridor. Many green infrastructure plans use minimum buffer widths of 200 to 300 meters (650 to 1,000 feet). Riparian buffers should include the floodplain plus some of the adjoining upland fringe. However, many water quality benefits can be obtained from riparian buffers that are much more narrow.

Sites. Important ecological, recreational, or cultural features may exist outside of protected core areas. These isolated features typically are smaller than cores but still deserving of protection. Isolated features are referred to as sites. If possible, sites should be connected to the overall network.

Green Infrastructure Planning

The Conservation Fund promotes planning of the green infrastructure network for a specific region. Ideally, the green infrastructure plan will become part of the region's long-range growth plan. Once the green infrastructure plan is in place, priority cores, hubs, and corridors can be maintained.

There is no set protocol for developing a green infrastructure plan. Fundamentally, green infrastructure planning is a stakeholder process. Stakeholders are persons or interest groups from the community that have some interest in the outcome of the plan. The stakeholder group is responsible for decisions regarding the plan. A technical or leadership group should be guiding the stakeholders through the process. The technical group compiles data from available sources and informs the stakeholders regarding the needs of the network. Through repetitive interaction, the groups will compile a plan satisfactory to all.

Components of the green infrastructure plan are:

- Stakeholder-derived vision statement
- Green infrastructure network design based on vision statement
- A plan outlining tools and resources to implement the network
- Information/education program
- Agency buy-in

Geographic information systems (GISs) are a valuable tool for compilation of the network of cores, hubs, and corridors. Some data layers that may be needed are:

- National Hydrography Dataset, United States Geological Survey
- Wetlands, U.S. Fish and Wildlife Service
- National Elevation Dataset, United States Geological Survey
- SSURGO Soils Survey, Natural Resources Conservation Service
- Land use and land cover, national or local sources
- Land ownership, national and local sources
- Gap Analysis Program (GAP), National Biological Information Infrastructure

Readers interested in preparing green infrastructure plans for a particular area should consult the materials of the Conservation Fund and the multiple case studies.

The Tools of Green Infrastructure

Green infrastructure planning is just the start of a green infrastructure program. The plan is worthless if there is no mechanism for implementation. In today's world, laws and regulations restricting the use of land are difficult to implement. Fortunately there are tools available for preservation of open space that don't resort to public taking of property.

Public Awareness and Education
Public education programs inform the community about what green infrastructure is and how they can help with the plan. Education programs include billboards, public service announcements, news releases, media interviews, and speakers bureaus. The program is not a one-time event. Efforts must be ongoing to keep the plan in the public mind.

Fee Simple Ownership
In some cases a core area may be so valuable from a conservation perspective that it should be held by a public entity. Fee simple ownership means that the holder of title to the land has absolute ownership of the property, subject only to governmental regulations and encumbrances that may have been placed on the deed by previous owners. It is not likely that any governmental or nonprofit agency can buy all of the land in a green infrastructure network.

Conservation Easements and Transfer of Development Rights
Property ownership can be thought of as owning a bundle of rights to property use. In a conservation easement, the landowner elects to give up some portion of that bundle of rights in return for just compensation. Compensation may be cash payment, tax incentives, or just the preservation of the property. The just compensation is provided by a public or nonprofit organization that desires to protect a natural resource. Frequently, a land trust is involved as the easement holder.

Conservation easements may be crafted for just about any conservation purpose. Easements have been acquired to preserve riparian corridors, woodlands, prairies, farmlands, and cultural and historic sites. The key to successful conservation easements is an agreement on the terms of the easement between the easement grantor or landowner and the easement purchaser. The easement agreement should clearly spell out what land is involved, what uses of that land are allowed and not allowed, what conservation value is protected, and the responsibilities of the lease grantor and grantee. Provision must also be made for ongoing inspection of the easement and enforcement of the agreed-on conditions. Easements may or may not allow public access, depending on the landowner's interests.

Transfer of development rights is similar to a conservation easement, except that land use zoning is involved. In a transfer of development rights, a public entity identifies areas where development is desired and other areas where it is not desirable. A developer with a project in a desired development area may want to develop to a higher density than the zone allows. If the government has a transfer of development rights ordinance, that developer may be able to buy the "right" to develop property from another landowner. In return, the government allows increased density on the developer's property. The net result is maintenance of the desired overall density, and maintenance of desired open space. The landowner selling the development rights retains the right to other uses of his/her property.

Conservation easements and transfer of development rights may be a valuable tool for preserving green infrastructure. Hubs and corridors may be priority areas for purchase of conservation easements by land trusts. Cities may also identify priority conservation areas as areas from which development rights may be purchased. The green infrastructure plan provides a guide as to where high-priority areas exist. Groups interested in conservation easements should make themselves familiar with their local laws and programs.

Cluster Development
Not all land in the green infrastructure system need be preserved in an undeveloped condition. Low impact development (LID) is a type of development that attempts to preserve pre-development ecological services. Primarily, LID strives to produce conditions such that the post-development stormwater hydrograph mimics the pre-development hydrograph. In LID, impervious surface is minimized, infiltration is enhanced, and runoff is minimized. Detail on LID design is provided in Chapter 15. Land use regulations frequently specify the density of homes or living units per acre or hectare. Traditional development spreads those living units somewhat evenly across the property. Conversely, lots can be clumped together into smaller areas, leaving larger contiguous areas open (Figure 14-3). Envision a golf course development without the golf course. Overall lot density is maintained at the required level, but lot size is reduced. This is the concept of cluster development.

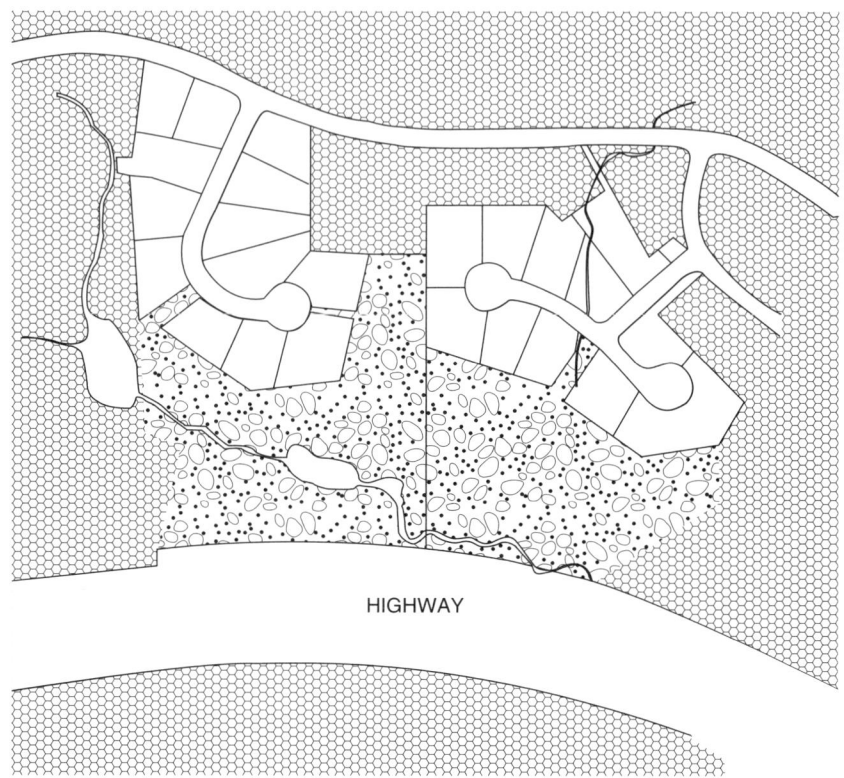

FIGURE 14-3. In a cluster development, the designer considers aspects of the land that provide common good to the community and the local property owners. Such aspects are preserved. Lots are clustered into smaller areas around the common property. The tradeoff is that smaller lots are allowed in return for preserving green space. (*Source*: Federal Highway Administration, *The Audible Landscape,* 1974.)

Cluster development offers advantages to both the community and the developer. The community benefits because open space (forest, meadow, wetland, park, or farmland) is preserved. The developer benefits from reduced gray infrastructure (roads, sewers, stormwater facilities), resulting in less investment, and frequently increased lot value.

Cluster developments that target preservation of priority conservation values are referred to as conservation subdivisions (Southwestern Illinois Resources Conservation & Development, 2006). For a community to effectively utilize conservation subdivisions in planning, it must provide clear conservation goals and priorities and review zoning and subdivision ordinances to allow cluster development. Incentives such as streamlined review and approval may increase developers' interest in conservation subdivisions.

Eminent Domain

The government may seize property without the owner's consent for public use. The power of eminent domain is most often used to secure rights of way for highways and utilities, or to protect critical infrastructure. Agencies exercising eminent domain on property must provide just compensation for the taking of that property. The use of eminent domain to preserve green infrastructure has not been tested in the United States.

Regulation

Local governments may regulate land use within their jurisdiction, in accordance with a land use plan. If the government adopts the green infrastructure network as part of its land use plan, then it may use regulations to establish elements of the network. Zoning regulations specify the type of use and the density of development. Zoning ordinances frequently specify setbacks (how far a structure has to be from a property line or physical feature) and minimum green space requirements. Overlay zones are additional restrictions placed on land for special purposes. Subdivision regulations provide standards to which developments must be built. These regulations include standards for streets, utilities, number of parking spaces, and preservation of trees.

No single tool will suffice for implementation of the green infrastructure network. A combination of tools, programs, and funding mechanisms will be required. This combination, referred to as the implementation quilt, is possibly the most important part of the green infrastructure plan, and it should be considered in the planning process. Implementation of a green infrastructure program will be a long-term commitment.

Scale Matters

Green infrastructure exists across multiple jurisdictions and at different scales, from continental to site. At the continental scale, planners are concerned with migratory patterns of species and global climate or water regulation. At the site scale, the planner may only be concerned with local water management and providing habitat for a specific species.

THE SUSTAINABLE CITIES INITIATIVE

There are numerous initiatives in various stages of development for assessing and designing sustainable cities. These initiatives will undoubtedly continue to be in rapid flux, so detailing each of them in a book would be pointless. The reader is advised to perform Web searches under the topic of sustainable cities in order to gauge the activity in this area. There are a couple of initiatives that are adequately established to discuss. These include the United Nations World Urban Forum (UNWUF) and the International Community for Leadership in Environmental Initiatives (ICLEI).

United Nations World Urban Forum

In 2008, human beings became a predominantly urban species. By 2050, based on current trajectories, more than 70 percent of the projected 9.25 billion people on earth will live in urban environments. Globally, we are not designing these urban areas; rather, they are being developed helter-skelter on the hillsides, in the drainage ditches, and on the dumps of existing cities. Currently, an estimated one billion people live in slums and barrios, cardboard and plastic cities that are unplanned, unmanaged, and unsustainable. Slums are defined as living areas that do not have one of the following four critical services: potable drinking water, sanitary waste disposal (liquid and solid), durable housing, and improved living spaces. The UNWUF recommends the following steps to address the immediate and impending crisis in urban systems (UN, 2009):

1. *Prioritize urban policy*: Develop formal commitments from governments to adhere to basic principles of social and environmental justice and sustainability. Global standards for an urban policy that responds to the local tensions between urban and rural needs must be developed, adopted, and implemented.
2. *Planning legislation*: Implementing urban planning will require national legislation to respond to rapidly expanding urban crises. Colonial-era policies that currently dominate urban planning are not adequate for this challenge.
3. *Decentralization of urban planning functions*: While global and national policies and legislation are necessary for empowering urban ecological design, decisions on urban planning issues should be made as close as possible to those affected by them.
4. *Urban planning function within municipalities*: Urban planning should not be separate from other decision making in urban systems, but rather should dictate criteria for all decisions to ensure that urban systems function in a sustainable manner.
5. *Urban research and data*: The process of designing urban ecosystems is evolving rapidly, and must be informed by local research and data relevant to the challenges of that municipality.

6. *Planning education*: Urban leaders around the world need common resources and frameworks to use for assessing their community needs.

This call for common strategy and action is resonating globally. The challenges for our global urban systems are in many ways very simple: Provide basic sanitation infrastructure to as many people as fast as possible, then work on sustainability. All other ecosystem services will follow, but cannot be sustainable without this first step. Potable water delivery systems and sewage waste collection and treatment are the top priorities.

The global community is making significant progress on this front. In 1990, almost half of urban dwellers globally lived in slums. That number was reduced to 36 percent by 2005 (UN, 2009). The challenge for the next 5 years will be greater than the accomplishments of the past 15 years (Figure 14-4). Municipal waste treatment gaps in Asia alone could exceed one billion people by 2015.

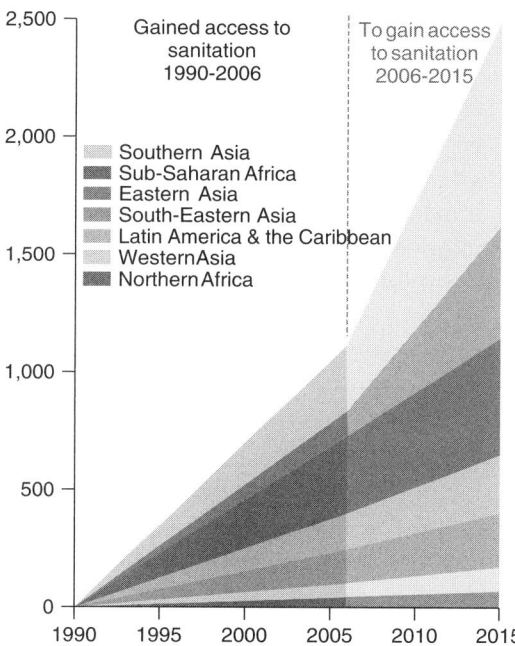

FIGURE 14-4. Population that gained access to an improved sanitation facility 1990–2006 (millions) and population that needs to gain access to an improved sanitation facility to meet the Millennium Development Goals (MDG) target, 2006–2015 (millions). (*Source*: UN, 2009.)

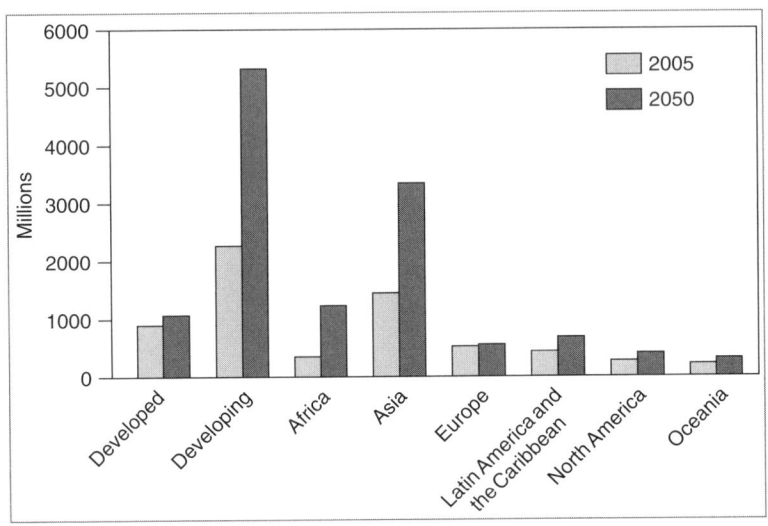

FIGURE 14-5. Urban population by region. (*Source*: UN, 2008.)

Global challenges for urban resources will include energy (and thus greenhouse gas emissions), water resources, housing, and transportation. These challenges are not dissimilar to those in the developed world, but are more extreme (Figure 14-5). The largest growth in urban demand will be in developing countries, especially those in Asia (not including Japan) and Africa. The urban population of 3.3 billion reside predominantly in smaller towns and villages—55 percent live in towns of less than 500,000 people (UN, 2010). Urban sprawl is resulting in the merging of previously separated urban areas, and creating even more discord between traditional governance structures and pressures for common criteria for managing urban systems. The ecological engineering opportunities in these dynamic situations are significant. Expanding cities have economic and social pressures that can be addressed using ecosystem services design, and can reduce infrastructure costs as well. The UNWUF platform does not currently address ecosystem services, but the opportunities for informing and enhancing urban design with ecological engineering principles are clear.

ICLEI: Local Governments for Sustainability

The International Community for Leadership in Environmental Initiatives (ICLEI) is an association of over 1,100 cities from 68 countries committed to sustainable development. The member cities incorporate a five-milestone structure to move city policies in sustainable directions: (1) establish a baseline, (2) set a target, (3) develop a local action plan, (4) implement the local action plan, and (5) measure results. These simple steps are not particularly advanced design, but can be powerful if informed by the proper metrics (ICLEI, 2008).

The first emphasis area for ICLEI has been greenhouse gas (GHG) emissions reduction from urban areas. ICLEI proposed a Local Government Operations Protocol (the Protocol) as a standardized guideline to assist local governments in quantifying and reporting GHG emissions associated with government operations. The Protocol requires a reporting process to identify emissions to be included in inventory, quantify emissions, and report emissions.

The Protocol is organized into five overarching accounting and reporting principles:

1. *Relevance*: GHG inventory should appropriately reflect the GHG emissions of the local government and should be organized to reflect areas over which local governments exert control and hold responsibility.
2. *Completeness*: All GHG emission sources and emissions-causing activities within the chosen inventory boundary should be accounted for (specific exclusions should be justified and disclosed).
3. *Consistency*: Consistent methodologies should be used in the identification of boundaries, analysis of data, and quantification of emissions to enable meaningful trend analysis over time, demonstration of reductions, and comparisons of emissions (changes should be disclosed).
4. *Transparency*: All relevant issues should be addressed and documented to provide a trail for future review and replication (sources and assumptions should be disclosed).
5. *Accuracy*: The quantification of GHG emissions should reflect the actual emissions, with minimum bias and adequate accuracy to enable decisions on priorities for reduction.

The Protocol requires participating local governments to assess emissions of all six internationally recognized GHG sources regulated under the Kyoto Protocol (other GHGs may also be reported):

1. Carbon dioxide (CO_2)
2. Methane (CH_4)
3. Nitrous oxide (N_2O)

4. Hydrofluorocarbons (HFCs)
5. Perfluorocarbons (PFCs)
6. Sulfur hexafluoride (SF_6)

The Protocol requires that cities account for emissions of each gas separately, report emissions in metric tonnes of each gas and metric tonnes of CO_2 equivalent (CO_2e), and report GHG inventories on a calendar year basis (not fiscal year). Cities establish a benchmark performance datum with which to compare current emissions. This base year should be a representative year for which accurate records of all key emission sources exist in sufficient detail for accurate inventory. Cities then develop a base-year emissions recalculation policy for defining methods to recalculate base-year emissions to reflect changes in the local government that would otherwise compromise consistency and relevance of the reported GHG emissions. Carbon stocks and estimations of project-specific GHG reductions may be reported optionally. Biological stocks and project-specific reductions should be reported separately from inventory emissions, and no line item adjustments should be made to the inventory based on these activities.

Participating cities set organizational boundaries for the GHG assessment. They select and apply an approach for consolidating GHG emissions (operational or financial control) and apply it consistently. Operational control is defined as those interests in which a local government has full authority to introduce and implement its operating policies at the operational level. Financial control is defined as those interests in which a local governmental operation is fully consolidated in financial accounts. ICLEI recommends operational control as an accounting method.

The local governments should account for 100 percent of direct GHG emissions from operations over which they have control (often referred to as Scope 1 emissions). Under the Protocol, Scope 2 emissions are indirect GHG emissions associated with the consumption of acquired electricity, heating, or cooling. These must also be accounted for in the ICLEI reporting structure. Cities under the Protocol do not have to report GHG emissions in which they have an interest but have no control (Scope 3).

Summary

Ecological services have real value in our economy. Replacing those services worldwide would literally cost trillions of dollars annually.

The green infrastructure network provides ecological services along with economic development and other human activity. The network of cores, hubs, and corridors provides habitat, refugia, and connectivity needed to preserve biodiversity. The network also provides for climate regulation, water supply and pollution control, and recreational opportunities.

A variety of tools are available for implementing the green infrastructure network while respecting private property ownership. Most importantly, the green infrastructure needs to be considered in local or regional land use planning and needs to be given priority by decision makers in the community. Green infrastructure provides the foundation for ecological sustainability in a community. It also provides priorities for ecological restoration projects. Sustainable building and development practices augment the foundation of green infrastructure. Sustainable practices are the subject of the remainder of this book.

> There can be no doubt that a society rooted in the soil is more stable than one rooted in pavements.
>
> —Aldo Leopold

Further Readings
Benedict, Mark A., and Edward T. McMahon, *Green Infrastructure: Linking Landscapes and Communities*, Island Press, Washington, DC, 2006.

References
Benedict, Mark A., and Edward T. McMahon, *Green Infrastructure: Linking Landscapes and Communities*, Island Press, Washington, DC, 2006.
Byres, Katherine, and Karin Marchetti Pointe, *The Conservation Easement Handbook*, The Land Trust Alliance, Washington DC, 2005.
Hutchinson, Ann, Monica Drewniany, Todd Stell, and Neil Kinsey, *Growing Greener: Conservation by Design*, The Natural Lands Trust, Media, PA, 2001.
ICLEI, Local Government Operations Protocol, Version 1.0., ICLEI, San Francisco, CA, 2008.
Southwestern Illinois Resource Conservation & Development, *Conservation Subdivision Design Handbook*, Mascoutah, IL, 2006.
UN, Planning Sustainable Cities, UN Habitat, Nairobi, Kenya, 2009, www.unhabitat.org/grhs/2009.
UN, Bridging the urban divide, *Urban World*, 1(5), UN Habitat, Nairobi, Kenya, 2010.
Welsch, David L., *Riparian Forest Buffers—Function for Protection and Enhancement of Water Resources*, U.S. Department of Agriculture, Forest Service, Northern Area State & Private Forestry, 1991.

15

Low Impact Development

> If we are to create a sustainable world—one in which we are accountable to the needs of all future generations and all living creatures—we must recognize that our present forms of agriculture, architecture, engineering, and technology are deeply flawed. To create a sustainable world, we must transform these practices. We must infuse the design of products, buildings, and landscapes with a rich and detailed understanding of ecology.
>
> —Sim Van Der Ryn, *Ecological Design*

INTRODUCTION

Real estate development has a profound impact on the ecosystem in almost all parts of the world. Development is the process of converting unused landscapes into commercial, industrial, or residential areas. Seldom are the needs of the ecosystem considered in developing landscapes, beyond the immediate safety of structures and their inhabitants. Innovative engineers, landscape architects, and developers are reevaluating standard practices and the adverse impact of development on the landscape. A new paradigm is emerging that accommodates desired development and preserves ecosystem functions. This paradigm is referred to as low impact development, or LID.

Perhaps the most obvious impact of conventional development on the ecosystem is the modification of the hydrologic cycle. The increase of impervious area during development greatly increases runoff from a site (Figure 15-1). The health of local streams and waterbodies is seriously compromised when as little as 10 percent of a watershed becomes impervious. At 25 percent imperviousness, streams become degraded (Center for Watershed Protection 1993). This impact is caused by hydrologic modification of streams (channelization), increased erosion due to increased peak discharges, loss of riparian vegetation, and degraded water quality from nonpoint sources of pollution.

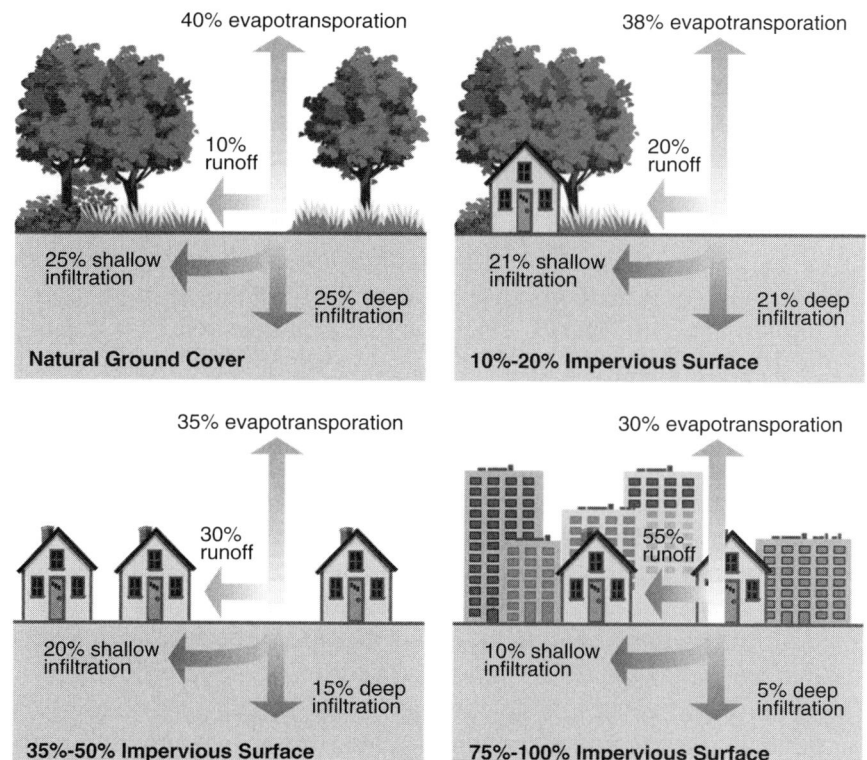

FIGURE 15-1. The impact of increasing imperviousness on the ratio of rainfall to runoff. As imperviousness increases, the percentage of rain lost to abstractions reduces while runoff increases.

The conventional practice of stormwater engineering dates to the time when open sewers were used for public sanitation. The primary function of drainage was protection of human health (Butler and Davies, 2000). The best way to protect health was to remove human and other wastes from habitation as quickly as possible. Society soon realized that this practice was polluting downstream receiving waters. London, England, installed separated storm and sanitary sewers in the mid- to late 1800s. Stormwater engineering continued to evolve with the philosophy of quickly removing water from the site. By the mid-twentieth century, conventional development practice used curbed and guttered streets as primary drainage conveyance to storm sewers and drainage ditches. This system was efficient at moving water off-site. However, the increased discharge rates and volume created liability for damage to downstream property owners and overtaxed the natural drainage system.

Conventional stormwater practice is not sustainable because it degrades ecological services provided by natural systems. LID is the evolution of stormwater engineering to utilize the natural systems. Since sanitary engineering has reduced potential contact with waterborne diseases, it is now possible to manage stormwater in a distributed fashion that mimics natural ecological processes. Mimicking ecological systems is the key to low impact development (LID). The goal of LID is to create hydrologic conditions post-development that match pre-development conditions. Carefully planned development that implements integrated management practices (IMPs) manages stormwater. Distributed stormwater management reduces the need for expensive end-of-pipe stormwater practices.

The LID approach uses five key components to accomplish the objective of matching pre-development conditions (Prince George's County, 1999 a), including:

- Site planning: Define development envelope, minimize imperviousness, disconnect impervious areas, consider drainage flow path.
- Hydrologic analysis: Delineate sub-watersheds, define design storm, model pre- and post-development conditions, compare developed to baseline, evaluate supplemental needs.
- Integrated management practice: Determine hydrologic control, define constraints, screen and evaluate IMPs, evaluate additional needs.
- Erosion and sediment control: Plan, schedule, and implement erosion control, sediment control, and maintenance.
- Public outreach and education: Generate knowledge of LID practices and responsibilities.

In this chapter, these key components are discussed in relation to their impact on the site's hydrology, water quality, and refugia. Design concepts are discussed in a final section on assessing ecological services and design.

HYDROLOGY

The hydrologic cycle (Figure 5-7) continually circulates water from the ocean to the atmosphere, back to the earth, and ultimately back into the ocean. Runoff is the portion of precipitation that is not abstracted by other processes. Abstractions from rainfall include interception by vegetation, surface storage, evapotranspiration, and infiltration. Increasing impervious area in a watershed causes increase in both runoff volume and peak discharge (Figure 15-1). Increase in runoff volume is caused by the reduction of precipitation abstraction (infiltration, interception, and surface storage). Water also moves more quickly and directly from remote points in the watershed to the outlet. Reducing the travel time means that shorter, more intense storms contribute to the flow, and more flow accumulates at the outlet simultaneously. The result is an increase in peak discharge. In the terms used in the NRCS's curve number technique for computing stormwater runoff, the curve number (CN) increases and the time of concentration (t_c) decreases.

The increased volume and peak discharge stress the receiving stream because bankfull discharge, the channel-forming flow, is experienced more frequently and for longer periods. The stream performs more work and moves more material downstream. As a result, erosion enlarges the stream to a point where it accommodates the new flow regime. Usually, the result is a wider channel with little channel diversity and riparian cover. Along with the loss of channel diversity is loss of habitat and in-stream refugia.

A less obvious impact of increasing impervious area is the loss of baseflow in the receiving stream. Because less water infiltrates and more runs off, the mass of water moves on downstream. With less infiltration, there is less water stored in the soil and aquifer to recharge the stream between storms. Lower baseflow also reduces habitat and allows the stream water to become warmer. The result is loss of fish species.

Conventional stormwater engineering practice treats rainfall runoff as a liability that has to be removed from the site as quickly as possible. LID, in contrast to conventional engineering, uses site hydrology as the framework for design. By utilizing the hydrology, the LID design team attempts to mimic the natural hydrology of the site.

To visualize the impact of development, consider a hypothetical stormwater discharge hydrograph (Figure 15-2). A hydrograph shows the instantaneous discharge rate at any point in time. The area under the hydrograph is the total volume of water discharged during the storm. In a natural system with a low runoff curve number (CN) and long time of concentration (t_c), the hydrograph rises slowly to a peak, then tapers off back to baseflow after rainfall ceases. In the developed condition, the CN is increased and the t_c decreased. The developed hydrograph rises quickly to a higher peak and then stays higher for a longer period.

A full description of site hydrology is provided in Chapter 6. To review, using the NRCS's Technical Release No. 55 (TR55) model, stormwater runoff, Q, is a function of precipitation, P, and a

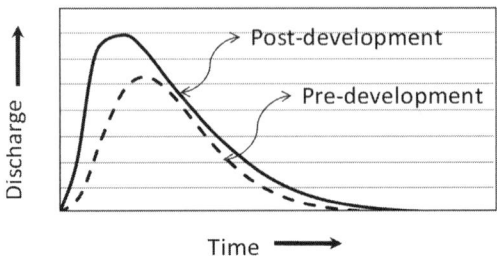

FIGURE 15-2. The impact of increasing impervious area on a hypothetical storm discharge hydrograph. Because of the increased curve number and decreased t_c, both the peak discharge and the volume of discharge from the storm increase.

coefficient S. The coefficient S is inversely related to the watershed's composite curve number (CN).

$$Q = \frac{(P - 0.2S)^2}{P + 0.8S}, \quad P > 0.2S \tag{15.1}$$

$$S = \frac{1000}{CN} - 10, \quad (Q, P \text{ and } S, \text{in.})$$

$$S = \frac{25400}{CN} - 254 \quad (Q, P \text{ and } S, \text{mm}) \tag{15.2}$$

The CN accounts for all abstractions to rainfall and computes runoff directly.

Peak discharge, Q_p, is the product of the unit peak discharge, q_u, and the watershed area. The unit peak discharge is a function of the time of concentration, t_c, and three coefficients related to S. Time of concentration is the time that it takes for water to travel from the hydrologically most remote point in the watershed to the point of discharge.

$$\log(q_u) = C_o + C_1 \log t_c + C_2 (\log t_c)^2 \tag{15.3}$$

The CN and t_c, along with the properties of soil, give us the tools necessary to design for LID. The objective is to minimize changes in the composite watershed CN, maintain or increase t_c, and make up for any changes with integrated management practices (IMPs) that enhance infiltration and evapotranspiration.

INITIAL STEPS

Collect site data including topographic maps, soil type, existing facilities, and land cover. Determine the existing drainage paths. For each drainageway that exits the site, delineate the watershed. In each of these watersheds, delineate areas of uniform land cover and soil type. Find the *CN* for each combination, and compute an area weighted *CN*. These numbers will be the pre-development *CN*s for the site. Determine the t_c for each drainageway; these are the pre-development times of concentration.

On a site map, delineate all hydrologic and other natural features that need to be maintained post-development. Features may include wetlands, floodplains, sinkholes, springs, seeps, rock outcrops, and steep slopes.

MINIMIZING CHANGE TO PRE-DEVELOPMENT *CN*

For residential subdivisions, a substantial part of the increase in *CN* comes from the presence of streets and alleys. By minimizing road distance, the developer not only reduces the change in *CN*, but also realizes savings from reduced paving and utility extensions. One approach to minimizing road length is to develop cluster housing (Figure 15-3). Zoning requirements typically specify the number of living units allowed per hectare of development. If regulations permit, cluster lots can be smaller than conventional lots, provided that compensating open space is maintained to meet the zoning requirement. The open space can be high-infiltration areas or other sensitive natural features. Locating the developed areas on less permeable soils will also help to reduce the difference between pre- and post-development conditions.

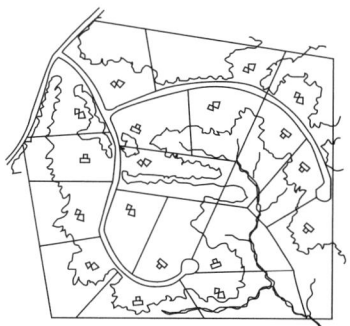
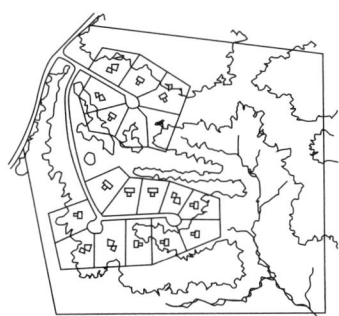

FIGURE 15-3. Minimizing roadway by following terrain and using cluster development.

FIGURE 15-4. Porous concrete pavement abutting standard asphalt. Runoff from the impervious asphalt only travels a few inches onto the porous concrete before it soaks into the pavement.

Streets need be only as wide as required for public safety. Frequently, road widths are set to accommodate fire trucks and other emergency vehicles. Eliminating on-street parking on one or both sides of the road may permit narrower streets and also provide adequate passage. Using narrower streets reduces imperviousness and helps maintain pre-development CN. If regulations allow, sidewalks may also be eliminated from one side of the street and still maintain function.

Permeable pavers and porous pavement may also be used to enhance infiltration and maintain pre-development CN (Figure 15-4). These paving materials may be used wherever traffic is light or parking infrequent. In residential areas, driveways may be made of permeable material.

Disconnecting impervious areas from the drainageway provides opportunity for infiltration through natural soils. Transition zones along paved lots and/or streets increase infiltration and disconnect the parking lot hydrologically. On homes and buildings, downspouts should be turned onto vegetated areas to disconnect them from the drainage system.

Other methods for reducing CN include using vegetated swales for drainage instead of paved ditches, preserving natural depressions, and reducing driveway length and width. Table 15-1 shows several

TABLE 15-1 Low Impact Development Planning Techniques to Reduce Post-Development *CN*

Practice	Process
Limit use of sidewalks	Reduces impervious surface, allows disconnectivity
Reduce length and width of roads	Reduces impervious surface, allows disconnectivity
Reduce driveway length and width	Reduces impervious surface, allows disconnectivity
Conserve natural resource areas	Maintains existing land cover, hydrologic soil group (HSG), and infiltration capacity
Minimize disturbance	Maintains existing land cover and hydrologic condition
Preserve infiltratable soils	Maintains existing HSG and infiltration capacity
Preserve natural depressions	Maintains existing land cover, percent imperviousness, and surface storage
Create transition zones	Maintains permeable land cover, reduces percent imperviousness
Use vegetated swales	Maintains permeable land cover
Preserve vegetation	Maintains permeable land cover, maintains evapotranspiration, maintains surface storage

Source of data: Prince George's County, Maryland.

practices to reduce *CN* and how those practices relate to physical processes.

Reducing *CN* helps to reduce the volume of runoff that must be managed, but often not enough to meet predevelopment conditions. The next step is to attempt to match or increase the site's t_c. Peak discharge rate also is dependent on the t_c of the watershed.

MAINTAINING OR INCREASING T_C

Rainfall intensity is inversely proportional to rainfall duration. A basic assumption in storm hydrology is that the most intense storm occurs over a time equal to the time of concentration. The increased rainfall intensity results in a steeper hydrograph and increased peak discharge. Practices that increase t_c include (DoD, 2004):

- Using flatter slopes
- Reducing the height of slopes
- Increasing the length of flow paths
- Minimizing the use of curb and gutters
- Minimizing storm sewers
- Using grassed swales
- Minimizing impervious areas
- Disconnecting impervious areas

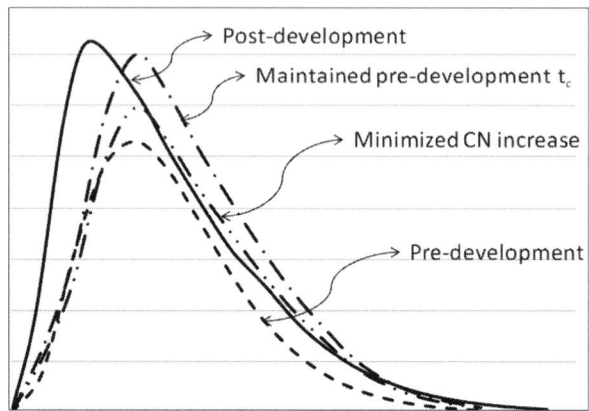

FIGURE 15-5. Hypothetical discharge hydrograph showing the benefit of reducing *CN* and maintaining or increasing t_c. In this case, additional measures will be necessary to maintain the pre-development hydrograph.

- Using weirs and check dams in swales
- Preserving naturally vegetated areas and existing topography

Minimizing the site's composite *CN* reduces runoff volume. Maintaining or increasing t_c reduces Q_p and further reduces runoff volume (Figure 15-5), but in most cases there will still be some increase in runoff. In those cases, IMPs can be selected to meet design objectives.

INTEGRATED MANAGEMENT PRACTICES

IMPs are distributed, multifunctional, small-scale controls, selected for their ability to achieve the site design water quantity and quality objectives in a cost-effective manner (DoD, 2004). Several of the most commonly used IMPS are discussed in the following paragraphs.

Bioretention

Bioretention areas are depressions with porous backfill and vegetated surface (Figure 15-6). If natural soil permeability is not adequate, an underdrain may be provided to encourage drainage. Bioretention provides both water quality through filtering and sequestering in vegetation, and also provides retention of stormwater to reduce runoff volume. A key consideration of bioretention is that surface ponding not extend past 48 hours. This draining period prevents conditions that encourage mosquito breeding. Bioretention is effective for treating

FIGURE 15-6. This bioretention facility at Beaver Water District's Administration Building in Lowell, Arkansas, captures runoff from parking lots and sidewalks.

stormwater that has run across commercial, industrial, or residential sites. Bioretention is not appropriate along high-speed roadways. Bioretention provides multiple functionality by providing habitat for native insects and a source of food for native wildlife.

Dry Wells

A dry well is a pit filled with aggregate that is located to catch water from roof downspouts or paved areas. Dry wells are small-scale facilities not appropriate for large volumes of water. Dry wells should also not be used where high sediment loads are expected, because the sediment will quickly fill the storage capacity of the well.

Rain Barrels and Cisterns

Rain barrels and cisterns, like dry wells, are placed to catch water from roof downspouts (Figure 15-7). The advantage of rain barrels and cisterns over dry wells is that the water can be reused. Rain barrels are often placed on homes to collect small volumes of water for gardening. Cisterns are larger in volume and are frequently used in commercial or industrial developments, where the water can be used for landscape irrigation or toilet flushing. Anytime water is reused, treatment must be provided to meet the minimum requirements for use. Cisterns should

FIGURE 15-7. A rain barrel installed on a residential downspout. Rain barrels reduce runoff volume and provide an alternate source of nonpotable water for the residence.

not be interconnected with potable water systems because of the risk of cross-connections.

Vegetated or Grassed Swales

Vegetated or grassed swales (Figure 15-8) are shallow grass-covered ditches for water conveyance. Grassed swales slow runoff, facilitate infiltration, and remove pollutants through filtration. Grassed swales can be used wherever runoff velocity can be maintained within acceptable ranges, and runoff volume is within the capacity of the swale.

Infiltration Trenches

Infiltration trenches are trenches that have been backfilled with stone. Infiltration trenches collect runoff during storm events and release it into the ground through infiltration. Infiltration trenches are frequently used behind other management practices that remove oil and grease. Trenches could become clogged if these pollutants were not removed in advance.

Tree Box Filters

Tree box filters are in-ground containers containing street trees. Tree box filters are used in urban areas to intercept runoff from streets and sidewalks and provide temporary storage upstream from storm drain

FIGURE 15-8. This vegetated filter strip at the Missouri Department of Conservation Discovery Center in Kansas City slows runoff velocity, provides infiltration, and traps sediment and pollutants.

inlets. The filters provide some pollution control and provide aesthetic relief in the urban landscape.

Vegetated or Green Roofs

Green roofs incorporate a vegetation layer into the structural component of the roofing system. Green roofs help reduce runoff by slowing runoff velocity across the rooftop and providing storage of water in the soil media. Green roofs also provide a cooling effect for the building and the surrounding area. Green roofs may also provide some pollutant removal.

Filter Strips

Filter strips are bands of dense vegetation planted downstream from a runoff source such as a parking lot. Filter strips slow runoff, provide infiltration, and trap pollutants. To achieve maximum efficiency, take care to ensure sheet flow onto the filter strips.

Rain Gardens

Rain gardens (Figure 15-9) are the smaller brother to the bioretention area. A rain garden is a shallow depression planted with native plants that receives runoff from a rooftop or other impervious area. Rain

FIGURE 15-9. A rain garden placed to collect runoff from a residential rooftop reduces runoff volume and provides aesthetic value to the home.

gardens are typically about 6 inches deep and are designed to pond water for up to 48 hours. Rain gardens require homeowner maintenance such as regular weeding and pruning. Rain gardens are not wetlands. They are designed to become completely dry between storms. Watering of the garden may be required until native plants are mature.

By managing the modification of the site's CN and t_c and distributed use of IMPs, the LID goal of matching the pre-development hydrograph can be achieved. Local regulations may require control of storms of a specific return frequency, typically the 5-, 10-, 25-, 50- and 100-year storms. If the LID does not provide adequate peak flow reduction, then conventional end-of-pipe management practices are required. Typical end-of-pipe practices include stormwater detention and constructed wetlands. Hydrologic analysis is required to verify the effectiveness of the LID design.

Water Quality

Polluted stormwater runoff or nonpoint source pollution (NPS) is now the leading cause of water quality impairment in the United States (EPA, 2009). Runoff from urban and suburban areas is contaminated with suspended sediment, oxygen-demanding substances, copper, zinc, phosphorus, and nitrogen. These pollutants have deleterious impact on our streams and lakes. Sediment may fill interstitial spaces in streambed gravel, causing loss of spawning area and refugia for macroinvertebrates and small fish. Copper and zinc can be directly toxic to aquatic organisms. Phosphorus and nitrogen are two macronutrients that are responsible for accelerated eutrophication of waterbodies. The algae resulting from eutrophication may also cause extreme diurnal dissolved oxygen swings in streams and lakes, stressing aquatic life.

Much of the pollution in stormwater runoff is contained in the first flush. That means that the majority of pollutants on pavement or lawns are entrained in the first half-inch or so of rainfall. The pollutograph for a particular pollutant precedes the hydrograph.

In natural or agricultural systems where infiltration is a major abstraction to rainfall, rains of one-half inch or less do not produce runoff. Urban systems, because of the dominance of impervious surfaces, produce runoff even in small rainfalls. In most watersheds, these small rainfalls account for the vast majority of events during a year. Many municipalities have adopted ordinances requiring treatment of the first flush to avoid the potential water quality impacts. Frequently, the requirement is to retain the first one-half inch of runoff on-site. Some areas may require retaining a larger storm event.

In LID developments, water quality is maintained through reducing runoff and through the distributed IMPs. In their low impact development manual, the Department of Defense (DoD, 2004) recommends using a treatment train approach for water quality management. Components of the treatment train include: minimization, natural filtration, constructed filtration, evaporation, and pollution prevention. A short discussion of each element is given in the following sections.

Minimization

Water quality is more easily managed in small treatable units. Small treatment units located close to the source of pollution are more effective than larger end-of-pipe facilities. By micromanaging site design, the LID designer can distribute treatment units at each source. An example of distributed treatment is the rain garden. Runoff from a portion of a roof is drained into the garden. The garden is designed to retain the first one-half inch of rain that falls on the roof. Any runoff beyond the first half-inch would overflow the rain garden into the larger drainage system.

Natural Filtration

The soil and vegetation growing on the soil provide physical, chemical, and biological treatment of pollutants in runoff. LID should use these natural processes to effectively treat polluted runoff. Vegetated filters, rain gardens, and grassed swales are IMPs that allow the designer to take advantage of those processes.

Constructed Filtration

If natural filtration cannot meet the water quality need, then filtration facilities that mimic natural processes may be installed at the facility. Examples are bioretention areas, filter trenches, and dry wells. On hardscapes, tree box filters and catch basin inserts can help filter runoff. When possible, constructed filtration should be protected by upstream natural filtration elements, to prevent rapid clogging of the filtration media.

Evaporation

By retaining water in shallow depressions, the evapotranspiration process can be increased. When the water evaporates, solids—and pollutants sorbed to those solids—are left behind. Native vegetation can then remediate the pollutants through biological processes. Rain gardens and bioretention are IMPs that provide retention of water in shallow depressions.

POLLUTION PREVENTION

Pollution can be prevented through occupant education, nutrient management on lawn surfaces, and street sweeping. All of the water quality elements in the treatment train require some occupant attention, but pollution prevention is mostly their responsibility. A strong occupant and public education/awareness program should accompany all LIDs.

The water quality treatment train is integral to the LID design. If the stormwater retention facilities incorporated into the LID design are not adequate to meet local ordinances, then conventional end-of-pipe treatment will be required. Even in these cases, LID minimizes the need for end-of-pipe treatment, preserving land for the developer as well as capital.

HYDROLOGIC ANALYSIS

In LID, as opposed to conventional development, the designer is thinking in terms of micromanagement and distributed control. Conventional stormwater engineering utilizes end-of-pipe treatment based on flow from the entire site. The LID designer must employ micromanagement in the hydrologic analysis of the project. The following steps in analysis are taken from the Prince George's County (1999 b) LID manual.

The initial step in hydrologic analysis for LID is the same as for conventional development. Collect the necessary site data to determine pre-development CN, t_c, and discharge volumes and rates, as discussed in Chapter 6.

LID Step 1: Determine the LID runoff curve number. Divide the site into elements of homogeneous land cover and hydrologic soil group. Find the CN for each design element from standard tables. Compute the area weighted composite CN for the project:

$$CN_p = \frac{CN_1 A_1 + CN_2 A_2 + \cdots CN_n A_n}{A_1 + A_2 + \cdots A_n} \quad (15.4)$$

where

CN_P = composite CN for site based on pervious area
CN_n = CN of the n element
A_n = area of the n element.

Calculate LID curve number based on connectivity of impervious area: Disconnecting impervious area effectively reduces the curve number. To find the effective CN based on connectivity, apply the following formula to the composite CN:

$$CN_c = CN_p + \frac{P_{imp}}{100} * (98 - CN_p)(1 - 0.5R) \quad (15.5)$$

where

CN_c = composite CN based on connectivity
P_{imp} = percent impervious cover
R = ratio of disconnected impervious area to total impervious area

Disconnected impervious area is the impervious area that drains across a vegetated area before entering the drainage system.

LID Step 2: Determine pre- and post-development time of concentration using any standard hydrological procedure. Every effort should be made to maintain post-development t_c the same as or longer than pre-development.

LID Step 3: Determine the design storm: The LID design storm may or may not be the same as the regulatory design storm. Regulatory requirements may specify a longer return frequency design storm for flood control issues. Frequently, flood management regulations require analysis to ensure no increase in peak discharge from the 2- and 10-year storms. Some areas may require no increase in the 100-yr peak discharge as well. For LID purposes only, a storm that occurs several times per year may be selected. The design storm will have to be selected based on local conditions and requirements.

LID Step 4: Determine LID stormwater management requirements for the design storm: Four conditions must be analyzed: volume required to maintain pre-development runoff volume with retention, volume required to maintain pre-development peak discharge with 100 percent retention, volume required to maintain detention pre-development peak discharge with 100 percent detention, and the water quality volume. Retention storage does not release water except through infiltration and/or evaporation. Detention storage has a controlled release that maintains some specified peak discharge.

The retention required to maintain pre-development runoff volume is the difference in between runoff post-construction and runoff pre-construction. For complicated sites, one of the available computer models may be used to compute runoff volume.

For illustration, consider a simple system with a pre-development CN of 50 and a LID CN of 60, and an area of 5 acres. Also assume that the design storm 24-hour rainfall is 5 inches. Using the NRCS method:

$$Q(inches) = \frac{(P - 0.2S)^2}{(P + 0.8S)}$$

$$S = \frac{1000}{CN} - 10$$

Pre-development

$$S = \frac{1000}{50} - 10 = 10$$

$$Q = \frac{(5 - 0.2 * 10)^2}{(5 + 0.8 * 10)} = 0.7 \text{ inches}$$

post-development

$$S = \frac{1000}{55} - 10 = 8.18$$

$$Q = \frac{(5 - 0.2 * 8.18)^2}{(5 + 0.8 * 8.18)} = 0.98 \text{ inches}$$

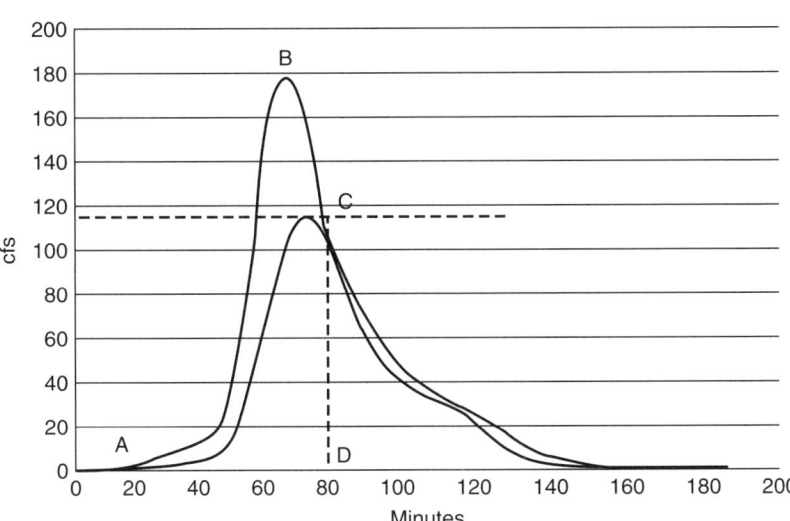

FIGURE 15-10. The retention and detention volume requirements to maintain pre-development peak discharge. Retention volume required is the area inside figure A-B-C-D-A. Detention volume required is the area inside figure A-B-C-A.

The retention volume is therefore:

$$0.98 \text{ inches} - 0.7 \text{ inches} = 0.28 \text{ inches}$$

To convert this volume to cubic feet, multiply 0.28 inches by the area of the site in square feet, and divide by 12. For this 5-acre site, the retention volume for maintenance of pre-development runoff volume is 5,082 cubic feet.

The requirement to maintain peak discharge using 100 percent detention can be computed from the site discharge hydrographs. In Figure 15-10, the requirement is the total area under the post-development hydrograph until the receding limb intersects with the pre-development peak discharge rate (area A-B-C-D-A). A close approximation of the area can be computed using a simplified triangular hydrograph or by integrating to find the area under the post-development hydrograph to point C. The requirement to maintain peak discharge using detention is the area between the post- and pre-development hydrograph, area A-B-C-A.

The hydrograph shows that the volume required for detention is always less than the volume required for retention. An intermediate volume may be found by using a hybrid approach with some retention and some detention, if desired.

The volume required to maintain pre-development peak discharge with retention may or may not be less than the volume required to maintain pre-development runoff volume. Both calculations must be made to verify the largest volume requirement.

The water quality volume is frequently specified by local ordinance as the volume equal to one-half inch of runoff over all impervious area on the site. To find this volume, multiply the impervious area in square feet by 0.042 feet ($\frac{1}{2}$ inch converted to feet).

LID Step 5: Find the required area for distributed retention IMPs (retention areas, rain gardens, etc.). From Step 4, the required retention volume for the site is the largest volume requirement (runoff volume, peak discharge management, or water quality). This volume is the total volume of distributed retention on the site. The surface area of retention IMPs is the required volume divided by the average depth of water in the elements.

$$A_r = \frac{V_r}{d} \tag{15.6}$$

where
A_r = area required for disturbed retention design elements,
V_r = volume required for retention,
d = depth of water allowed in retention elements.

The retention IMPs should be uniformly distributed around the site. The soil's saturation infiltration rate should also be checked to ensure that the retention IMPs drain completely within 48 hours of filling.

REFUGIA

LID is primarily a hydrologic-based approach to development. Even so, the distributed treatment unit's overall design can provide other ecosystem services. Thinking ecologically and looking for multifunctionality can build habitat and refugia into the development. Preserving sensitive landscapes such as wetlands, riparian zones, and hillsides simultaneously maintains *CN* and adds natural habitat refugia and corridor functions to the site.

An ecotone is a zone where two ecosystems overlap. In an ecotone, characteristics of both ecosystems intertwine. Biodiversity is high because species from both ecosystems inhabit the space. In LID projects, ecotones occur when natural areas such as floodplains, wetlands, or hillsides are preserved. This is an example of thinking multifunctionally. Preserving sensitive landscapes provides storage of runoff, reduces the watershed's CN, increases the t_c, and provides an ecotone to increase biodiversity.

Edges are sharp transitions between ecosystems. A few species benefit from edges, but in general, species avoid edges for the comfort of their normal habitat. Transitions in ecotones should be gradual, rather than sharp, to minimize the edge effect.

On a regional scale, connectivity is important to maintaining biodiversity. Both plant and animal species thrive better in connected landscapes. Local disturbance may wipe a particular species out of a section of the landscape, but if it is connected to other sections, replacements can be recruited to fill the niche. Genetic diversity is also maintained if individual animals can move between communities. Corridors that connect intact patches in the landscape allow movement of both individuals and genetic material between patches. Both riparian and upland corridors can be preserved in LID with careful site planning. Corridors, if preserved for movement of local fauna, should be wide enough to provide a gradual transition between developed land and the interior habitat of the corridor.

ECOSYSTEM SYSTEM SERVICES ASSESSMENT/DESIGN

Low impact development is based on maintaining natural hydrologic conditions. Ecological services related to the hydrology that are protected or enhanced are: conveyance of water and sediment, habitat and refugia, water temperature regulation, water quality regulation, and aesthetic and spiritual values. The LID design process differs from conventional design in that site hydrology is integral to design, rather than fitted in to accommodate development. The LID design process can be summarized into five steps (DoD): define project goals and objectives, perform site evaluation and analysis, develop LID control strategies, design the LID site, and develop operation and maintenance procedures. While the step-by-step process is somewhat rigid, it provides a very efficient way to plan a LID project.

Step 1: Define Project Objectives and Goals

The four aspects of stormwater control are runoff volume, peak discharge rate, flow frequency and duration, and water quality. Using the procedures outlined earlier, evaluate the existing stormwater system with respect to each of the four aspects. Based on the evaluation, determine goals and the feasibility of control of volume, peak discharge, flow frequency and duration, and water quality. Consider the possibility of reuse of stormwater while setting goals. Next, prioritize and rank the goals and objectives. Finally, define the hydrologic controls necessary to meet the objectives.

Step 2: Perform Site Evaluation and Analysis

A site evaluation provides details that assist in development of a LID project. Compile the available data on the site, including topographic maps and aerial photographs, drainage maps, utilities, soils maps, and land use plans. Make an on-site investigation to find LID opportunities such as pollutant-generating areas, disconnects from combined sewers, potential green corridors, and areas where water quality and quantity controls can be installed. Determine site constraints, including available space, infiltration characteristics, water table, slope, drainage patterns, sun and shade, wind, critical habitat, and underground utilities. Identify areas for protection including wetlands, Karst features, floodplains, and steep slopes. Evaluate deed restrictions such as easements, as well as regulatory restrictions such as building setbacks. Delineate the watershed and micro-watershed areas, taking into account previously modified drainage patterns. If available, compile baseline hydrologic and water quality data from stream gages, sampling sites, local averages, or modeling results. Review local regulations and codes with respect to land development.

Step 3: Develop LID Control Strategies

To minimize the runoff potential of the development, the hydrologic evaluation is made an ongoing part of the design process. Site drainage can suggest locations for both green areas and building sites. Open drainage systems help integrate the site with natural features and create an aesthetically pleasing landscape. Substeps for developing control strategies are:

1. *Determine the design storms*: Local regulations may stipulate design storms that must be considered in design. Those requirements may limit or constrain the use of LID techniques, or necessitate structural controls employed in conjunction with LID.
2. *Select a stormwater modeling technique*: The model selected depends on the type of development, complexity of planning goals, and designer's preference. For most LID projects, the

NRCS's TR55 or TR20 hydrologic model is adequate, and both are available in the public domain. Most project reviewers will be familiar with the TR55 and TR20 protocols. More complex developments may require more complex models.
3. Evaluate current conditions: Using the results of the modeling, estimate baseline values for the four key aspects: volume, peak discharge, frequency and duration, and water quality.
4. *Select nonstructural site-planning techniques*: Minimize total impervious area with alternative roadway layouts, reduced on-street parking, permeable paving materials, and, where possible, reduced road widths and sidewalks limited to one side of roads. Minimize directly connected impervious areas. Modify drainage flow paths to increase time of concentration. Maximize overland flow, use open swales, and increase vegetation. Define the development envelope. Restrict ground disturbance to the extent possible. Phase site grading to maintain ground cover. Locate development on areas with low hydrologic function, such as tight soils.
5. *Evaluate site-planning benefits, and compare with baseline values.* Use the site hydrologic model to evaluate each of the four key aspects in light of the site-planning efforts. This step helps determine if site planning meets the goals and objectives, or if additional practices are needed.
6. *Evaluate the need for IMPs*: If site planning does not meet the LID objectives, then IMPs need to be selected to ameliorate the hydrologic changes. Distribute the IMPs around the site to treat runoff close to its source. Conduct modeling scenarios to evaluate the benefit of the added IMPs.
7. *Evaluate additional needs*: If IMPs cannot provide the needed control for either volume or peak discharge, additional management practices may be required. Conventional end-of-pipe methods, such as stormwater detention and constructed wetlands, can be added if necessary. Following the seven substeps described may eliminate the need for these conventional methods. The result can be savings to the developer.

Step 4: Design LID Site or Master Plan

Sketch a design concept that distributes the LID practices around the project. Use a variety of surfaces. Look for opportunities for multifunctional capabilities such as combining habitat or refugia with stormwater features. Develop a site master plan identifying all key control issues and implementation areas. Specify LID technologies.

Step 5: Develop Operation and Maintenance Procedures

As with drainage systems, proper operation and maintenance of LID practices is required. Unlike those of conventional development, however, many LID drainage practices belong to the property owners. Those owners must be made aware of their responsibility. In a well-designed LID facility, operation and maintenance should be much like normal yard maintenance and be obvious to the owner. Different integrated management practices have different requirements. Some general principles are: Keep IMPs and flow ways clear of debris, water vegetation as required, maintain grassed areas, and prune plantings as needed.

The Department of Defense's five-step approach leaves out one very important step: construction management. In LID, it is essential that the designer be on-site to verify that IMPs are installed, specifications are met, and plans are followed. General contractors may or may not be familiar with LID practices. Proper functioning of IMPs is not always intuitively obvious. If left to their own ingenuity, contractors without LID experience will revert to what has worked successfully for them in the past. That probably will not be what the designer planned.

Even though LID's primary objective is maintenance of pre-development hydrologic conditions, because of LID's dependence on vegetative practices it also provides other ecological services (Figure 15-11). Terrestrial ecosystem services provided include nutrient and carbon cycling, terrestrial habitat and refugia for terrestrial species, climate control, and biodiversity. Sites developed with LID also tend to have greater aesthetic value than other developments, and frequently provide higher property value.

SUMMARY

As the world's population grows, anthropogenic use of landscapes is inevitable. The way we use those landscapes will to a large extent determine how well we maintain the ecosystem services provided to us by nature. A particular threat in areas of rapid population growth is impact to the water resource through chemical, biological, and physical alteration of the system. Low impact development recognizes the three axioms of ecosystem services design: (1) everything is connected, (2) everything is changing, and (3) we are all in this together. With its emphasis on maintaining pre-development runoff volume, peak discharge, frequency and duration of flow, and water quality, LID is one tool to help meet the needs for human development and the needs of

FIGURE 15-11. Architect's model of Habitat Trails, Rogers, Arkansas. The Habitat Trails development is a LID low-income development. Drainage is via grassed swales between houses and along the streets into a wet meadow retention area. Providing pervious pavers along the sides to meet fire department criteria for width minimizes the impervious paved area of streets. Drives and parking spaces are crushed brick to maintain perviousness. In this development, green space is retained to provide a common lawn/playground. (University of Arkansas Community Design Center)

the ecosystem. The fact that LID also provides higher property values and improved aesthetic values is an added benefit.

Looking to the future, professionals will need to understand ecological functions, use those functions to meet the joint needs of mankind and the ecosystem, work across disciplines, and communicate effectively to the public. Today's education system recognizes the need to train professionals to meet these needs. Certifying ecological design professionals can ensure that minimum criteria are met. Ecological certification is the topic of the last section of this book.

> If in a city we had six vacant lots available to the youngsters of a certain neighborhood for playing ball, in might be "development" to build houses on the first, and the second, and the third, and the fourth, and even the fifth, but when we build houses on the last one, we forget what houses are for.
>
> —Aldo Leopold, *A Plea for Wilderness Hunting Grounds*

Further Readings

Department of Defense, *Unified Facilities Criteria, Design: Low Impact Development*, Manual UFC 3-210-10, United States Department of Defense, Washington, DC, 2004.

Prince George's County, *Low Impact Development Hydrologic Analysis*, Prince George's County, Maryland, Department of Environmental Resources Programs and Planning Division, Largo, MD, 1999.

References

Butler, David, and John W. Davies. *Urban Drainage*, Spon Press, London, UK, 2000.

Center for Watershed Protection, *Water Protection Research monograph 1; The Impact of Impervious Cover on Aquatic Ecosystems*, Center for Watershed Protection, Ellicott City, Md, 2003.

DoD, Unified Facilities Criteria, *Design: Low Impact Development*, Manual UFC 3-210-10, United States Department of Defense, Washington, DC, 2004.

EPA, *Water Quality Inventory: Report to Congress*, 2004 Reporting Cycle, EPA 841-R-008-01, United States Environmental Protection Agency, Office of Water, Washington, DC, 2009.

Prince George's County, *Low Impact Development Design Strategies: An Integrated Design Approach*, Prince George's County, Maryland, Department of Environmental Resources Programs and Planning Division, Largo, MD, 1999a.

Prince George's County, *Low Impact Development Hydrologic Analysis, Prince George's County*, Maryland, Department of Environmental Resources Programs and Planning Division, Largo, MD, 1999b.

16

Ecosystem Services Design in Agriculture and Industry

Until man duplicates a blade of grass, nature can laugh at his so-called scientific knowledge.

—Thomas Edison

INTRODUCTION

Assessing ecosystem services is the first step in any ecological engineering project. The difficulty with assessment is that most ecosystem services cannot be measured directly. Thus, surrogate metrics, or indicators, must be utilized to quantify these processes. Several criteria for developing those indicators have been discussed (see Chapter 13). To date, the methods for measuring ecosystem services across the landscape have not been standardized.

There is movement towards standardization, however, in both the agricultural and industrial sectors of the economy. The Food, Conservation, and Energy Act of 2008 (Public Law 110-234, 122 Stat. 923, enacted May 22, 2008, H.R. 2419, the so-called Farm Bill) specifically requires assessment of "environmental services" from agricultural systems:

SEC. 2709. ENVIRONMENTAL SERVICES MARKETS

Subtitle E of title XII of the Food Security Act of 1985 is amended by inserting after section 1244 (16 U.S.C. 3844) the following new section:

SEC. 1245. ENVIRONMENTAL SERVICES MARKETS

(a) TECHNICAL GUIDELINES REQUIRED.—The Secretary shall establish technical guidelines that outline science-based methods to measure the environmental services benefits from conservation and land management activities in order to facilitate the participation of farmers, ranchers, and forest landowners in emerging environmental services markets. The Secretary shall

give priority to the establishment of guidelines related to farmer, rancher, and forest landowner participation in carbon markets.

(b) ESTABLISHMENT. — The Secretary shall establish guidelines under subsection (a) for use in developing the following:

1. *A procedure to measure environmental services benefits.*
2. *A protocol to report environmental services benefits.*
3. *A registry to collect, record, and maintain the benefits measured.*

This law calls for a framework for assessment of agricultural sustainability from an ecosystem services perspective, with measures of specific benefits. Those metrics or indicators are not presently available; however, there are some basic indicators that can be applied to all agricultural systems.

Similar efforts are under way in industries, especially the consumer packaged goods sector, where consumers are asking industries to inventory and reduce their contribution to ecosystem service degradation. As with agriculture, no standard for assessing ecosystem services impacts has been developed for industries. The World Resources Institute (WRI) has developed the Corporate Ecosystem Services Review, the most comprehensive approach available for industries to evaluate and reduce their negative impacts on ecosystem services.

Development of these indicators is the front line for ecological engineering. We manage what we measure. We do not measure ecosystem services well. This chapter provides a review of agricultural indicators for sustainable management of ecosystem services, and WRI's Corporate Ecosystem Services Review methodology. Ecological engineers, working with technology teams, can use these indicators in benchmarking and designing ecosystem services restoration and management strategies.

AGRICULTURAL SUSTAINABILITY INDICATORS

Farm- or field-level indicators of agricultural sustainability should reflect the character and constraining conditions of place. Very few indicators are comprehensive enough to be universally applied across all agriculture in the United States (or the world). Place matters. Therefore, the indicators of agricultural sustainability for assessing farm- or field-level status should be very specific to the locale. There are, however, common variables that can be used to measure field- and farm-level sustainability across almost all agricultural production systems.

Developing a set of indicators for sustainability of agricultural production at the field level is attractive for many reasons: they would provide producers with an explicit, practice-based measure of their local impact; they can be integrated across landscape units (watersheds, ecoregions) to evaluate cumulative impacts and trends; and they can direct interventions and innovations to the source of impact. Current technological innovations in telemetered data acquisition, remote/real time environmental and crop production monitoring, Internet-based reporting, and geographic information system (GIS) database interfaces provide real-time platforms for developing and implementing these innovations. The potential value to agricultural producers includes a site-customized recordkeeping system for production, the ability to model and project practice benefits (pesticide application, irrigation schedules, and crop rotations), and centralization of producer certification applications and records.

The primary indicators at field scale of agricultural sustainability are those that reflect the conditions and character of place. These indicators are most commonly considered by sustainability media: environmental (soil, water, and habitat), social/cultural, and economic. Sources of information about these indicators at varying scales are available (Table 16-1), but the most effective indicators will be those measured and managed at the production unit (field) level contemporary to production practice (Dale and Polasky, 2007).

Summary of Sustainability Indicators

Sustainability indicators by category are summarized in Table 16-1. Each indicator represents a measurable phenomenon, and thus a monitorable and assessable instrument in managing agricultural production for sustainable outcomes. Sources of information aggregated to the national level are available for each indicator.

Environmental Indicators for Soil

Soil is a living matrix, and the foundation for human civilization. Soil evolves over millennia in a location from weathered parent material (underlying geology) and is transported across the globe by wind, water, gravity, and ice. The primary concerns with soil fertility (the ability of a soil to produce crops) include erosion, loss of organic matter, acidification, salinization, and tertiary structure. Field-scale indicators should include measures of each of these variables.

Erosion
All soils erode. It is the nature of soil. However, human activities accelerate this process dramatically. Sustainable soil management would

TABLE 16-1 Field-Scale Indicators for Sustainable Agricultural Production

		Indicator	Source
Environmental	Soil	Erosion	RUSLE2 model, National Resources Conservation Service (NRCS)
		Organic matter loss	NRCS, USGS STATSGO and SSURGO, Conservation Technology Information Center (CTIC)
		Soil acidification	USDA NASS fertilizer, based on local conditions
		Soil salinization	Salt balance index, based on local conditions
	Water	Water quantity	USDA Census of Agriculture
		Water quality	EPA, USGS National Water Quality Assessment (NAWQA) Program
	Habitat	Landform	USGS Seamless GIS, National Land Cover Dataset
		Sensitive ecoforms	USGS Seamless GIS
		Corridor nodes	USGS Seamless GIS, Google Earth
		Refugia	Based on local conditions
Social/Cultural		Educational attainment	USDA Census of Agriculture, US Census Bureau
		Health and well-being	US GAO, US BLS, US DOL National Agricultural Workers Survey
		Farmworker income	US DOL National Agricultural Workers Survey
		Demographics distribution	USDA Census of Agriculture, US Census Bureau
Economic		Output prices	USDA National Agricultural Statistics Service (NASS)
		Input prices	USDA Economic Research Service (ERS)
		Yield efficiency	USDA NASS and ERS
		Risk management	USDA ERS, USDA NASS, USDA Census of Agriculture
		Farm and farmworker income	USDA ERS, state extension crop production budgets, Bureau of Labor Statistics

ideally prevent any erosion, and repair historic erosion by increasing the depth of the top horizon to 1.5 times deeper than the root zone of the crop(s) being produced. This is a rather high standard—justified by the concept that fertile soils should have root capacity greater than the crop being grown—and one that is rarely met. The most common approach is to determine the current or predicted loss rate, using the Universal Soil Loss Equation (USLE) or the Revised Universal Soil Loss Equation (RUSLE2) and comparing that with the tolerable soil erosion rate (T) (NRCS, 1999). Soil management systems that yield less erosion than T are presumed to be sustainable. The difficulty is determining a defensible value of T. This approach takes into account place-specific variables such as soil rainfall, soil erosivity, length and slope, tillage practices, and conservation measures employed at the field level.

Organic Matter Loss
Soil is a living matrix, with bacteria, fungi, viruses, nematodes, and other life forms processing organic matter into biomass, energy, and mineral components. Organic matter in soil is much more than just the root mass and residue left from crop cycles. As much as 40 percent of organic matter in soils is active biomass, competing for nutrients, water, carbon, and oxygen in soil's three-dimensional matrix. Soil-weathering processes associated with cultivation often reduce organic content in the soil matrix. Organic matter increases water- and nutrient-holding capacity of the soil, enhances reaction sites for degradation of biomass, buffers pH and other chemical characteristics, and creates micropores for gas and liquid exchange throughout the soil horizon. Sustainable agricultural production systems must at least maintain, and in most cases should increase, organic content of the soil. Loss of organic content represents the imminent death of the soil matrix.

Soil Acidification
The process of fertilization, especially with ammonia, can result in soil pH shifting to below 5.5 (acidification). In some regions, acid deposition (rain) can accelerate this process. Soil acidification occurs when the acidifying agent exceeds the buffering capacity of the soil, resulting in rapid shift of soil pH. Soil acidification is destructive because it leads to breakdown in the pore structure of the soil, and in some cases solubilization of toxic metals. Acidification can be ameliorated through the application of lime or other basic soil amendments. However, over time, these amendments can lead to salinization. The best protection is to preserve the organic matter of the soil to keep soil-buffering capacity high.

Soil Salinization
Salt content in soils, especially when composed of chloride salts of sodium, potassium, sometimes calcium and magnesium can result in loss of critical tertiary structure; the soil breaks down. The process of salt buildup in soil (salinization) contributes to loss of water infiltration, nutrient processing and exchange, and oxygen transfer. Some salt

occurs naturally in soils, and can be affected by climate, geography, topography, and other variables. Anthropogenic salinization is caused by irrigation with saline water and use of fertilizers that build up salts in the soil. This process can be reduced or even reversed by managing the salt added to soils, and by enhancing drainage. The salt balance index (SBI) is a simple and reasonable indicator of soil salinization. This index is soil- and crop-specific, providing explicit management guidelines to producers at the field scale (Thayalakumaran et al., 2007).

Environmental Indicators for Water

Water resource management is second only to soil management, in importance in sustainable agricultural production. Water resources are not uniformly distributed across the planet. Agricultural production practices have been adapted to the resources that are available in a given place. However, in many places, technology has been employed to enhance water availability in a manner that requires energy, depletes water from other areas, and, in some cases, causes degradation of local water resources. Sustainable agricultural practices must include sustainable use of water quantity and sustainable quality of discharge water.

Water Quantity

Water quantity, or, more specifically, the contemporary availability of water, is central to any agricultural production strategy. Crop production practices that use rainwater directly or through on-site capture and irrigation are clearly the most sustainable. Use of supplemental irrigation water from surface and groundwater sources creates several challenges to sustainable production. Surface waters can contain relatively high salt concentrations, resulting in salinization. Poor irrigation management can increase nutrient and chemical loading to streams, rivers, lakes, and estuaries. Inter-basin transfer of water for irrigation creates water deficits in the source basins and hydrologic regime changes in the receiving basins. Groundwater sources for irrigation that are withdrawn in balance with recharge are potentially sustainable. Irrigation using groundwater that is withdrawn at rates in excess of recharge is unsustainable. Water use allocation conflicts, such as urban versus agricultural, indicate the increasingly unsustainable character of current water resources management practices. In this conflict, water scarcity will become the norm for agricultural producers. Water conservation practices should be place-specific, but the conservation ethic should be universal in sustainable agriculture.

Water Quality
The ultimate indicator of the effectiveness of sustainable agricultural production strategies in a basin is water quality in that basin. Water quality criteria have been established for every waterbody in the U.S. by the U.S. Environmental Protection Agency (USEPA) and each state. These criteria include narrative and numeric standards for parameters such as sediment, biological oxygen demand (BOD), indicator bacteria (fecal and total coliforms), and pH. Criteria for nutrient concentrations (nitrogen and phosphorus) are being developed for each ecoregion. These criteria acknowledge that aquatic ecosystems have an inherent capacity to treat pollutants, and attempt to limit the loads of pollutants to those systems to below their treatment capacity. This strategy has not worked very well, because nonpoint sources of pollution such as agricultural runoff have not been monitored or managed in this process. A sustainable agricultural production system must have no deleterious impact on water quality in the region. Nutrient, sediment, and agricultural chemical loads must be controlled at the edge of the field, and waterbodies must be protected from impact through riparian zones, streamside buffer areas, and other best management practices. Agricultural producers in watersheds with impaired water quality have an increased performance burden with regards to sustainable practices. State and federal water quality management criteria often assess Total Maximum Daily Loads (TMDLs) for pollutants of concern, raising issues about point versus nonpoint sources of sediments, nutrients, and other agriculture-related pollutants.

Environmental Indicators for Habitat

Habitat
The single biggest impact and potential ecological service provided by agricultural production besides feed, food, fiber, and fuel could be enhanced habitat. Habitat measures range from very simple to very complex. The most reasonable measures are those that producers can make themselves and that measure parameters that landholders can affect. These include landform, sensitive ecoforms, corridor nodes, and refugia.

Landforms
The simplest measures of habitat are landforms: forest, meadow, riparian area, pasture, field, and ecotones. Proportional indices of landform provide a reasonable and easy way to measure, monitor, and enhance desirable landforms. Producers should inventory percent of landforms

over time, monitor trajectories, and set explicit goals for maintaining thresholds based on biomes. Ecotones can be measured over time using simple Google Earth or similar imagery and enhanced as part of sustainable agricultural practice strategies by ecoregion. Landforms are surrogates for ecosystem services within ecoregions. Changing landforms changes ecosystem services.

Sensitive Ecoforms
Human activities, especially urbanization, have destroyed critical habitat matrices (ecoforms) in almost every ecoregion on Earth. In the U.S., these are typically identified in association with endangered species (old growth forests, bottomland hardwood forests, and coastal fringe wetlands, for example). These critical ecoforms are inventoried by ecoregion, so agricultural producers have explicit knowledge of their place and proximity to these sensitive systems. Sustainable agricultural production in an area must seek to recover and protect these sensitive ecoforms.

Corridor Nodes
Fragmentation of the ecosystem has created loss of cohesive corridors that are critical for the successful life strategies of many species. The most common corridors are associated with terrestrial mammals, and include riparian and woodland trajectories. However, flight corridors for migratory songbirds and waterfowl are also critical for sustainable agricultural practices. Corridor nodes are the corridor landforms at the edge of property that connect one system with another. These can be measured using simple Google Earth images for parcels, and monitored, managed, and enhanced to achieve sustainable agricultural production goals for each ecoregion.

Refugia
Refugia are places set aside for nonhuman life. These are places designated as NOT for any human use, presence, or activity. They can be quite small and still be effective. Sustainable agricultural producers in intensively managed ecoregions should designate and protect a portion of their land for refugia. This proportion should be landform-specific and sensitive to the conditions of place—endangered or threatened species in the region, loss of critical habitat type, or knowledge of a population of indigenous species.

Social and Cultural Indicators

Social and cultural issues are one of the three categories of sustainability outcome indicators. Social factors include the health and well-being of farmers and their families. Indicators include educational attainment, maintenance of rural culture, and heritage practices. Social and cultural sustainability is often more difficult to measure than economic or environmental sustainability, in part because it has a less direct impact on monetary expenditures or income. Nevertheless, maintaining the social and cultural fabric of the communities who rely on an agricultural livelihood will help ensure a safe, healthful, and enduring food system. Clearly, communities will vary widely in these indicators by region and type of cropping system, but continuous improvement will be a strong signal for sustainability.

Educational Attainment
Education is important not only for farmers' children who may or may not remain on the farm, but also for farmers themselves, to remain knowledgeable and competitive. Education provides the opportunity for children of farmers to either move to new opportunities or remain on the farm and enrich their own farm with their new knowledge and understanding of agriculture, science, and technology. Farm management is becoming increasingly complex as a result of new environmental demands and requirements, new technologies to increase efficiency and productivity, new demands from segmented consumer groups, and global competition. Education is extremely important for maintaining a viable system that keeps up with the quick pace of the global economy.

Health and Well-Being
Farming relies more heavily on physical labor than most industries today. Maintaining the health and well-being of farmers is critical to maintaining a healthy farm and a healthy farm system. In light of the escalating costs of health care and health insurance, maintaining health is a critical factor in economic prosperity. The National Agricultural Workers Survey provides some data for the health of farm owners and farmworkers. Good indicators for the health of those employed on the farm are the level of those with some sort of health insurance, as well as the level of farms that practice best management with regard to toxic chemicals such as pesticides.

Farmer Numbers and Age/Gender/Race Distribution
Although a good number and good distribution of farmers does not guarantee cultural sustainability and prosperity, it generally is a good indicator. Maintaining a good demographic distribution and a healthy population size will not only help sustain a lasting farming system,

but will benefit society as well. Transparency and food awareness will enhance the health of the consumer and the prosperity of the farmer. Many farming communities are growing significantly older because the younger generation is not following in their parents' footsteps. Sustainable communities need both a robust population and a healthy influx of people with new energy and ideas.

Farmworker Income
Income must be sufficient to provide for housing, food, and clothing, at a minimum, for non-owner farmworkers. Benefits should also include sufficient compensation to allow farmworkers to afford education and health services for their families.

Economic Indicators

Economic indicators, the third category of sustainability along with environmental and social/cultural, are equally important. Without economic sustainability, there will be no long-term system. Food will instead be produced in other places, where the economics are more favorable. Economic factors are easier to measure because they have direct monetary impacts. However, economic conditions are fairly volatile, because global competition for both inputs and outputs affect prices. Therefore, risk management, in terms of diversification of income source and debt management, will help maintain an economically sustainable farm system.

Output Prices
Output prices depend not only on supply and demand of the crop itself, but also on prices of other competing crops. These prices can be easily measured, but can change rapidly based on global supply.

Input Prices
Farmers require numerous inputs, many of which are decreasing in availability and/or increasing in cost. These include fuel and energy sources, water, and energy-intensive fertilizers and pesticides. In addition, as a result of changing demographics and immigration patterns, labor costs are increasing for many agricultural sectors.

Yield Efficiency
Farmers can optimize their income and minimize their environmental impacts by increasing their efficiency of input to outputs. Using less fertilizer to produce the same yield will reduce both the cost to

the farmer and the cost to the environment in terms of energy usage and greenhouse gas emissions from production of fertilizers and soil nitrous oxide emissions. Reduced-tillage and no-tillage practices can also decrease fuel usage, maintain organic matter, and minimize soil carbon dioxide emissions.

Risk Management
Because of a growing global economy, there is an increasing fluctuation of input and output prices. In addition, generally, long-term trends of input prices are going up, and output prices are going down. Therefore, farmers must begin or continue to manage these price risks, in addition to yield risk due to weather or infestation. This means management of cash flow to stay afloat when short-term supply and demand impacts are negative. It also means the diversification of income sources, whether by producing multiple crops, specialized crops with market segmentation and high value-added (e.g., organic or biodynamic) appeal, nontraditional farm income (e.g., agri-tourism), or off-farm income.

Farm and Ag Sector Income
Output prices and input prices in large part determine the incomes of farmers. This can be measured in many ways, including farm income, average household income, agricultural GDP, and economic impact.

Field-Scale Indicators

Development of thresholds for indicators of sustainable agricultural ecosystem services should be a collaborative effort of stakeholders in the Ecosystem Services Advisory Committee (ESAC). This list of field-scale indicators was developed in order to provide a starting point in assessing ecosystem services on agricultural lands by indicator (Table 16-2). The indicators are expressed as guidelines for unsustainable boundaries in agricultural production practices. Some indicators are absolute benchmarks, and can be measured at the field edge. Many of the proposed indicators are directional trends, rather than absolute benchmarks. Sustainable ecosystem services management is a process of iterative improvement, rather than absolute attainment.

Agricultural sustainability categories (environmental, social/cultural, and economic) are natural divisions across a continuum that in total represents agricultural production. Any consideration of a single category or weighting of categories within an index or scorecard represents an explicit value judgment. Caution should be exercised

TABLE 16-2 Field-Scale Indicators for Sustainable Agricultural Production

Category		Indicator	Threshold
Environmental	Soil	Erosion	Soil erosion above T is unsustainable; ultimately, the goal is no net loss of soil from any agricultural production activity.
		Organic matter loss	Soil organic matter that is declining indicates unsustainable agricultural practices. Soil with organic matter below regional soil-type benchmark is marginal, and should be remediated as part of a sustainable agricultural management strategy.
		Soil acidification	Soil pH that deviates from soil type characteristic pH could potentially result in unsustainable soil characteristics, and should be remediated as part of a sustainable agricultural management strategy.
		Soil salinization	Soil salinity that exceeds the soil type characteristic salinity indicates a potential problem with agricultural production practices and should be remediated as part of a sustainable agricultural management strategy.

	Water	Water quantity	The following water use practices are inherently unsustainable:
			    Irrigation practices that use groundwater sources withdrawn in greater quantities than recharge rates
			    Agricultural production using water from inter-basin transfers resulting in degraded ecosystem services from the source basin
			    Irrigation practices that create water resource demands in excess of rainfall and surface water storage supplies in water-stressed areas

		Water quality	Use of water in any way that results in the discharge of water from an agricultural production facility with contaminants significantly above background levels in the receiving body is unsustainable. Primary pollutants of concern, in order of concern, are: sediment, nutrients (nitrogen and phosphorus), bacteria (pathogens and other coliforms), salinity, pH, toxics (pesticides, herbicides, etc.), and organic matter.
	Habitat	Landform	Sustainable landforms are those that are not decreasing ecological services. Agricultural practices that change landforms to reduce critical ecological services are unsustainable.
		Sensitive ecoregions	Agricultural production strategies that reduce or in any way negatively impact sensitive ecoregions are unsustainable.
		Corridor nodes	Agricultural production strategies that reduce or in any way negatively impact corridor nodes are unsustainable.
		Refugia	Agricultural production strategies that reduce or in any way negatively impact refugia are unsustainable. In addition, any production strategy that does not incorporate a habitat management plan into the overall agricultural production plan is unsustainable.

(continues)

TABLE 16-2 (*continued*)

Category	Indicator	Threshold
Social/Cultural	Educational attainment	Agricultural production strategies that prohibit or in any way impede educational opportunities for children are unsustainable.
	Health and well-being	Agricultural production strategies that do not explicitly manage the health and well-being of workers are unsustainable. Sustainable agricultural production strategies must incorporate worker safety management plans and monitoring into the overall production strategy.
	Demographics distribution	Producer demographics that represent the dominant population are sustainable; producer demographics that are controlled by a subset of the population are not sustainable.
	Farmworker income	Agricultural production practices that generate farmworker incomes less than a living wage for the worker are unsustainable.
Economic	Output prices	Output prices that vary more than 25% in a cropping season create unsustainable stresses on producers.
	Input prices	Input prices that vary more than 25% in a cropping season create unsustainable stresses on producers.
	Yield efficiency	Yield efficiencies that decline are unsustainable.
	Risk management	Agricultural production strategies that do not provide economic risk management, including access to credit, are unsustainable.
	Farm and farmworker income	Agricultural production practices that generate farm producer or worker incomes less than a living wage are unsustainable.

in developing such normative aggregators, as they will be inherently biased. Such indicators should be developed by the ESAC to ensure stakeholder representation in normative judgments. Social and economic indicators are particularly prone to cultural bias; for example, the use of child labor in crop harvest may be seen as exploitative in some cultures, and as part of the norms of another. The pastoral family farm in American folklore was only possible with compulsory child labor (chores). The distinction between economic opportunity and exploitation is a tricky one. Clearly, any compulsory labor that creates unhealthy or unsafe conditions for a child is unethical. Similarly, any situation that denies a child an opportunity for an education is unethical and unsustainable. Finally, sustainability in this context is an iterative process, dependent upon incremental categorical improvements. Innovation, communication, and exploration are critical foundations for sustainable agricultural production. Investment in these foundations is critical if agricultural production practices are to continue to evolve to sustainable thresholds.

INDUSTRIAL SUSTAINABILITY METRICS

The loss of ecosystem services at a global scale affects more than common resources. The supply chains of industries can be affected in dramatic ways, leading to loss in productivity and profitability. Many industries are engaged in mapping and inventorying their supply chain impacts on ecosystem services. The World Resources Institute (WRI) developed the Corporate Ecosystem Services Review (ESR) as a tool for supporting informed decisions regarding supply chains and ecosystem services impacts (WRI, 2008).

The ESR is organized into five discrete stages, or steps (Figure 16-1). These steps lead a company through the process of evaluating ecosystem services impacts across their supply chain. The process can take as long as four months to complete, and represents a comprehensive evaluation of corporate activities. Much of the work is associated with supply chain inventory assessment, which supports other corporate needs, and can enhance opportunities in efficiency beyond the assessment of ecosystem services.

Step One: Scope Selection

Identifying the scope of the ESR is critical for defining boundaries for the processes under consideration. Typical scopes can be product-level assessments (following all impacts from Product A from cradle to grave in a life-cycle approach) or process-level assessments (evaluating all agricultural products in the company supply chain for impacts on water resources, for example). The process of selecting the scope for an ESR usually involves answering three key questions (Figure 16-2):

1. Which stage of the value chain is being considered? This can be limited to just the company, can extend back to the suppliers, and can even reach forward to the consumers.
2. Specifically, who will perform the ESR, and where will it be performed? If suppliers are included in the scope, which suppliers? Which sectors? Which geographic areas? The ecological engineer will also need to define the aspects of the company being evaluated, and if downstream (customer) impacts are in the scope, these will also need to be explicitly defined.

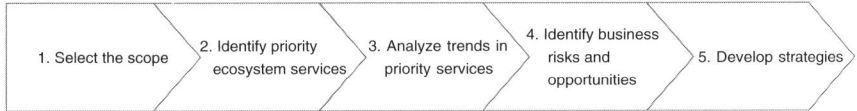

FIGURE 16-1. Steps in a corporate ecosystem services review. (*Source*: WRI, 2008.)

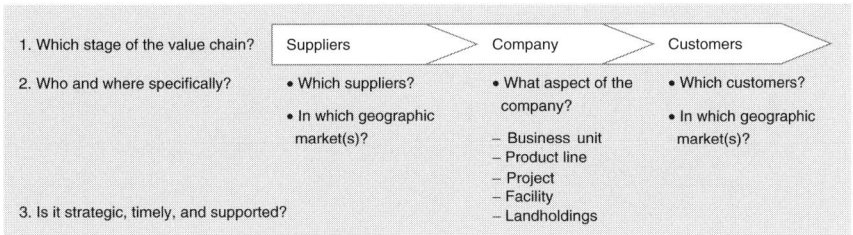

FIGURE 16-2. Scope selection considerations in a corporate ecosystem services review. (*Source*: WRI, 2008.)

3. Is the ESR scope strategic, timely, and supported? This is self-evident, but the scope should be consistent with other initiatives within the company, and comply with the structure and culture of the company. The purpose of an ESR is to inform and motivate decisions; this process requires buy-in of management up and down the supply chain.

Step Two: Identify Priority Ecosystem Services

The ecosystem services that are priorities for the company should be identified using a rational strategy for inventorying and assessing impacts of products and activities. The WRI protocol uses a sequential ranking process (WRI, 2008). However, the most common way for a company to evaluate company impacts is using life-cycle assessments (LCAs). Life-cycle assessments provide quantitative process models to evaluate production processes, analyze options for innovation, and improve understanding of complexity in agricultural production systems. An LCA can identify processes and areas where process changes—potentially enabled by new research and development—can significantly reduce the associated impacts. Broadly, LCAs consist of four stages (Figure 16-3):

1. Define the goal and scope.
2. Conduct life-cycle inventories (collection of data needed to perform the necessary calculations).

322   Ecological Engineering Design

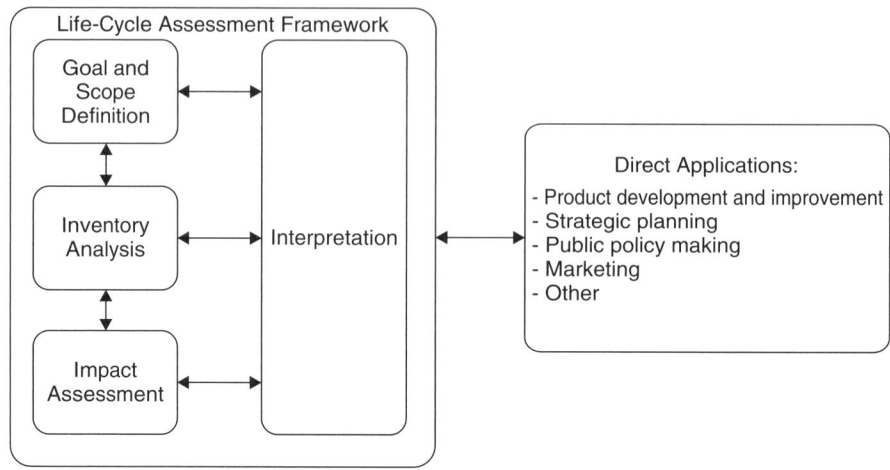

FIGURE 16-3. Schematic diagram of life-cycle assessment protocols.

3. Perform impact assessment.
4. Analyze and interpret the results.

The structure of an LCA is determined by its purpose. The scope of LCAs can be as broad as *"all material and energy inputs and outputs of a process or product"* or as specific as *"water use in production of a crop."* The goal and scope definition phase is Step One in the ESR, which involves defining and describing the product, process, or activity; establishing the aims of the LCA and the context in which it is to be performed; and identifying the life-cycle stages and environmental impact categories to be reviewed for the assessment. The depth and breadth of LCAs can differ considerably, depending on their goals.

The life-cycle inventory (LCI) analysis phase is an inventory of input/output data with regard to the system being studied; it involves identifying and quantifying energy, water, materials, and environmental releases (e.g., air emissions, solid wastes, wastewater discharge) during each life-cycle stage. The life-cycle impact assessment (LCIA) phase evaluates human and ecological effects of material consumption and environmental releases identified during the inventory analysis. Life-cycle interpretation is the final phase of the LCA procedure, in which the results of an LCI or an LCIA, or both, are summarized and discussed as a basis for conclusions. This phase is aimed at identifying the most significant environmental impact category and the life-cycle stage. An LCA for comparison of processes should be structured following ISO 14040:2006 and ISO 14044:2006. These standards provide

Key: ● High ○ Medium Low + Positive impact − Negative impact ? Don't know						
	Suppliers		Company operations		Customers	
Ecosystem service	Dependence	Impact	Dependence	Impact	Dependence	Impact
Provisioning						
Crops				○ −		
Livestock				● −		
Capture sheries						
Aquaculture						
Wild foods				○ +		
Timber and other wood ber				● +		
Other bers (e.g., cotton, hemp, silk)						
Biomass fuel			○	● +		
Freshwater			●	● −		
Genetic resources			○	○ ?		
Biochemicals, natural medicines, and pharmaceuticals				○ +		
Regulating						
Air quality regulation				? ?		
Global climate regulation			○	● +		
Regional/local climate regulation			○	○ +		
Water regulation			●	● −		
Erosion regulation			○	○ −		
Water puri cation and waste treatment				○ −		
Disease regulation						
Pest regulation						
Pollination						
Natural hazard regulation						
Cultural						
Recreation and ecotourism				● +		
Ethical values				○ +		

FIGURE 16-4. Example priority ranking of ecosystem services in a corporate ecosystem services review. (*Source*: WRI, 2008.)

an internationally agreed-upon method of conducting the LCA, but leave significant degrees of flexibility in methodology for customizing individual projects. The prioritization process in ecosystem services assessment should include a process-dependent assessment of dependence and impact across the supply chain (Figure 16-4).

Step Three: Analyze Trends in Priority Ecosystem Services

The ecosystem services that are priorities must be evaluated based on trends and impacts from company activities. The analysis should address the state of knowledge regarding the condition of ecosystem services at each scale in the supply chain, trends in each ecosystem service, the direct and indirect drivers for those trends, and the impacts of each actor on those services (company and others) (Figure 16-5).

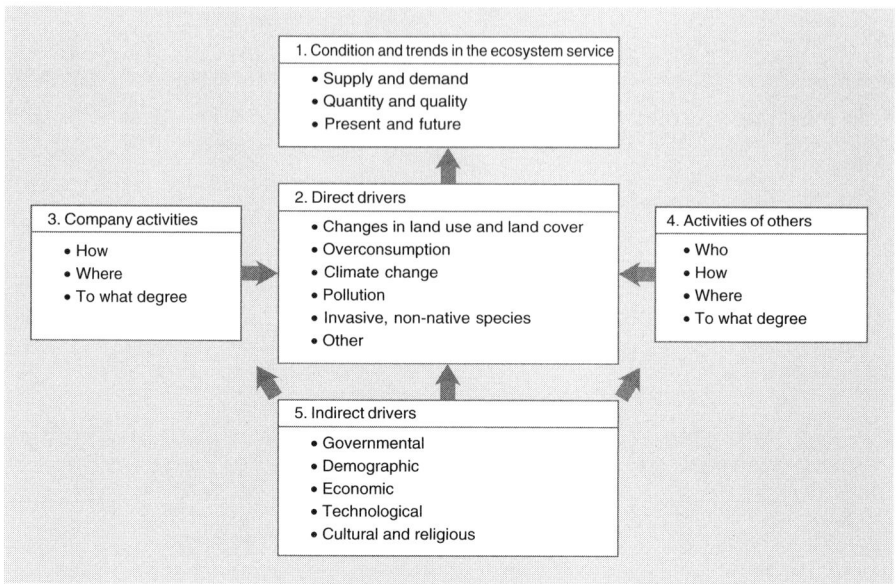

FIGURE 16-5. Ecosystem services trends and drivers in a corporate ecosystem services review. (*Source*: WRI, 2008.)

This process will be challenging because of the lack of data associated with almost all supply chain processes. The LCA can help inform this process to a certain degree, but this is where the ESAC can be most helpful in providing local knowledge and expert judgment. As ecosystem services inventories are developed and databases expanded, the utility function of this step will increase. The ecotechnology team should be able to anticipate the stresses of a given ecosystem service based on common stresses such as land use change, water use, overexploitation, pollution loads, and others. The ecosystem services design process described in Chapter 3 should be used, whereby the ecotechnology team develops conceptual models of impacts across ecosystem services.

Step Four: Identify Business Risks and Opportunities

In Step Four of an ESR, the implications for the company of the findings from Steps One, Two, and Three are integrated into an analysis of risks and opportunities (Figure 16-6). Industries face many risks from loss of ecosystem services. These include operational risks (loss of production capacity), regulatory and legal risks (fines, restrictions in operations), reputational risks (being associated with unpopular

Type	Risk	Opportunity
Operational	• Increased scarcity or cost of inputs • Reduced output or productivity • Disruption to business operations	• Increased efficiency • Low-impact industrial processes
Regulatory and legal	• Extraction moratoria • Lower quotas • Fines • User fees • Permit or license suspension • Permit denial • Lawsuits	• Formal license to expand operations • New products to meet new regulations • Opportunity to shape government policy
Reputational	• Damage to brand or image • Challenge to social "license to operate"	• Improved or differentiated brand
Market and product	• Changes in customer preferences (public sector, private sector)	• New products or services • Markets for certified products • Markets for ecosystem services • New revenue streams from company-owned or managed ecosystems
Financing	• Higher cost of capital • More rigorous lending requirements	• Increased investment by progressive lenders and socially responsible investment funds

FIGURE 16-6. Risks and opportunities inventory in a corporate ecosystem services review. (*Source*: WRI, 2008.)

or illegal activities), market and product risks (brand competition), and financing risks (lending restrictions based on corporate behavior) (WRI, 2008). These risks also represent opportunities for enhanced management and process control. The ESR was designed by WRI to provide a structured methodology for companies to develop strategies for managing risks and identifying opportunities associated with ecosystem services.

Step Five: Develop Strategies for Addressing Risks and Opportunities

In Step Five of an ESR, the company develops strategies for addressing risks and opportunities associated with ecosystem services in their supply chain. The objective is to provide managers within the company with explicit options for the company in regards to ecosystem services management. These strategies fall into three categories: internal changes, sector or stakeholder engagement, and policy-maker engagement (Figure 16-7) (WRI, 2008). Internal changes are the things companies can do within operations, product or market strategies, procurement, and other areas of direct control. The LCA can inform this process, and the ESAC can provide critical feedback to the company about opportunities for increased efficiency. Ultimately, this process should improve a company's competitive standing in the marketplace through improved supply chain efficiency, increased quality control,

FIGURE 16-7. Categories of strategies for addressing risks and opportunities in a corporate ecosystem services review. (*Source*: WRI, 2008.)

and improved relationships with direct and indirect suppliers and customers.

Sector engagement includes industry groups, cross-sector collaboration, engagement with nongovernmental organizations (NGOs), and other stakeholders to develop collaborative strategies for managing ecosystem services. Identifying the pre-competitive space is important for this strategy to be effective. If management of ecosystem services is a product differentiator within the marketplace, then the ESR could become a competitive vehicle, and sector engagement would likely not be effective. If the process of managing ecosystem services is pre-competitive, then companies can collaborate through NGOs and trade groups to address common problems and concerns, as well as to enhance common opportunities for supply chain safety and security.

Policy-maker engagement often results when a compelling common good requires government engagement in changing policy. Government policy tends to be broad, and thus can be challenged in addressing specific issues. Policies regarding resource allocation can be very effective at motivating behavioral change. Those policies include tax incentives, subsidies, and other economic benefits to support ecosystem services management. Conservation easements, protected areas, and zoning can also be very effective at protecting critical land areas. Organizations such as The Nature Conservancy have been very successful in developing priority areas for conservation.

> Only he who has planted a pine grove with his own hands, or built a terrace, or tried to raise a better crop of birds can appreciate how easy it is to fail; how

futile it is passively to follow a recipe without understanding the mechanism behind it.

—Aldo Leopold

References

The Corporate Ecosystem Services Review, World Resources Institute, Washington, DC, 2008.

Dale, V., and S. Polasky, Measures of the effects of agricultural practices on ecosystem services, *Ecological Economics*, 64(2): 286–296, 2007.

Falk, D., M. Palmer, and J. Zedler, eds., *Foundations of Restoration Ecology*, Island Press, Washington, DC, 2006.

NRCS, *National Soil Survey Handbook*, title 430-VI, Natural Resources Conservation Service, Washington, DC, 1999.

Thayalakumaran, T., M. Bethune, and T. McMahan, Achieving a salt balance – Should it be a management objective? *Agricultural Water Management*, 92(2): 1–12, 2007.

Index

A

Abiotic material:
　habitat impacts of, 198
　in soil, 140, 141
Accounting units (watersheds), 86
A (Equatorial) climate classification, 66, 67
A/D, *see* Analysis and deliberation
Adaptation, 150–152
Aeration, soil, 137–138
Afrotropical realm, 65
Agricultural sustainability indicators, 308–320
　economic indicators, 310, 316–317, 319
　environmental indicators, 309–314
　field-scale indicators, 310, 317–320
　social and cultural indicators, 310, 314–316, 319
Agriculture, 13–14
　ecosystem services for, 247–251
　effects of, on watershed, 101
　land used for, 9, 12
　as largest biome, 80
　and workforce, 7, 33
AIDS, 6, 7
Altitudinal belts, 68
American Society of Civil Engineers (ASCE), 44, 47–48
Analysis and deliberation (A/D), 56, 216, 262
Antarctic realm, 65
Area:
　of site, 106
　species-area relationship, 193–194
　of watershed, 84–87
Arid (B) climate classification, 66, 67
ASCE (American Society of Civil Engineers), 44, 47–48
Aspect, of site, 107
Assemblages, 190
Assessment. *See also* Indicators
　of ecosystem services design, 61–62
　in stream restoration, 223–227
Assimilation efficiency, 171
Atmospheric feedback loops, 211–212
Augmentation designs, 189–190
Australian realm, 65
Autonomy (legitimacy), 49
Average ecological efficiency (EFF), 173, 174
Average energy flux, 173

B

Bank erosion, 236–238
Bankfull, 229, 231–232, 235–236
Base flow, 97, 228
Basins (watersheds), 86
B (Arid) climate classification, 66, 67
BDI (biomass density index), 120, 121
Bed load (sedimentology), 233–235
Bequest uses (ecological services), 246
Best management practices (BMPs), 101, 102
Biodiversity:
　changes in, 91
　and ecosystem services, 29, 31, 35–39
　effect of land use change on, 12
　and refugia, 300
　and wetlands, 259
Biofuels, 9
Biogeochemical cycling, 176
Biogeographical realms, 65–66
Biomass accumulation ratio, 172
Biomass density index (BDI), 120, 121

329

Biomes:
    classification of, 66–71
    and value of ecosystem services, 24
Bioretention, 290–291
Biosphere, 3, 162
Biotemperature, 67
Biotic material:
    in community structure, 190–193
    interconnectedness of, 145
    in soil, 140, 141
BMPs (best management practices), 101, 102
Bogs, 258
Boreal Forests/Taiga biome, 70
Bureau of Reclamation, 234–235

C

CAD (computer-aided design), 85, 106
Carbon:
    and nutrients, 208, 209
    in soil, 142–143
Carbon cycle, 178–181
Catchment, *see* Watershed
C (Warm temperate) climate classification, 66, 67
Chemical characteristics, of soil, 123–124
Cisterns, 291–292
Clean Water Act, 99, 100
Climate:
    and biomes, 66–68
    and hydrologic cycle, 177–178
    impact of, on ecosystem changes, 202–203
    of sites, 124–125
Climate change, 79, 91
Climax community, 199
Cluster development, 273–274
*CN*s, *see* Curve numbers
Colonization sequence, 197
Commensalism, 192–193
Community control processes, 207–210
Community structure, 187–201
    abiotic filters in, 198
    biotic interactions in, 190–193
    disturbance regimes in, 199–200
    habitat heterogeneity in, 200–201
    hierarchical processes in, 187–188
    metapopulations in, 193–195

regional processes in, 195–198
types of restoration design, 188–190
Competition, 191–192, 209–210, 217
Computer-aided design (CAD), 85, 106
Conceptual mapping, 56–58
Condensation, 92
Connections, 155
Connectivity, 240–241, 301
Conservation easements, 272–273
Conservation Fund, 267, 271
Conservation movement, 23
Conservation Units (CUs), 37
Constructed filtration (water quality), 295–296
Consumer feedback loops, 214–215
Consumption, 192
Continental climate zone, 69
Continuum concept, 190
Convective precipitation, 93
Cores (green infrastructure), 269–270
Core species, 196
Corporate Ecosystem Services Review (ESR), 308, 320–326
Corridors, 224, 270
    and dispersal, 155–156
    as environmental indicators, 314
    and land use, 92
    in low impact development, 301
    riparian, 156, 241–242
Croplands, 12, 253. *See also* Agriculture
Cultural indicators, 314–316, 319
Cultural services:
    from forests, 254
    from grasslands, 255
    overview, 26, 29, 32, 34–35
    of soil, 132
    from streams, 222
    from urban areas, 263
    from watersheds, 84
    from wetlands, 257
Culverts, 240–241
Curve numbers (*CN*s), 111–114, 285–289, 297
CUs (Conservation Units), 37

D

D (Snow) climate classification, 66, 67
Deforestation, 11, 252. *See also* Forests

Demographics:
  changes in, 6–8
  of farmer populations, 315–316
Demographic Units (DUs), 37
Deserts and Xeric Shrublands biome, 71
Design teams, 54–55
  and assessment, 61
  designing ecosystem services by landform, 246
  and land use change, 80
  and legitimacy, 48
  and place, 64
Development, low impact, *see* Low impact development (LID)
Development rights, transfer of, 272–273
Direct uses (ecological services), 246
Discharge hydrograph, 117–119
Disease, 14
Dispersal, 196–198
Dispersion, 197–198
Disturbance, 199–200, 240
Divisions (ecoregions), 72
DoD (United States Department of Defense), 295, 304
Domains (ecoregions), 72
Drainage, 89, 283, 302–303
Dryland areas, 11, 12
Dry wells, 291
DUs (Demographic Units), 37

E

E (Polar) climate classification, 66, 67
Ecological design:
  ecological engineering vs., 2
  principles of, 51
Ecological efficiency, 171, 174
Ecological engineering:
  axioms of, 2–3, 50
  defined, 1
  ecological design vs., 2
Ecology, principles of, 148–149
Economic indicators, 310, 316–317, 319
Economic policy, 37, 38
Ecoregions:
  classification of, 72–78
  and climate change, 79
  and land use change, 80–81

Ecosystem control, 202–210
  of community, 207–210
  of population, 204–207
Ecosystem services, 22–39. *See also specific services*
  and biodiversity, 35–39
  classification of, 24–35
  defined, 1, 22
  demands for, 8–9
  and land use, 37
  origin of, 22–24
  value of, 24
Ecosystem Services Advisory Committee (ESAC):
  and agricultural sustainability indicators, 317, 319
  and ecoregion classification, 79
  and industrial sustainability indicators, 324
  role of, in designing ecosystem services, 54–56, 58, 59, 246
  in urban areas, 262
Ecosystem services design, 42–62
  analysis and deliberation in, 56
  assessment in, 61–62, 301–304
  axioms of, 50–51
  elements of, 147
  ethics for, 46–49
  goals in, 59–61
  and legitimacy, 48–50
  management structure for, 55–56
  mapping processes in, 56–58
  priorities in, 58–59
  problems with current methods, 43–46
  process of, 245–247
  role of design team in, 54–55
  synthesis in, 53–54, 62
Ecosystem services indicators, 250–251, 256
Ecosystem services loss index (ESLI), 60
Ecosystem services sustainability index (ESSI), 60, 246
Ecosystem stability, 174
Ecotones, 156, 158, 300–301, 314
Ecotourism, 35, 102
Edge effects, 156, 158, 301
EFF (average ecological efficiency), 173, 174

Effective rainfall, 111–114. *See also* Runoff
Emergent properties, 215
Emergy, 168–169
Eminent domain, 275
Endangered species, 16, 31, 160
Energy balance, 164–168
Energy density, 147–148, 169–170
Energy flow, 164–169. *See also* Trophic levels
  energy as unit of analysis, 168–169
  energy balance, 164–168
Environmental indicators, 309–314
Environmental variation, 152–154
Equatorial (A) climate classification, 66, 67
Equatorial climate zone, 69
Erosion, soil, 309–311
Erosion control, 284
ESAC, *see* Ecosystem Services Advisory Committee
ESLI (ecosystem services loss index), 60
ESR (Corporate Ecosystem Services Review), 308, 320–326
ESS (evolutionarily stable strategy), 153
ESSI (ecosystem services sustainability index), 60, 246
Ethics:
  for ecosystem services design, 46–49
  of sustainability, 3
Ethiopian realm, 65
EUs (Evolutionary Units), 37
Evaporation, 296
Evapotranspiration, 67, 68, 95, 178
Evolutionarily stable strategy (ESS), 153
Evolutionary Units (EUs), 37
Existential uses (ecological services), 246
Exploitation efficiency, 171
Extinctions, 199, 240

F

Farmers, 315–317
Feedback loops (ecosystem element), 162–163
Feedback processes, 210–215
  atmospheric feedback loops, 211–212
  consumer feedback loops, 214–215
  and population growth, 207
  soil feedback loops, 212–214

Fee simple ownership, 272
Fertility, of soil, 139–141
Fertility rates, human:
  and prosperity, 17
  trajectory of, 4–7
Field-scale indicators, 310, 317–320
Filter strips, 293
First Green Revolution (GR1), 16–17, 32
Fitness, of populations, 151
Floods, 43–46, 227–230, 232
Flooded Grasslands and Savannas biome, 70
Floodplains, 99
Flow pathways (ecosystem element), 162–163
Food:
  human demand for, 8–9
  increases in production of, 13
Food, Conservation, and Energy Act, 307–308
Forces (ecosystem element), 162–163
Forests, 11, 251–254
Forestry, 101–102
Framework design approach, 38, 39
Fresh water:
  demand for, 15
  and nitrogen cycle, 182
Frontal precipitation, 92
Froude's number, 239

G

Gadugi cultural idea, 3, 46–47
Genetic differentiation, 36–37
Geographic information system (GIS) software:
  for green infrastructure planning, 271
  and site, 106, 120
  and watershed, 84, 85, 91
Geology:
  of site, 107
  watershed, 89
Geomorphology:
  and stream restoration, 235–238
  and wetlands, 261
GHGs, *see* Greenhouse gases
GIS software, *see* Geographic information system software
Government Land Office (GLO), 91, 120

Government regulation, 275
GPP (gross primary production), 170, 178
Grassed swales, 292, 293
Grasslands, 214, 253–256
Gravelius index, 88
Gray infrastructure, 267, 268, 274
Grazing, 101, 214
Greenhouse gases (GHGs), 183, 211, 261, 279–280
Green infrastructure design, 267–281
  infrastructure network, 268–271
  planning for, 271
  programs for, 263
  and sustainable cities, 275–280
  tools in, 272–275
Green roofs, 293
GR1 (first Green Revolution), 16–17, 32
Gross heterotrophic production efficiency, 171
Gross primary production (GPP), 170, 178
Growth rates, of populations, 206–207
Guilds, 190

H

Habitat:
  defined, 66
  environmental indicators for, 313–314
  field-scale indicators for, 318
  and stream restoration, 238–239
Habitat diversity, 202
Habitat functions, 26
Habitat heterogeneity, 200–201
H (Highlands) climate classification, 66, 67
Heat island effects, 10
HEC-RAS (Hydrologic Engineering Center River Analysis System), 232–234
Heritability, of traits, 151–152
Highlands (H) climate classification, 66, 67
HIV, 6, 7
Holarchical organization, 215–216
HSGs (hydrologic soil groups), 112
Hubs (green infrastructure), 270
HUCs (hydrologic unit codes), 85–87
Humans:
  activities of, in rainforests, 252
  and coastal and estuary species, 175
  and ecosystem, 30–32
  and land use change, 80

population issues, 4–8, 36, 204–207
  and watershed, 100–103
Hurricane Katrina, 43–46, 48
Hydrographs, 96–97, 114, 117–119
Hydrologic analysis, 284, 296–300
Hydrologic cycle, 176–178, 284
Hydrologic Engineering Center River Analysis System (HEC-RAS), 232–234
Hydrologic soil groups (HSGs), 112
Hydrologic unit codes (HUCs), 85–87
Hydrology. *See also* Water
  and low impact development, 284–286, 296–300
  of sites, 107–119
  and soil, 123
  and stream restoration, 227–233
  and wetlands, 258–259
Hypoxia, 214

I

ICLEI (International Community for Leadership in Environment Initiatives), 278–280
Ideal free distribution, 197
Immigration, 7–8
IMPs, *see* Integrated management practices
Indicators:
  agricultural sustainability, *see* Agricultural sustainability indicators
  ecosystem services, 250–251, 256
  industrial sustainability, 320–326
Indirect uses (ecological services), 246
Industrial sustainability indicators, 320–326
Infiltration, 110–111
Infiltration trenches, 292
Information functions, 26
Input prices (agricultural sustainability indicator), 316
Integrated management practices (IMPs), 284, 290–296, 300, 303
Interactions (ecosystem element), 162–163
Interception, of precipitation, 95
Intergovernmental Panel on Climate Change (IPCC), 178, 180
Internal changes, 325

International Community for Leadership in Environment Initiatives (ICLEI), 278–280
Introduction designs, 188–189
Invasive species resistance, 33
IPCC (Intergovernmental Panel on Climate Change), 178, 180

J
Justice, 50

K
Karst terrain, 89, 90
Kirpich, P.Z., 116
Köppen-Geiger climate classification, 66, 67
Kyoto Protocol, 279

L
Land cover, 90–92
Landforms, 245–264
 agricultural lands, 247–251
 and ecosystem function, 154–160
 forests, 11, 251–254
 grasslands, 214, 253–256
 inventories by, 59
 as measures of habitat, 313–314
 metrics, 158–160
 in urban areas, 260–264
 wetlands, 24, 256–260
Land matrix, 119–120
Landscape structure, 91–92
Land use change, 8, 9, 11–12, 17
 and ecoregions, 80–81
 and ecosystem services, 37
 and watershed, 90–92, 101, 223, 224
 zoning issues with, 273
Land use/land cover (LULC) datasets, 120
Latitudinal belts, 68
LCAs (life-cycle assessments), 321–325
Legislation, 276
Legitimacy, 48–50, 55
LID, see Low impact development
Life-cycle assessments (LCAs), 321–325
Life zones, 67–68

Longitudinal stream profile, 225, 226
Low impact development (LID), 282–305
 adoption of, 264
 and assessment, 301–304
 and hydrology, 284–286, 296–300
 increasing t_c in, 289–290
 integrated management practices for, 284, 290–296, 300, 303
 pollution prevention in, 296
 preserving ecological services with, 273
 refugia in, 300–301
 and runoff curve numbers, 285–289, 297
LULC (land use/land cover) datasets, 120

M
Maintenance procedures, 304
Major land resource areas (MLRAs), 78
Malnourishment, 5, 6, 16–17
Management, 55–56, 83
Mangroves biome, 71
Mapping, conceptual, 56–58
Marine coastal systems, 24
Marshes, 258
MASH (minimum amount of suitable habitat), 195, 196
Mass flow, 175–184
 carbon cycle, 178–181
 hydrologic cycle, 176–178
 nitrogen cycle, 181–183
 phosphorus cycle, 183–184
Matrix, of landscape, 92
MDG (Millennium Development Goals), 277
MEA, see Millennium Ecosystem Assessment
Mean annual precipitation, 67, 68, 94
Mediterranean climate zone, 69
Mediterranean Forests, Woodlands and Scrub biome, 71
Metapopulations:
 in community structure, 193–195
 growth rates of, 207
Microbial priming effect, 211
Microhabitats, 238
Microorganisms, 90, 182, 192
Millennium Development Goals (MDG), 277

Index   335

Millennium Ecosystem Assessment (MEA), 24, 28–35, 66
Minimal viable metapopulation (MVM), 195
Minimal viable population (MVP), 194–195, 246
Minimum amount of suitable habitat (MASH), 195, 196
Mining, impacts of, 102
MLRAs (major land resource areas), 78
Montane Grasslands and Shrublands biome, 70
Mutualism, 193
MVM (minimal viable metapopulation), 195
MVP (minimal viable population), 194–195, 246

N

National Agricultural Workers Survey, 315
National Hydrologic Dataset, 87
National Weather Service, 95, 108, 125
Natural filtration (water quality), 295–296
Natural Resources Conservation Service (NRCS):
  curve number method, 112, 113
  generation of synthetic storm patterns by, 109
  and hydrographs, 118
  land classification, 78
  peak runoff rate, 114–117
  and soils, 129
  and watershed, 85, 90, 101
NCP (net community production), 170–171
Nearctic realm, 65
Nemoral climate zone, 69
Neotropical realm, 65
Net community production (NCP), 170–171
Net heterotrophic production efficiency, 171
Net primary production (NPP), 170–171
Net primary productivity (NPP), 173–175, 178, 181, 190, 213
Niches, 153
Nitrogen cycle, 181–183, 208, 209, 211
Nitrous oxide (N_2O), 183
Nonpoint source pollution (NPS), 294

NPP, see Net primary production; Net primary productivity
NRCS, see Natural Resources Conservation Service
Nurseries, 78
Nutrients:
  cycling of, 34
  and plants, 208–209
  in primary production, 173
  from soil, 140

O

Oceanian realm, 65
Operation procedures, 304
Option uses (ecological services), 246
Organic matter loss, 311
Oriental realm, 65
Orographic precipitation, 92, 93
Output prices (agricultural sustainability indicator), 316

P

Palearctic realm, 65
Parasitism, 192
Patches, 92, 154–155, 159
Peak runoff rate, 114–117
Permafrost thaw, 211
Phosphorus cycle, 183–184, 208, 209
Photosynthesis, 34, 98
Place. *See also specific components of place, e.g.:* Sites
  and cultural identity, 34–35
  defined, 64
  and organisms, 149–154
Plant hardiness zones, 78–79, 125
Polar (E) climate classification, 66, 67
Polar climate zone, 69
Policy-maker engagement, 325, 326
Pollination, 33–34
Pollution:
  and land use, 91
  prevention of, 296
  and prosperity, 17
  in runoff, 294–296
  of water, 14, 99–100, 313
Potential evapotranspiration ratio, 67, 68
Potential support ratio (PSR), 7

Poverty, 5, 7
PPR (primary productivity required), 174, 175
Precipitation:
    and hydrologic cycle, 176–178
    mean annual, 67, 68, 94
    at site, 108–110
    and watershed, 92–97
Precipitation frequency analysis, 94, 95
Predator interactions, 192, 214
Primary design, 53
Primary production, 34, 170–173
Primary productivity required (PPR), 174, 175
Procedural legitimacy, 49–50
Production functions, 26
Properties (ecosystem element), 162–163
Prosperity, 17, 32
Provinces (ecoregions), 72
Provisioning services:
    from agricultural lands, 247, 248
    from forests, 254
    from grasslands, 255
    overview, 25, 29, 32–33
    of soil, 131
    from streams, 222
    from urban areas, 263
    from watersheds, 84
    from wetlands, 257
PSR (potential support ratio), 7
Public awareness/education, 272, 284, 296

Q

Quantitative systems ecology, 1

R

Rain barrels, 291–292
Rainfall, effective, 111–114. *See also* Runoff
Rainfall intensity, 289. *See also* Precipitation
Rainforests, 251–252
Rain gardens, 293–294
Recreational tourism, 35, 102
Refugia:
    as environmental indicator, 314
    in low impact development, 300–301

Regional processes (community structure), 195–198
Regulating services:
    from agricultural lands, 247, 248
    from forests, 254
    from grasslands, 255
    overview, 25, 29, 33–34
    from streams, 222
    from urban areas, 263
    from watersheds, 84
    from wetlands, 257
Regulation functions, 26
Reintroduction designs, 189
Relief, 107
Reproductive strategies, of species, 204–205
Resource competition, 209–210
Restoration design, 187–190
Revised Universal Soil Loss Equation (RUSLE2), 310
Reynold's number, 239
Riparian corridors, 156, 241–242
Riparian vegetation, 98, 102–103
Risks, business, 324–326
Risk management, 317
Roads, 287–289
Roofs, vegetated, 293
Runoff:
    and low impact development, 284–290, 294–296
    peak runoff rate, 114–117
    from site, 111–117
    and watershed, 95–97, 102
Rural communities, 261–262
RUSLE2 (Revised Universal Soil Loss Equation), 310

S

Safety, 58
Salt balance index (SBI), 312
Satellite species, 196
Scale:
    in ecosystem services design, 59
    and green infrastructure, 275
SCS (Soil Conservation Service), 109, 111. *See also* Natural Resources Conservation Service (NRCS)
Sector engagement, 325, 326

Sedimentation and River Hydraulics – One Dimension (SRH-1D) model, 235
Sediment control, 284
Sedimentology, 233–235
Self-organization, 216–220
Service Providing Units (SPUs), 37
Sites, 106–127
   biological characteristics of, 119–124
   climatological characteristics of, 124–125
   evaluation and analysis of, 302
   in green infrastructure, 270–271
   hydrological characteristics of, 107–119
   physical characteristics of, 106–107
Site planning, 284
Slope, 107
Slums, 276, 277
Snow (D) climate classification, 66, 67
Social indicators, 310, 314–316, 319
Soil(s), 129–143
   on agricultural lands, 249
   and carbon cycle, 181
   chemical characteristics, 123–124
   ecology of, 141–143
   engineering characteristics, 121–123
   environmental indicators for, 309–312
   feedback loops, 212–214
   fertility of, 139–141
   field-scale indicators for, 318
   hydrologic characteristics, 123
   infiltration of, 110–111
   morphology, 130, 132–135
   and nitrogen cycle, 182
   physics of, 136–139
   roles of, in environment, 129–132
   and site, 120–124
   textural class, 121
   and watershed, 89, 90
Soil Conservation Service (SCS), 109, 111. *See also* Natural Resources Conservation Service (NRCS)
Soil Science Society of America, 129
Soil Survey Geographic (SSURGO) Database, 90
Solar energy, 125, 168–170
Solar radiation, 10
Sørensen similarity index, 195–196
Spatial distribution, 36
Species-area relationship, 193–194
Species diversity, 202
Species pool, 196
Species richness, 36
Spiritual ecosystem services, 35
SPUs (Service Providing Units), 37
SRH-1D (Sedimentation and River Hydraulics – One Dimension) model, 235
SSURGO (Soil Survey Geographic) Database, 90
Stakeholder participation, 56, 58, 271, 317, 319, 325
State Soil Geographic (STATSGO) Database, 90
Storm flow, 97, 227–228
Stormwater engineering, 283–285, 296–297
Stormwater runoff, *see* Runoff
Stream density, 89
Stream evolution model, 225–227
Stream order, 97–99
Stream restoration, 222–243
   assessment in, 223–227
   connectivity in, 240–241
   and geomorphology, 235–238
   and habitat, 238–239
   and hydrology, 227–233
   and riparian corridor, 241–242
   and sedimentology, 233–235
Substantive legitimacy, 50
Subtropical climate zone, 69
Succession, 199
Supporting services:
   from agricultural lands, 247, 248
   from forests, 254
   from grasslands, 255
   overview, 25, 29, 34
   of soil, 132
   from streams, 222
   from urban areas, 263
   from watersheds, 84
   from wetlands, 257
Surface storage, 95
Survivorship, 205–206
Suspended load (sedimentology), 233
Sustainability. *See also* Agricultural sustainability indicators; Industrial sustainability indicators
   in cities, 275–280
   defined, 3
   principles of, 3

Sustainable prosperity, 3
Synthesis (ecosystem services design), 53–54, 62

T

T (tolerable soil erosion rate), 310–311
T_c (time of concentration), 285–286, 289–290
Temperate Broadleaf and Mixed Forests biome, 70
Temperate Coniferous Forests biome, 70
Temperature:
    classification of life zones by, 67–68
    and site climate, 125
    soil, 136–137
Terrestrial landscapes, 55, 68–70, 198, 304
Thermodynamics, 164–165
Time of concentration (t_c), 285–286, 289–290
TMDLs (Total Maximum Daily Loads), 100
Tolerable soil erosion rate (T), 310–311
Topographic heterogeneity, 200–201
Topography:
    defined, 107
    watershed, 88–89
Total Maximum Daily Loads (TMDLs), 100
Tree box filters, 292, 293
Trenches, infiltration, 292
TR55 method, 117
Trophic levels, 169–175
    design of, 173–175
    and energy density, 169–170
    and primary production, 170–173
Tropical and Subtropical Coniferous Forests biome, 70
Tropical and Subtropical Dry Broadleaf Forests biome, 70
Tropical and Subtropical Grasslands, Savannas, and Shrublands biome, 70
Tropical and Subtropical Moist Broadleaf Forests biome, 70
Tropical climate zone, 69
Tundra biome, 71

U

United Nations, 9
United Nations World Urban Forum (UNWUF), 276–278
U.S. Army Corp of Engineers, 258
U.S. Forest Service, 241
United States Department of Agriculture (USDA), 78, 79, 125
United States Department of Defense (DoD), 295, 304
United States Environmental Protection Agency (USEPA), 99, 102, 268, 313
United States Geological Survey (USGS), 85, 86
Universal Soil Loss Equation (USLE), 310
UNWUF (United Nations World Urban Forum), 276–278
Urbanization and urban areas, 9–11
    impacts of, 102
    landforms in, 260–264
    sustainable cities, 275–280
USDA, *see* United States Department of Agriculture
USEPA, *see* United States Environmental Protection Agency
USGS (United States Geological Survey), 85, 86
USLE (Universal Soil Loss Equation), 310

V

Variation, environmental, 152–154
Vegetated roofs, 293
Vegetated swales, 292, 293
Vegetation, riparian, 98, 102–103
Vegetation types, 68

W

Warm temperate (C) climate classification, 66, 67
Warm temperate climate zone, 69
Water. *See also* Watershed
    environmental indicators for, 312–313
    field-scale indicators for, 318
    fresh, 15, 182
    human demand for, 14–16

in primary production, 173
quality of, 14, 99–100, 241, 259, 294–296, 313
quantity of, 14
in soil, 138–139
Watershed, 83–103
defined, 83
human impacts on, 100–103
hydrologic characteristics of, 92–99
impact of land use changes on, 37
physical characteristics of, 84–92
services, 84
and stream restoration, 223–225, 228–229
water quality characteristics of, 99–100
WCED (World Commission on Environment and Development), 3
Weather processes, 10
Wetlands, 24, 256–260
WFR (world fertility rate), 5
Wind direction/speed, 124–125
World Commission on Environment and Development (WCED), 3
World fertility rate (WFR), 5
World Resources Institute (WRI), 308, 320
World Wildlife Fund (WWF):
biomes classification, 69–71
ecoregions classification, 76, 78
WRI (World Resources Institute), 308, 320

Y

Yield efficiency (agricultural sustainability indicator), 316–317

Z

Zoning, 273, 275, 287